NUMERICAL ANALYSIS

NUMERICAL ANALYSIS

Kalyan Kr Mukherjee
Formerly Head, Department of Statistics
Asutosh College, Kolkata

New Central Book Agency (P) Ltd

Co-Published by

Bijay Publications

E-mail : bijaypublications@gmail.com

Exclusively Distributed by

Bijay Pustak Bhandar (P) Ltd

90/6A M. G. Road, Bhawani Dutta Lane, Kolkata-700007

✆ : (033) 4605-7856, 8777898576, 8274916104

E-mail : bijaypustakbhandar@gmail.com

NCBA

REGD OFFICE
8/1 Chintamoni Das Lane, Kolkata 700 009, India
email: ncbakolkata@gmail.com

OVERSEAS
NCBA (UK) Ltd, 149 Park Avenue North, Northampton, NN3 2HY, UK
email: ncbauk@yahoo.com

CORPORATE OFFICE
212 Shahpur Jat, New Delhi 110 049
email: ncbaexports@ncbapvtltd.com

BIJAY PUBLICATIONS
90/6A M. G. Rd., Bhawani Dutta Lane, Kolkata-7
E-mail : bijaypublications@gmail.com

BIJAY PUSTAK BHANDAR (P) LTD
90/6A M. G. Rd., Bhawani Dutta Lane, Kolkata-7
E-mail : bijaypustakbhandar@gmail.com

NUMERICAL ANALYSIS • Kalyan Kr Mukherjee

First Published: 1990
Second Edition: 1998
Reprinted: 2002, 2006, 2011, 2015, 2018, 2024

PUBLISHER • TYPESETTER • PRINTER
New Central Book Agency (P) Ltd
8/1 Chintamoni Das Lane, Kolkata 700 009

CO-PUBLISHED BY
Bijay Publications
90/6A M G Road, Bhawani Dutta Lane, Kolkata 700007

PROJECT TEAM
Abhijit Sen and Salil Das

ISBN: 81 7381 520 8

Price ₹ 495.00

Dedicated to all my students:
Past and Present of
Ashutosh College, Kolkata

Preface on First Edition

The present book on Numerical Analysis is intended to cover the syllabi of different Indian Universities in Statistics (Honours) and Mathematics (Honours) courses. It may also serve as a good help-book for the engineering students. Every endeavour has been made to present the underlying principles in a lucid manner without compromising rigour. As for mathematical and statistical pre-requisites, only the elementary knowledge of College Algebra and Probability is assumed.

In writing a book of this kind, I have freely consulted the works of the well-known authors in this area (e.g. J.B. Scarborough, H. Freeman, E.T. Whittaker and G. Robinson). I wish to record my indebtedness to those authors of their works. I shall deem my labour to be amply rewarded, if the present book proves to be of some help to the students for whom it is intended. I shall always welcome any suggestion for further improvement of the book from my friends, fellow teachers and past students working in this area, which may be incorporated in the subsequent editions.

I am much indebted to my wife, Smt. Sharmila Mukherjee, whose inspiration and help has enabled me to write out this volume.

Lastly, I wish to express my sincere thanks to my publisher Mr Arunabha Sen of New Central Book Agency (P) Ltd, who took personal care to bring out the book in its present get up in a short time.

15 August, 1990 — Kalyan Kr Mukherjee

Department of Statistics
Ashutosh College
Calcutta 700026

Preface on Second Edition

I wish to express my gratitude to my enumerable friends and colleagues for their warm reception extended to the first edition of the present text.

There were a number of misprints and computational errors in the earlier edition. This was a constant source of irritation to all serious readers. I have taken this opportunity to rectify those misprints and errors in the present revision. Some of the chapters, specially chapters 2 and 3, have been enlarged to cater for the needs of the students of the engineering colleges and other universities in the Eastern India and the South. I shall regard my labour to be amply rewarded, if the present edition receives the same appreciation from the present students in statistics and other disciplines.

15 April, 1998 Kalyan Kr Mukherjee
Calcutta 700084

Contents

Chapter 1

The Accuracy of Approximate Calculations

Summary

1.1 Introduction

Generally, in any computational problem in statistics, specially in the area of numerical analysis, the workers will invariably encounter all kinds of numbers—rational or irrational, real or complex. Their ultimate objective will be to obtain result with greatest possible accuracy. For example, numerical data for solving our everyday problems, are not usually exact. They are just approximations, correct to two, three, four or more figures. For example, we find the solution of the equation as follows $x^2 = 8 \Rightarrow x = \pm\sqrt{8} = \pm 2.8284$.

The result is not exact, it is correct to 4 decimal places. If we desire greater accuracy, we could write $x = \pm 2.828427$.

It is correct up to 6 decimal places. It is an approximate calculation. One can easily appreciate that the error in a computed result may be due to the following two sources: (a) *errors in the data* and (b) *errors in the computation.*

There is no remedy at our disposal, after the collection and recording of the data for the errors in (a).

But we may minimise the second type of error as much as we may please. We shall therefore assume in our present discussion that the computations are carried out by the workers in such a manner as to reduce the error of computations to negligible. The object of the present chapter is to set forth some basic rules and methods relating to approximate calculations and to give methods for estimating the accuracy of the results computed.

1.2 Approximate numbers and significant figures

1.2.1 Approximate numbers

It will be appropriate for our discussion if we make a distinction between numbers which are exact (in the absolute sense), and those which express approximate values. The numbers like 3, 1/3, 1/2, 50, etc. are exact numbers because there is no approximation or uncertainty associated with them. However, we have numbers such as π, $\sqrt{3}$, e, etc. which are exact but cannot be expressed exactly by finite number of digits. When written in digital form, π must be written as 3.1416, e as 2.7183, etc. Such numbers are therefore approximations to the true values, and in such cases, they are called approximate numbers. Approximate numbers may therefore be defined as a number which is used as an approximation to an exact number and differs only slightly from the exact number for which it stands.

1.2.2 Significant figures

A significant figure is any one of the digits 1, 2, 3, ..., 9 and 0 is a significant figure except when it is used to fix the decimal point or to fill the places of unknown or discarded digits. Thus, in the number 0.00134, the significant figures are 1, 3, 4. The zeros are used here merely to fix the decimal point and therefore not significant. But in the number 0.1204, the significant figures are 1, 2, 0, 4. In a number 26300, there is nothing in the number, as written, to show whether or not the zeroes are significant figures. However, let us rewrite the same number 26300 in any one of the following ways: (a) 2.63×10^4, (b) 2.630×10^4, and (c) 2.6300×10^4.

In (a), the significant figures are 2, 6, 3.

In (b), the significant figures are 2, 6, 3, 0.

In (c), the significant figures are 2, 6, 3, 0, 0.

Hence the correct answer should be: *The number of significant figures in* 26300 *will be* 3 *or* 4 *or* 5.

1.3 Rounding of numbers

If we try to divide 3 by 7 we get 3/7 = 0.42857142....

Here the quotient never terminates. If we have to use such numbers in a practical computation, we must cut it down to manageable form, such as 0.429, or 0.4286 or 0.42857, etc. This process of cutting off superfluous digits and retaining as many as we desire, is called rounding off.

To *round off* a number is to retain a certain number of digits, *counted from the left* and drop the others. Thus, to round off e to three, four, five and six figures, respectively, we have 2.72, 2.718, 2.7183, 2.71828.

Numbers are rounded off so as to cause the *least possible error.*

We may state the general rule for rounding off a number to n significant digits: "Discard all digits to the right of the nth place. If the discarded number is less then half a unit in the *n*th place, leave the *n*th digit unchanged; if the discarded number is greater than half a unit in the *n*th place, add 1 to the *n*th digit. If the discarded digit is *exactly* half a unit in the *n*th place, leave the *n*th digit unaltered if it is an even number, but increase it by 1 if it is an odd number."

When a number has been rounded off according to the rule stated above, it is said to be *correct to n significant figures*.

EXAMPLE 1.3.1. Round off the following numbers correct to four significant figures:

(a) 27.1345 becomes 27.13
(b) 27.8793 " 27.88
(c) 27.3993 " 27.40
(d) 27.3550 " 27.36
(e) 27.3650 " 27.37

When the rule given above is followed strictly, the errors due to rounding are largely cancelled by one another.

1.4 Absolute, relative and percentage errors

Absolute error of a number is the numerical difference between the true value of the quantity and its approximate value as written.

The relative error is the absolute error divided by the true value.

The percentage error is 100 times the relative error. Thus, if P be the approximate value corresponding to the true value $P + \Delta P$, then

$E_a \equiv$ Absolute error = True value − approximate value = $P + \Delta P - P = \Delta P$.

$$E_r \equiv \text{Relative error} = \frac{\text{Absolute error}}{\text{True value}} = \frac{\Delta P}{P + \Delta P} = \frac{E_a}{P}.$$

$$E_p \equiv \text{Percentage error} = \frac{\Delta P}{P + \Delta P} \times 100 = E_r \times 100.$$

NOTE 1.4.1. The absolute error in a numerical computation correct to n significant figures cannot be greater than half a unit in the nth place.

EXAMPLE 1.4.1. The number 27.1345 rounded to 27.13 correct to 4 significant figure has absolute error $E_a = 27.1345 - 27.13 = 0.0045$. According to the rule given in note 1, E_a will be less than 1/2 × (a unit at the 4th place), i.e. 1/2 ×0.01 = 0.005.

NOTE 1.4.2. Relative and percentage errors are independent of units of measurement. But absolute error depends on the unit of measurement.

The relation between the relative error and the number of correct figures is given by the following theorem:

THEOREM 1.4.1. If the first significant figure of a number is k and the number is correct to n significant figures, then the relative error is less than $1/(k \times 10^{n-1})$. (Before we give a formal proof of the theorem, we shall first verify the statement with regard to a number of examples.)

EXAMPLE 1.4.2. Let the number 271.37 be correct to five significant figures. Here the first significant figure k is 2 and the number of significant figures n is 5. Then $E_r \leq 1/(2\times 10^{5-1})$ $= 0.5/10^4 = 0.00005$.

Solution: Again according to note 1, $E_a \leq 0.01 \times 1/2 = 0.005$.

For relative error, we have

$$E_r \leq \frac{0.005}{271.37 - 0.005} = \frac{5}{271370 - 5} = \frac{1}{54274 - 1}$$
$$= \frac{1}{2 \times 27136.5} < \frac{1}{2 \times 27136} < \frac{1}{2 \times 10^4}$$

Hence the theorem holds.

EXAMPLE 1.4.3. Let us consider the number 356.320. We assume that the last digit 0 is introduced to fill up the position of a discarded digit and therefore not significant.

Solution: Here $k = 3$, $n = 5$ and $E_a \leq 10 \times 1/2 = 5$. Then

$$E_r \leq \frac{5}{356320 - 5} = \frac{1}{2 \times 35632 - 1} = \frac{1}{2\left(35632 - \frac{1}{2}\right)} < \frac{1}{2 \times 3 \times 10^4} < \frac{1}{3 \times 10^4}$$

[*Note*: E_r is not certainly less than $1/\left(2k \times 10^{n-1}\right)$.]

Let us now give the Proof of the Theorem 1.

Proof. Let N be the exact value of the number, n be the number of significant figures, m be the number of correct decimal places.

Three cases must be distinguished, namely, (a) $m < n$, (b) $m = n$, and (c) $m > n$.

CASE 1.1. $m < n$.

Here the number of digits in the integral part of N is $n - m$. Denoting the first significant digit of N by k, we have $E_a \leq \frac{1}{2} \cdot 10^{-m}$, $N \geq k \cdot 10^{n-m-1} - \frac{1}{2} \cdot 10^{-m}$.

Hence

$$E_r \leq \frac{\frac{1}{2} \cdot 10^{-m}}{k \cdot 10^{n-m-1} - \frac{1}{2} \cdot 10^{-m}} = \frac{10^{-m}}{2k \cdot 10^{n-1} \cdot 10^{-m} - 10^{-m}}$$
$$= \frac{1}{2k \cdot 10^{n-1} - 1} = \frac{1}{2\left(k \cdot 10^{n-1} - \frac{1}{2}\right)}$$

Noting that n is a *positive integer* and that k is any one of the digits from 1 to 9 inclusive, we may see that $2k \cdot 10^{n-1} - \frac{1}{2} > k \cdot 10^{n-1}$ in all cases except $k = 1, n = 1$, which is a trivial case, where $N = 1, 0.01$, etc. (i.e. where N has just one significant digit which is 1, a case which never occurs in practice).

Hence for all other cases, $2k \cdot 10^{n-1} - 1 > k \cdot 10^{n-1}$. $\therefore E_r < \frac{1}{k \cdot 10^{n-1}}$.

CASE 1.2. $m = n$.

N is a pure fraction (decimal) and k is the first digit after the decimal point.

Then $E_a \leq \frac{1}{2} \cdot 10^{-m}$, $N \geq k \cdot 10^{-1} - \frac{1}{2} \cdot 10^{-m}$.

$$\therefore E_r \leq \frac{\frac{1}{2} \cdot 10^{-m}}{k \cdot 10^{-1} - \frac{1}{2} \cdot 10^{-m}} = \frac{10^{-m}}{2k \cdot 10^{-1} - 10^{-m}}$$
$$= \frac{1}{2k \cdot 10^{m-1} - 1} = \frac{1}{2k \cdot 10^{n-1} - 1} < \frac{1}{k \cdot 10^{n-1}}$$

CASE 1.3. $m > n$.

In this case k occupies the $(m - n + 1)$th decimal place and therefore $E_a \leq \frac{1}{2} \cdot 10^{-m}$, $N \geq k \cdot 10^{-(m-n+1)} - \frac{1}{2}10^{-m}$.

$$\therefore E_r \leq \frac{\frac{1}{2} \cdot 10^{-m}}{k \cdot 10^{-m+n-1} - \frac{1}{2}10^{-m}} = \frac{10^{-m}}{2k \cdot 10^{-m+n-1} - 10^{-m}}$$

$$= \frac{1}{2k \cdot 10^{n-1} - 1} < \frac{1}{k \cdot 10^{n-1}}$$

The theorem therefore holds in all the cases.

EXAMPLE 1.4.4. Let $N = 453.16852$ be correct to 8 significant figures.

Solution: Here $k = 4$, $n = 8$ and $E_a \leq \frac{1}{2} \times 10^{-5} = \frac{1}{2} \times 0.00001 = 0.000005$.

$$E_r < \frac{0.000005}{453.16852 - 0.000005} = \frac{5}{453168520 - 5} = \frac{1}{2 \times 45316852 - 1}$$

$$= \frac{1}{2\left(45316852 - \frac{1}{2}\right)} < \frac{1}{2 \times 4 \times 10^7} < \frac{1}{4 \cdot 10^7}$$

1.5 The general formula for errors

Let N be a function of several independent quantities $u_1, u_2, \ldots, u_n$ which are subject to errors of magnitude $\Delta u_1, \Delta u_2, \ldots, \Delta u_n$, respectively. These errors in u's will result in an error ΔN in N.

Accordingly,

$$N + \Delta N = f(u_1 + \Delta u_1, u_2 + \Delta u_2, \ldots, u_n + \Delta u_n)$$

Using Taylor's theorem for a function of several variables, we have

$$N + \Delta N = f(u_1, u_2, \ldots, u_n) + \Delta u_1 \cdot \frac{\partial f}{\partial u_1} + \Delta u_2 \cdot \frac{\partial f}{\partial u_2} + \cdots + \Delta u_n \cdot \frac{\partial f}{\partial u_n}$$

[Since the errors $\Delta u_1, \Delta u_2, \ldots, \Delta u_n$ are always very small quantities, we have neglected the squares, products and higher powers of Δu_i's in the above expansion].

$$\therefore \Delta N = \Delta u_1 \cdot \frac{\partial f}{\partial u_1} + \Delta u_2 \cdot \frac{\partial f}{\partial u_2} + \cdots + \Delta u_n \cdot \frac{\partial f}{\partial u_n} \tag{1.5.1}$$

This is the general formula for computing the error of a function and it includes all possible cases. We may note that the right-hand member of (1.5.1) is merely the total differential of the function N.

For the relative error of the function N, we have

$$E_r = \frac{\Delta N}{N} = \frac{\Delta u_1}{N} \cdot \frac{\partial f}{\partial u_1} + \frac{\Delta u_2}{N} \cdot \frac{\partial f}{\partial u_2} + \cdots + \frac{\Delta u_n}{N} \cdot \frac{\partial f}{\partial u_n} \tag{1.5.2}$$

EXAMPLE 1.5.1. Let

$$N = \frac{kx^m y^n z^p}{v^q \omega^r} \quad (k, m, n, p, \text{ and } q, r \text{ are constants}) \tag{1.5.3}$$

Solution: Then (1.5.2) gives the relative error

$$E_r = \frac{\Delta N}{N} = m\left(\frac{\Delta x}{x}\right) + n\left(\frac{\Delta y}{y}\right) + p\left(\frac{\Delta z}{z}\right) - q\left(\frac{\Delta v}{v}\right) - r\left(\frac{\Delta \omega}{\omega}\right)$$

Since errors $\Delta x, \Delta y, \Delta z, \Delta v, \Delta \omega$ are just as likely to be positive as negative, we take all the terms with positive sign.

$$E_r \leq m\left|\frac{\Delta x}{x}\right| + n\left|\frac{\Delta y}{y}\right| + p\left|\frac{\Delta z}{z}\right| + q\left|\frac{\Delta v}{v}\right| + r\left|\frac{\Delta \omega}{\omega}\right| \quad (1.5.4)$$

1.6 Application of error formulae to the fundamental operations of arithmetic and to logarithms

1.6.1 Addition

Let $N = u_1 + u_2 + \ldots u_n$.

$$\Delta N = E_a = \Delta u_1 + \Delta u_2 + \cdots + \Delta u_n \quad (1.6.1)$$

Thus, the absolute error of a sum of approximate numbers is therefore equal to the algebraic sum of their absolute error (Rule 1).

1.6.2 Subtraction

In this case,

$$N = u_1 - u_2$$

and $\Delta N = \Delta u_1 - \Delta u_2$ (1.6.2)

Since Δu_1, Δu_2 may be positive or negative, in order to obtain the maximum error, we must take the sum of the absolute values of the errors.

$\therefore\ E_r \leq |\Delta u_1| + |\Delta u_2|$

[*Note:* When one approximate number is subtracted from another, both of them must be rounded off to the same place before subtracting.]

1.6.3 Loss of significant figures by subtraction

In the subtraction of two approximate numbers, a serious type of error may be present when the numbers are nearly equal. For example, suppose we are to subtract 632.993 from 633.358. The numbers are correct to six significant figures, but no more. Their difference 0.365 is correct only to 3 significant figures.

Errors arising out of the disappearance of the most important significant numbers from the left may be frequently found in practice. The inaccuracy introduced by virtue of this type of computational work, can be lessened in one of the *two* ways:

1. By approximating each of the two numbers with sufficient accuracy *before* subtraction. Thus, to find the difference $\sqrt{3.02}-\sqrt{3}$ correct to five significant figures, we take $\sqrt{3.02}$ = 1.737815 and $\sqrt{3}$ = 1.732051.
   Then $\sqrt{3.02} - \sqrt{3} \simeq 1.737815 - 1.732051 = 0.005764$ (correct to four significant figures only—the last one doubtful).
   This method is limited when the two given numbers are approximate and true to only a few digits.
2. By transforming the expression whose value is desired.

Thus, to find the value of $(1 - \cos x)$, where x is small and no extended mathematical table is at hand, we may use Taylor's expansion.

$$1 - \cos x = 1 - \left(1 - \frac{x^2}{2!} + \frac{x^4}{4!} \cdots\right) = \frac{x^2}{2} - \frac{x^4}{24} + \ldots$$

1.6.4 Multiplication

In this case, $N = u_1, u_2, \ldots, u_n$.

$$\frac{\Delta N}{N} = \frac{\Delta u_1}{u_1} + \frac{\Delta u_2}{u_2} + \cdots + \frac{\Delta u_n}{u_n} \tag{1.6.3}$$

Thus, the relative error of a product of n approximate numbers is equal to the algebraic sum of the relative errors of the separate numbers. E_a will be $E_r \times N$.

1.6.5 Division

Here $N = u_1/u_2$.

$$\therefore\ E_r < \left|\frac{\Delta u_1}{u_1}\right| + \left|\frac{\Delta u_2}{u_2}\right|$$

The relative error of a quotient is therefore equal to the algebraic sum of the relative errors of divisor and dividend.

A simple formula for the absolute error of a quotient can be found as shown under: Let ΔQ be the absolute error of u_1/u_2. Then

$$\Delta Q = \frac{u_1 + \Delta u_1}{u_2 + \Delta u_2} - \frac{u_1}{u_2} = \frac{u_1 u_2 + u_2 \Delta u_1 - u_1 u_2 - u_1 \Delta u_2}{u_2(u_2 + \Delta u_2)} = \frac{u_2 \Delta u_1 - u_1 \Delta u_2}{u_2(u_2 + \Delta u_2)}$$

$$= \frac{u_1\left(\dfrac{u_2 \Delta u_1 - u_1 \Delta u_2}{u_1 u_2}\right)}{u_2 + \Delta u_2} = \frac{u_1\left(\dfrac{\Delta u_1}{u_1} - \dfrac{\Delta u_2}{u_2}\right)}{u_2 + \Delta u_2}$$

Let ω be the greatest absolute value of

$$\frac{\Delta u_1}{u_1}, \quad \text{or,} \quad \frac{\Delta u_2}{u_2}$$

If u_1 and u_2 are both subject to errors of the *same order of magnitude*

$$\Delta Q \le \frac{u_1(\omega + \omega)}{u_2 - \omega u_2}, \quad \text{i.e.} \quad \Delta Q \le \frac{2u_1\omega}{u_2(1 - \omega)}$$

If ω is negligible in comparison with 1, we have

$$\Delta Q \le 2\left(\frac{u_1}{u_2}\right)\omega, \quad \text{if only } u_1 \text{ and } u_2 \text{ are both subject to error} \tag{1.6.4}$$

However, if only u_1 or u_2 is subject to error

$$\Delta Q \le \left(\frac{u_1}{u_2}\right)\omega \tag{1.6.5}$$

[*Note:* Accuracy of a quotient should be examined by means of the relative error. The statement made in regard to products holds for quotient also. If either divisor or dividend is more accurate in comparison to the other, the more accurate number should be rounded off. This may result in *one more* significant figure than the less accurate one. The result is correct to the number of significant figures as given in the less accurate number.]

EXAMPLE 1.6.1. Find the number of trustworthy figures in the quotient 672.6/293.7, assuming that both numerator and denominator are approximate and true only to the number of digits given.

Solution: Here the largest relative error is

$$\omega = \frac{0.05}{293.7} = 0.0001702$$

The quotient u_1/u_2 is 672.6/293.7 = 2.29009.

Hence by (1.6.4)

$$\Delta Q \leq 2(2.29009)(0.0001702) = 0.000780$$

Since this error affects the fourth decimal place and not the third, we may write the quotient 672.6/293.7 = 2.290.

NOTE 1.6.1. Greatest value of the quotient will be

$$\frac{672.6 - 0.05}{293.7 + 0.05} = \frac{672.65}{293.65} = 2.2907 \simeq 2.29$$

Least value of the quotient

$$\frac{672.6 + 0.05}{293.7 - 0.05} = \frac{672.65}{293.65} = 2.28953 \simeq 2.29$$ (both correct to 3 significant figures; the two results agree).

NOTE 1.6.2. In case of quotient, relative error, instead of absolute error ΔQ, may be interesting.

$$E_r = \frac{\Delta Q}{u_1/u_2} = \frac{\left(\frac{\Delta u_1}{u_1} - \frac{\Delta u_2}{u_2}\right)}{1 + \frac{\Delta u_2}{u_2}} \simeq \frac{\Delta u_1}{u_1} - \frac{\Delta u_2}{u_2}$$

provided $\Delta u_2/u_2$ is small in comparison to 1.

Relative error in this case is equal to the difference of the relative errors in the individual numbers.

1.7 Propagated error in evaluating a function $f(x)$ of a single variable x having an error Δx

We expand

$$f(x + \Delta x) = f(x) + \Delta x f'(x) + \frac{(\Delta x)^2}{2!} f''(x) + \ldots$$

Since Δx is usually small, we may neglect terms from $(\Delta x)^2$ onwards and write

$$f(x + \Delta x) - f(x) = \Delta x f'(x) \Rightarrow \Delta y = \Delta x f'(x), \quad \text{or,} \quad \Delta x = \frac{\Delta y}{f'(x)} \tag{1.7.1}$$

This is a fundamental equation for computing error in arguments taken from a table. Δy as written above is the error in the computed function whose values are tabulated, and Δx is the corresponding error in argument.

EXAMPLE 1.7.1. $f(x) = \sin x$; $f'(x) = \cos x$ and $\dfrac{\Delta y}{\Delta x} \simeq \cos x$; $\Delta x \simeq \Delta y \sec x$ radians.

EXAMPLE 1.7.2. $f(x) = e^x$; $f'(x) = e^x$; $\dfrac{\Delta y}{\Delta x} \simeq e^x$; and $\Delta x = e^{-x}\Delta y$.

EXAMPLE 1.7.3. $f(x) = \log_a x = \log_e x \times \log_a e$; $f'(x) = \frac{1}{x} \times \log_a e$; $\frac{\Delta y}{\Delta x} \simeq \frac{1}{x} \times \log_a e$; $\Delta x \simeq \frac{x\Delta y}{\log_a e}$.

EXERCISES

1. Round-off the following numbers correct to three significant figures:
   (a) 32.3526. [*Ans:* 32.4; E_a : −0.0474.]
   (b) 1.0532. [*Ans:* 1.05; E_a : 0.0032.]
   (c) 0.005736. [*Ans:* 0.00574; E_a : −0.000004.]
   (d) 0.0003058. [*Ans:* 0.000306; E_a : −0.0000002.]
   (e) 0.0028561. [*Ans:* 0.00286; E_a : −0.0000039.]
   Find the absolute error in each case.
2. Find the number of significant figures in the following numbers:
   (a) 1.0236. [*Ans:* 5.]
   (b) 0.002406. [*Ans:* 4.]
   (c) 261000. [*Ans:* 3 or 4 or 5 or 6.]
   (d) 0.1053. [*Ans:* 4.]
   (e) 278.03. [*Ans:* 5.]
   (f) 37.1002. [*Ans:* 6.]
3. The following numbers are all approximate and correct up to their last digits: 256.786, 27.48, 321.0, 56.482, 16.82. Find their sum and the absolute error of the sum. [*Ans:* 678.1; −0.032.]
4. Determine the absolute error E_a of the following *approximate* numbers, given their relative errors:
   (a) $x_a = 13146$, $E_r = 0.1\%$. [*Ans:* 0.13×10^{-2}.]
   (b) $x_a = 3.48$, $E_r = 0.7\%$. [*Ans:* 0.024.]
   (c) $x_a = 0.864$, $E_r = 5\%$. [*Ans:* 0.045.]
5. Determine the number of correct digits in the number x, given its absolute error.
   (a) $x = 0.3681$, $E_a = 0.23 \times 10^{-2}$. [*Ans:* One.]
   (b) $x = 48.2461$, $E_a = 0.21 \times 10^{-2}$. [*Ans:* Four.]
   (c) $x = 841.256$, $E_a = 0.1$. [*Ans:* Three.]
   (d) $x = 000985$, $E_a = 0.1 \times 10^{-4}$. [*Ans:* Two.]
   (e) $x = -33.783$, $E_a = 0.3 \times 10^{-2}$. [*Ans:* Four.]
6. Determine the number of correct digits in the number x, given its relative error E_r.
   (a) $x = 0.4785$, $E_r = 0.2 \times 10^{-1}$. [*Ans:* One.]
   (b) $x = 86.34$, $E_r = 0.1$. [*Ans:* Zero.]
   (c) $x = 0.1857$, $E_r = 0.1$. [*Ans:* Zero.]
   (d) $x = 39461$, $E_r = 0.1$. [*Ans:* Two.]
   (e) $x = 386.4$, $E_r = 0.3$. [*Ans:* One.]
7. Find the relative error for the evaluation of $u = x_1x_2$ with $x_1 = 4.51$, $x_2 = 8.32$ having error E_a in $x_1 = 0.01$ and E_a in $x_2 = 0.01$. [*Ans:* ±0.0034.]
8. Compute relative error in computing $u = x^3 + 2x^2 - x$ for $x = \sqrt{3}$, taking $\sqrt{3} \simeq 1.73$. [*Ans:* 0.0016.]

   [*Hints:* $\frac{\Delta u}{u} = \frac{3x^2 + 4x - 1}{x^3 + 2x^2 - x}\Delta x$; here $\Delta x \leq 0.005$.]

Chapter 2

Difference Operators Δ and E

Summary

2.1 Finite differences

Consider the function $y = f(x)$. Here x is the independent variable and y, the dependent variable. Suppose the values of the independent variable are given at equispaced interval h as

$x_0, \;\; x_0 + h, \;\; x_0 + 2h, \;\; x_0 + 3h, \;\; \ldots$

The corresponding values of y will be

$$\begin{aligned} y_0 &= f(x_0), \\ y_1 &= f(x_0 + h), \\ y_2 &= f(x_0 + 2h), \\ y_3 &= f(x_0 + 3h), \\ &\ldots\ldots\ldots\ldots \end{aligned}$$

The value of the independent variable x is called the argument and the value of the dependent variable y is called the entry. So we may say that the following entries:

$y_0,\ y_1,\ y_2,\ y_3,\ \ldots$

are obtained corresponding to the values of the argument

$x_0,\ x_0 + h,\ x_0 + 2h,\ x_0 + 3h,\ \ldots$

The difference between the consecutive values of the independent variable is called the interval of differencing (*here h*). Unless otherwise stated explicitly, the interval of differencing will be unity.

The differences

$y_1 - y_0$ [or, $f(x_0 + h) - f(x_0)$]

$y_2 - y_1$

$y_3 - y_2$

......

are called the first differences of the function $y = f(x)$, and are denoted respectively by

$\Delta y_0 = y_1 - y_0 \equiv f(x_0 + h) - f(x_0)$

$\Delta y_1 = y_2 - y_2 \equiv f(x_0 + 2h) - f(x_0 + h)$

$\Delta y_2 = y_3 - y_2 \equiv f(x_0 + 3h) - f(x_0 + 2h)$

In general,

$$\begin{aligned}\Delta y_r &= y_{r+1} - y_r \\ &= f(x_0 + \overline{r+1}h) - f(x_0 + rh)\end{aligned}$$

where Δ is not a quantity, but a symbol representing an *operation*.

The differences of the above first differences are called the second differences of $f(x)$ and are denoted by

$\Delta^2 y_0 \equiv \Delta^2 f(x_0) \equiv \Delta y_1 - \Delta y_0$

$\Delta^2 y_1 \equiv \Delta^2 f(x_0 + h) \equiv \Delta y_1 - \Delta y_1$

$\Delta^2 y_2 \equiv \Delta^2 f(x_0 + 2h) \equiv \Delta y_3 - \Delta y_2$

and so on.

Here it must be emphasised that the symbol Δ^2 does not represent the square of a quantity Δ but denotes repetition of the operation Δ. Thus, $\Delta^2 f(x_0)$ means (i.e. $\Delta^2 y_0$)

$$\begin{aligned}\Delta^2 f(x_0) &= \Delta\{\Delta f(x_0)\} \\ &= \Delta\{f(x_0 + h) - f(x_0)\} \\ &= \Delta f(x_0 + h) - \Delta f(x_0)\end{aligned}$$

The differences of these second differences are called the third differences. Denoting these by $\Delta^3 f(x_0)$, $\Delta^3 f(x_0 + h)$, $\Delta^3 f(x_0 + 2h)$, etc. we have

$$\begin{aligned}\Delta^3 f(x_0) &= \Delta^2\{\Delta f(x_0)\} \\ &= \Delta^2\{f(x_0 + h) - f(x_0)\} \\ &= \Delta^2 f(x_0 + h) - \Delta^2 f(x_0)\end{aligned}$$

THEOREM 2.1.1. Let a function $y = f(x)$ have the values $y_0, y_1, y_2, \ldots, y_n$ corresponding to the values of the argument $x_0, x_0 + h, x_0 + 2h, \ldots, x_0 + nh$; then for any non-negative integer m

$$\Delta^m y_r = \sum_{i=0}^{m} (-1)^{m-i} \binom{m}{i} y_{r+i}$$

Proof. We see that

$$\Delta y_r = y_{r+1} - y_r = \sum_{i=0}^{1} (-1)^{1-i} \binom{1}{i} y_{r+i}$$

$$\begin{aligned}\Delta^2 y_r &= \Delta y_{r+1} - \Delta y_r = (y_{r+2} - y_{r+1}) - (y_{r+1} - y_r)\\ &= y_{r+2} - 2y_{r+1} + y_r\\ &= \sum_{i=0}^{2} (-1)^{2-i} \binom{2}{i} y_{r+i}\end{aligned}$$

Thus, the theorem holds good for $m = 1$ and $m = 2$.

Let us assume that the result also holds for $m = k (k < m)$.

Then

$$\Delta^k y_r = \sum_{i=0}^{r=k} (-1)^{k-i} \binom{k}{i} y_{r+i}$$

$$\begin{aligned}\Delta^{k+1} y_r &= \sum_{i=0}^{k} (-1)^{k-i} \binom{k}{i} \Delta y_{r+i}\\ &= \sum_{i=0}^{k} (-1)^{k-i} \binom{k}{i} (y_{r+i+1} - y_{r+i})\\ &= \sum_{i=0}^{k} (-1)^{k-i} \binom{k}{i} y_{r+i+1} - \sum_{i=0}^{k} (-1)^{k-i} \binom{k}{i} y_{r+i}\\ &= \sum_{j=1}^{k+1} (-1)^{k+1-j} \binom{k}{j-1} y_{r+j} + \sum_{i=0}^{k} (-1)^{k+1-i} \binom{k}{i} y_{r+i}\\ &= y_{r+k+1} + \sum_{j=1}^{k} (-1)^{k+1-j} \binom{k}{j-1} y_{r+j} + \sum_{i=1}^{k} (-1)^{k+1-i} \binom{k}{i} y_{r+i} + (-1)^{k+1} y_r\\ &= y_{r+k+1} \sum_{i=1}^{k} (-1)^{k-i+1} \left\{ \binom{k}{i-1} + \binom{k}{i} \right\} y_{r+i} + (-1)^{k+1} y_r\\ &= y_{r+k+1} + \sum_{i=1}^{k} (-1)^{k+1-i} \binom{k+1}{i} y_{r+i} + (-1)^{k+1} y_r\\ &= \sum_{i=0}^{k+1} (-1)^{k+1-i} \binom{k+1}{i} y_{r+i}\end{aligned}$$

Therefore, the **result holds for** $m = k + 1$, **provided** it holds for $m = k$. But it has been established for $m = 2$ **and** $m = 3$. Hence by mathematical law of induction, we may write that the theorem **holds good for any integer (positive)** m.

2.2 Horizontal and diagonal difference tables

The difference Table given in Table 2.2 is called a *diagonal* difference table. The majority of difference tables are of diagonal type, but for many purposes, a more compact table (Table 2.1) called a *horizontal* difference table, is preferable. In this type of difference tables, the differences of different order are denoted by subscripts instead of exponents. Using the notations of horizontal differences, we can rewrite the difference table in the horizontal form as follows:

Table 2.1: *Horizontal difference table*

x	y_x	$\Delta_1 y_x$	$\Delta_2 y_x$	$\Delta_3 y_x$
x_0	y_0			
$x_0 + h$	y_1	$y_1 - y_0 = \Delta_1 y_1$		
$x_0 + 2h$	y_2	$y_2 - y_1 = \Delta_1 y_2$	$\Delta_1 y_2 - \Delta_1 y_1 = \Delta_2 y_2$	
$x_0 + 3h$	y_3	$y_3 - y_2 = \Delta_1 y_3$	$\Delta_1 y_3 - \Delta_1 y_2 = \Delta_2 y_3$	$\Delta_2 y_3 - \Delta_2 y_2 = \Delta_3 y_3$
$\vdots$	$\vdots$	$\vdots$	$\vdots$	$\vdots$

We may note the following differences in the two forms:

For example,

$$\Delta^3 y_0 = y_3 - 3y_2 + 3y_1 - y_0$$
$$\text{and } \Delta_3 y_3 = y_3 - 3y_2 + 3y_1 - y_0$$

Thus, $\Delta^3 y_0$ and $\Delta_3 y_3$ are numerically the same. In general

$$\Delta^n y_k = \Delta_n y_{n+k} \quad \text{(going forward from } y_k\text{)}$$
$$\text{and } \Delta_m y_n = \Delta^m y_{n-m} \quad \text{(going backward from } y_n\text{)}$$

2.3 The difference table for a polynomial y_x

The diagonal difference table for a polynomial function will reveal certain distinctive feature, which can be illustrated by taking y_k as a second degree polynomial of argument x.

Let

$$y_x = 2x^2 + 3x$$

The argument x is taken at equispaced interval $h = 2$.

Table 2.2: *Diagonal difference table of* $y_x = 2x^2 + 3x$

x	y_x	Δy_x	$\Delta^2 y_x$	$\Delta^3 y_x$
1	5			
		22		
3	27		16	
		38		0
5	65		16	
		54		0
7	119		16	
		70		
9	189			

We may note that second order finite differences are constant and equal to 16 in this case and the third order differences are zero. It may be shown, following the above lines, that if y_x be a polynomial of degree m in x; mth order finite differences will be constant and $(m+1)$th order differences and above will all be zero, a result which will be often used in interpolation problems (Chapter 3).

2.4 Propagation of an error in a finite difference table

Let us consider $(n + 1)$ values of the dependent variable y_x, given by $y_0, y_1, y_2, \ldots, y_n$. Suppose they are *true* values of a function y_x. We further suppose that the value y_5 is affected with an error ε, so that the erroneously tabulated value is $y_5 + \varepsilon$. We consider only the part of the difference table having the erroneous entry $y_5 + \varepsilon$ located at the middle. The successive differences of the y's are shown as under:

Table 2.3: *Propagation of error in a diagonal difference table*

y_x	Δy_x	$\Delta^2 y_x$	$\Delta^3 y_x$	$\Delta^4 y_x$
y_0				
	Δy_0			
y_1		$\Delta^2 y_0$		
	Δy_1		$\Delta^3 y_0$	
y_2		$\Delta^2 y_1$		$\Delta^4 y_0$
	Δy_2		$\Delta^3 y_1$	
y_3		$\Delta^2 y_2$		$\Delta^4 y_1 + \varepsilon$
	Δy_3		$\Delta^3 y_2 + \varepsilon$	
y_4		$\Delta^2 y_3 + \varepsilon$		$\Delta^4 y_2 - 4\varepsilon$
	$\Delta y_4 + \varepsilon$		$\Delta^3 y_3 - 3\varepsilon$	
$y_5 + \varepsilon$		$\Delta^2 y_4 - 2\varepsilon$		$\Delta^5 y_3 + 6\varepsilon$
	$\Delta y_5 - \varepsilon$		$\Delta^3 y_4 + 3\varepsilon$	
y_6		$\Delta^2 y_5 + \varepsilon$		$\Delta^4 y_4 - 4\varepsilon$
	Δy_6		$\Delta^3 y_5 - \varepsilon$	
y_7		$\Delta^2 y_6$		$\Delta^4 y_5 + \varepsilon$
	Δy_7		$\Delta^3 y_6$	
y_8		$\Delta^2 y_7$		$\Delta^4 y_6$
	Δy_8		$\Delta^3 y_7$	
y_9		$\Delta^2 y_8$		
	Δy_9			
y_{10}				

We may note the following peculiarities:

1. The effect of an error increases with the successive differences,
2. The coefficients of ε's are the binomial coefficients with alternating signs, positive and negative,
3. Algebraic sum of the errors in any difference column is zero,
4. Table also shows that the maximum error in the differences lies in the same *horizontal line* as the erroneous tabular value.

Similar type of situation will arise showing the propagation of error in a horizontal difference table.

Table 2.4: *Propagation of error in a horizontal difference table*

y_x	$\Delta_1 y_x$	$\Delta_2 y_x$	$\Delta_3 y_x$	$\Delta_4 y_x$
y_0				
y_1	$\Delta_1 y_1$			
y_2	$\Delta_1 y_2$	$\Delta_2 y_2$		
y_3	$\Delta_1 y_3$	$\Delta_2 y_3$	$\Delta_3 y_3$	
y_4	$\Delta_1 y_4$	$\Delta_2 y_4$	$\Delta_3 y_4$	$\Delta_4 y_4$
$y_{5+\varepsilon}$	$\Delta_1 y_5 + \varepsilon$	$\Delta_2 y_5 + \varepsilon$	$\Delta_3 y_5 + \varepsilon$	$\Delta_4 y_5 + \epsilon$
y_6	$\Delta_1 y_6 - \varepsilon$	$\Delta_2 y_6 - 2\varepsilon$	$\Delta_3 y_6 - 3\varepsilon$	$\Delta_4 y_6 - 4\varepsilon$
y_7	$\Delta_1 y_7$	$\Delta_2 y_7 + \varepsilon$	$\Delta_3 y_7 + 3\varepsilon$	$\Delta_4 y_7 + 6\varepsilon$
y_8	$\Delta_1 y_8$	$\Delta_2 y_8$	$\Delta_3 y_8 - \varepsilon$	$\Delta_4 y_8 - 4\varepsilon$
y_9	$\Delta_1 y_9$	$\Delta_2 y_9$	$\Delta_3 y_9$	$\Delta_4 y_9 + \varepsilon$
y_{10}	$\Delta_1 y_{10}$	$\Delta_2 y_{10}$	$\Delta_3 y_{10}$	$\Delta_4 y_{10}$

Here again, the effect of the error is being manifested in the same fashion as in the table 2.3. But in this table 2.4, the *first erroneous difference of any order in the same horizontal line* as the erroneous tabular value.

The law according to which an error is propagated in a difference table enables one to trace such an error to its source and correct it. As an illustrative example let us consider the following case: (J.B. Scarborough)

EXAMPLE 2.4.1. Find and correct by means of differences the error in the table[1]

y_x	$\Delta_1 y_x$	$\Delta_2 y_x$	$\Delta_3 y_x$	$\Delta_4 y_x$	$\Delta_5 y_x$
20736					
28561	7825				
38416	9855	2030			
50625	12209	2354	324		
65540	14915	2706	352	$28 + \varepsilon$	
83521	17981	3066	360	$8 - 4\varepsilon$	−20
104976	21455	3474	408	$48 + 6\varepsilon$	+40
130321	25345	3890	416	$8 - 4\varepsilon$	−40
160000	29679	4334	444	$28 + \varepsilon$	+20

[1] When writing numerical difference tables, it is customary to omit the zeros between the decimal point and the first significant figure to the right of it; that is to say, the differences are expressed in units of the last figure retained.

We may note in this connection that if an error is present in the data, the differences of some order (*here* 5) will become alternating in sign. One should therefore carry the differencing for enough to allow the error to be revealed in this manner.

Since the algebraic sum of the fifth differences is 0, the fifth differences found in this example are accumulated errors. Referring now to table 2.4, we have

$-5\varepsilon = -20$

$+10\varepsilon = 40$

$-10\varepsilon = -40$

$+5\varepsilon = +20$

suggesting $\varepsilon = 4$ to the nearest unit.

The distribution of differences in column $\Delta_4 y_x$ will be

	$28 - \varepsilon$,	$8 + 4\varepsilon$,	$48 - 6\varepsilon$,	$8 + 4\varepsilon$,	$28 - \varepsilon$
i.e.	24	24	24	24	24

(*constant fourth order differences*).

The first erroneous entry will have to be located against the entry obtained by the horizontal line through 28 in column under $\Delta_4 y_x$.

The corresponding entry, 65540 is erroneously recorded as true value $+\varepsilon$. Hence the correct entry will be 65540 − 4 = 65536.

2.5 Use of operator *E* (Shift Operator)

Let $y = f(x)$ be given for the following equispaced values of the argument

$x_0, \quad x_0 + h, \quad x_0 + 2h, \quad \ldots$

The corresponding entries will be

$f(x_0), f(x_0 + h), f(x_0 + 2h), \ldots$

The operator E on $f(x_0)$ will be given by

$Ef(x_0) = f(x_0 + h)$

In general,

$Ef(x) = f(x + h)$

if h is the unit of differencing. We may mention again that if not explicitly stated, the usual unit of differencing is unity.

Similarly,

$E^2 f(x_0) = f(x_0 + 2h)$,

and in general,

$E^m f(x_0) = f(x_0 + mh)$

where m is any *rational number*, positive or negative.

E^m means the operator E (called shift operator) is repeated m times on $f(x_0)$.

For example:

$$E^5 f(x_0) = f(x_0 + 5h)$$

$$E^{3/2} f(x_0) = f\left(x_0 + \frac{3}{2}h\right)$$

$$E^{-3} f(x_0) = f(x_0 - 3h)$$

2.6 Relation between operators Δ and *E*

We may note

$$\begin{aligned}\Delta f(x_0) &= f(x_0 + h) - f(x_0)\\ &= Ef(x_0) - f(x_0)\\ &= (E - 1)\cdot f(x_0) \qquad (2.6.1)\end{aligned}$$

Thus, the operator Δ is equivalent to the operator $E - 1$. Symbolically, we may write

$\Delta \equiv E - 1$, or, $E \equiv 1 + \Delta$

2.7 Relation between differences and derivatives

We shall assume that the unknown function $y = f(x)$ is continuous in a finite interval and is continuously differentiable.

Then

$$f(x+h) = f(x) + hf'(x) + \frac{h^2}{2!}f''(x) + \cdots$$

$$\Rightarrow \quad E\cdot f(x) = \left\{1 + hD + \frac{h^2D^2}{2!} + \frac{h^3D^3}{3!} + \cdots\right\} f(x) \text{ where } D \equiv \frac{d}{dx}$$

$$= e^{hD} f(x) \qquad (2.7.1)$$

Symbolically, the shift operator E is equivalent to operator e^{hD}.

$\therefore 1 + \Delta \equiv e^{hD}$

$$\therefore D \equiv \frac{1}{h}\log_e(1+\Delta), \text{ or, } D \equiv \frac{1}{h}\left\{\Delta - \frac{\Delta^2}{2} + \frac{\Delta^3}{3} - \frac{\Delta^4}{4} + \cdots\right\}$$

Thus,

$$D.f(x)\Big]_{x=x_0} = \frac{1}{h}\left\{\Delta - \frac{\Delta^2}{2} + \frac{\Delta^3}{3} - \frac{\Delta^4}{4} + \cdots\right\} f(x_0)$$

$$= \frac{1}{h}\left\{\Delta f(x_0) - \frac{1}{2}\Delta^2 f(x_0) + \frac{1}{3}\Delta^3 f(x_0) - \frac{1}{4}\Delta^4 f(x_0) + \cdots\right\}$$

If interval of differencing is unity,

$$D \equiv \left\{\Delta - \frac{1}{2}\Delta^2 + \frac{1}{3}\Delta^3 - \frac{1}{4}\Delta^4\right\}$$

(neglecting higher order differencing).

2.8 Properties of Δ operator

We must carefully note the difference between the expressions $E^2 y_x$ and $(Ey_x)^2$. Here $E^2 y_x$ means y_{x+2h} (if h is the interval of differencing), whereas $(Ey_x)^2$ means $(y_{x+h})^2 = y^2_{x+h}$ and certainly y_{x+2h} is not the same as y^2_{x+h}.

We may also note the following properties of Δ operator:

1. $\Delta\{f(x) \pm \phi(x)\} = \Delta f(x) \pm \Delta\phi(x)$.
2. $\Delta\{cf(x)\} = c\Delta f(x)$, where c is a constant.
3. $\Delta c = 0$.
4. $\Delta\{f(x)\cdot\phi(x)\} = f(x+h)\phi(x+h) - f(x)\phi(x)$

$$= f(x+h)\{\phi(x+h) - \phi(x)\} + \phi(x)\{f(x+h) - f(x)\}$$
$$= f(x+h)\Delta\phi(x) + \phi(x)\Delta f(x).$$

Also

$$\Delta\{f(x)\phi(x)\} = \phi(x+h)\Delta f(x) + f(x)\Delta\phi(x)$$

5. $\Delta\left\{\dfrac{f(x)}{\phi(x)}\right\} = \dfrac{f(x+h)}{\phi(x+h)} - \dfrac{f(x)}{\phi(x)}$

$$= \frac{f(x+h)\phi(x) - f(x)\phi(x+h)}{\phi(x+h)\phi(x)}$$
$$= \frac{\phi(x)\{f(x+h) - f(x)\} - f(x)\{\phi(x+h) - \phi(x)\}}{\phi(x+h)\phi(x)}$$
$$= \frac{\phi(x)\Delta f(x) - f(x)\Delta\phi(x)}{\phi(x+h)\phi(x)} \quad [\text{where } \phi(x) \neq 0].$$

2.9 Finite difference on polynomials

The function $y = a_0 + a_1x + a_2x^2 + \cdots + a_nx^n$, $a_n \neq 0$ is a rational integral function of the nth degree in x, where the indices are positive integers, n being the greatest, and $a_0, a_1, a_2, \ldots, a_n$ are constants of which $a_n \neq 0$ and others are unrestricted constants.

A rational integral function is also called a polynomial. The above expression may be called a polynomial of degree n in x and may be denoted by $\phi_n(x)$.

Let us consider a simple polynomial of degree 3 in x, $\phi_3(x) = x^3 + 2x^2 + x + 1$. The interval of differencing is h.

$$\begin{aligned}\Delta\phi_3 &= \phi(x+h) - \phi(x)\\ &= (x+h)^3 + 2(x+h)^2 + (x+h) + 1 - x^2 - 2x^2 - x - 1\\ &= 3hx^2 + (3h^2 + 4h)x + (h^3 + 2h^2 + h)\\ &\equiv b_0x^2 + b_1x + b_2\end{aligned}$$

where $b_0 = 3h$, $b_1 = 3h^2 + 4h$, $b_2 = h^3 + 2h^2 + h$.

See that $\Delta\phi_3(x)$ yields a polynomial of degree 2 in x, i.e. one less than the given polynomial $\phi_3(x)$.

$$\begin{aligned}\Delta^2\phi_3(x) &= b_0(x+h)^2 + b_1(x+h) + b_2 - b_0x^2 - b_1x - b_2\\ &= 2b_0\cdot h\cdot x + b_1h\\ &= 2h(3h)x + h(3h^2 + 4h) \quad [\text{substituting the values of } b_0 \text{ and } b_1]\\ &= 3!h^2x + h^2(3h + 4)\end{aligned}$$

$\therefore\ \Delta^2\phi_3(x)$ gives a polynomial of degree 1 in x. The degree of polynomial is reduced by 1 at each stage of differencing.

$$\Delta^3\phi_3(x) = 3!h^2(x+h) + h^2(3h+4) - 3!h^2x - h^2(3h+4)$$
$$= 3!h^3 \quad [= \text{constant}]$$

Thus, the third order finite difference on a polynomial of degree 3 will be a constant and the fourth order difference of $\phi_3(x)$ will therefore be zero. In example 2.3, Table 2.31, the entry in the difference table is $y_x = 2x^2 + 3x$, which is a polynomial of degree 2, in x. The second order difference $\Delta^2 y_x$ is therefore constant in the table of differences. We may note that in example 2.3, h is taken to be 2. Therefore, second order differences are $2(h^2).2! = 2.2^2.2! = 16$ throughout the column of $\Delta^2 y_x$.

We now propose to give a formal proof to the following theorem:

THEOREM 2.9.1. If $\phi_n(x)$ be a polynomial of nth degree in x, $\Delta^n\phi_n(x)$ will be a constant.

Proof. Let

$$\phi_n(x) = a_nx^n + a_{n-1}x^{n-1} + \cdots + a_1x + a_0$$

be a polynomial of degree n in x, the ($a_n \neq 0$) unit of differencing being h (say)

$$\Delta\phi_n(x) = a_n\{(x+h)^n - x^n\} + a_{n-1}\left\{(x+h)^{n-1} - x^{n-1}\right\} + \cdots$$
$$+ a_1\{(x+h) - x\} + (a_0 - a_0)$$
$$= a_n\binom{n}{1}hx^{n-1} + \left\{a_n\binom{n}{2}h^2 + a_{n-1}\binom{n-1}{1}h\right\}x^{n-2} + \cdots + a_1h$$
$$\equiv b_{n-1}x^{n-1} + b_{n-2}x^{n-2} + \cdots + b_0 \quad \text{(say)}$$

where

$$b_{n-1} = \binom{n}{1}ha_n$$
$$b_{n-2} = \binom{n}{2}h^2a_n + \binom{n-1}{1}ha_{n-1}, \quad \text{etc.}$$

Thus, the first order difference of $\phi_n(x)$ reduces the degree of polynomial $\phi(x)$ by 1. Let us assume that the mth order ($m < n$) difference of $\phi_n(x)$ reduces the degree of polynomial by m. In other words, $\Delta^m\phi_n(x)$ gives a polynomial of degree $(n-m)$ in x.

Thus,

$$\Delta^m\phi_n(x) = c_{n-m}x^{n-m} + c_{n-m-1}x^{n-m-1} + \cdots + c_0$$
$$\Delta^{m+1}\phi_n(x) = \Delta\{\Delta^m\phi_n(x)\}$$
$$= c_{n-m}\left\{(x+h)^{n-m} - x^{n-m}\right\}$$
$$+ c_{n-m-1}\left\{(x+h)^{n-m-1} - x^{n-m-1}\right\} + \cdots$$
$$+ c_1\{(x+h) - x\} + (c_0 - c_0)$$
$$= \binom{n-m}{1}hc_{n-m}x^{n-m-1}$$
$$+ \left\{\binom{n-m}{2}h^2c_{n-m} + \binom{n-m-1}{1}hc_{n-m-1}\right\}x^{n-m-2} + \cdots + c_1h$$
$$\equiv d_{n-m-1}x^{n-m-1} + d_{n-m-2}x^{n-m-2} + \cdots + d_0 \quad \text{(say)}$$

where

$$d_{n-m-1} = c_{n-m}\binom{n-m}{1}h$$

$$d_{n-m_1} = \left\{c_{n-m}\binom{n-m}{2}h^2 + c_{n-m-1}\binom{n-m-1}{1}h\right\}$$

....

$$d_0 = c_1 h$$

Thus, $\Delta^{m+1}\phi_n(x)$ gives a polynomial of degree $n-m-1$ in x. In other words, $\Delta^{m+1}\phi_n(x)$ reduces the degree of polynomial by $(m+1)$, provided $\Delta^m\phi_n(x)$ has already reduced the degree of polynomial by m. Since we have proved that $\Delta\phi_n(x)$ has produced a polynomial of degree $(n-1)$ in x and $\Delta^2\phi_n(x)$ has given another polynomial of degree $(n-2)$ in x, so will $\Delta^3\phi_n(x)$ generate a polynomial of degree $(n-3)$ in x. Hence by mathematical induction, we may say that $\Delta^n\phi_n(x)$ will produce a polynomial of degree $n-n$ (i.e. 0) in x, which is a constant and higher order differences of $\phi_n(x)$ will all be zero.

If, however, the function y_x be other than a polynomial, the nth order finite difference will not be constant, however great n may be.

EXAMPLE 2.9.1. Find the nth order difference of C^x.

$$\Delta C^x = C^{x+h} - C^x \quad \text{[interval of differencing being } h\text{]}$$

$$= C^x(C^h - 1), \quad [= \text{some constant} \times C^x]$$

$$\Delta^2 C^x = (C^h - 1)\Delta C^x$$

$$= (C^h - 1)^2 C^x$$

....

Proceeding in this way, and noting the symmetry, we may finally write

$$\Delta^n C^x = (c^h - 1)^n \cdot C^x,$$

which is still a function of x.

NOTE 2.9.1. Unless otherwise stated, the interval of differencing is unity.

THEOREM 2.9.2. Show that

$$f(x_0 + nh) = f(x_0) + \binom{n}{1}\Delta f(x_0) + \binom{n}{2}\Delta^2 f(x_0) + \cdots + \binom{n}{n}\Delta^n f(x_0)$$

Proof.

$$\text{LHS} = f(x_0 + nh) = E^n f(x_0) = (1+\Delta)^n f(x_0)$$

$$= \left\{1 + \binom{n}{1}\Delta + \binom{n}{2}\Delta^2 + \cdots + \binom{n}{r}\Delta^r + \cdots + \binom{n}{n}\Delta^n\right\} f(x_0)$$

$$= f(x_0) + \binom{n}{1}\Delta f(x_0) + \binom{n}{2}\Delta^2 f(x_0) + \cdots + \binom{n}{r}\Delta^r f(x_0) + \cdots + \binom{n}{n}\Delta^n f(x_0)$$

where n is any positive integer.

2.10 Separation of symbols

In obtaining the value of $E^n f(x_0)$ in terms of $f(x_0)$ and its differences, we have used the symbolic relation $E \equiv 1 + \Delta$ and have expanded $(1+\Delta)^n$ by the binomial theorem without introducing the function $f(x_0)$ from both sides of the identity

$$E^n f(x_0) = (1 + \Delta)^n f(x_0)$$

is known as the method of separation of symbols. This enables us to establish readily many relations involving $f(x)$ and differences of $f(x)$.

EXAMPLE 2.10.1. Use the method of separation of symbols to prove the following identities:

(a) $u_0 + \dfrac{u_1 x}{1!} + \dfrac{u_2 x^2}{2!} + \cdots \infty = e^x \left[u_0 + x\Delta u_0 + \dfrac{x^2}{2!}\Delta^2 u_0 + \dfrac{x^3}{3!}\Delta^3 u_0 + \cdots \right].$

(b) $u_0 - u_1 + u_2 - u_3 + \cdots \infty = \dfrac{1}{2}u_0 - \dfrac{1}{4}\Delta u_0 + \dfrac{1}{8}\Delta^2 u_0 - \dfrac{1}{16}\Delta^3 u_0 + \cdots \infty.$

(c) $u_1 x + u_2 x^2 + u_3 x^3 + \cdots = \dfrac{x}{1-x}u_1 + \dfrac{x^2}{(1-x)^2}\Delta u_1 + \dfrac{x^3}{(1-x)^3}\Delta^2 u_1 + \cdots.$

Solution:

(a)
$$\begin{aligned}
\text{LHS} &= u_0 + \frac{xEu_0}{1!} + \frac{x^2E^2u_0}{2!} + \cdots \\
&= \left\{1 + \frac{xE}{1!} + \frac{x^2E^2}{2!} + \cdots\right\} u_0 \\
&= e^{xE} u_0 \\
&= e^{x(1+\Delta)} u_0 \\
&= e^x \cdot e^{x\Delta} u_0 \\
&= e^x \left\{1 + \frac{x\Delta}{1!} + \frac{x^2\Delta^2}{2!} + \frac{x^3\Delta^3}{3!} + \cdots\right\} u_0 \\
&= e^x \left\{1 + \frac{x}{1!}\Delta u_0 + \frac{x^2}{2!}\Delta^2 u_0 + \frac{x^3}{3!}\Delta^3 u_0 + \cdots\right\} \\
&= \text{RHS}
\end{aligned}$$

(b)
$$\begin{aligned}
\text{LHS} &= u_0 - Eu_0 + E^2 u_0 - \cdots \\
&= (1 - E + E^2 - E^3 + \cdots) u_0 \\
&= (1 + E)^{-1} u_0 \\
&= (2 + \Delta)^{-1} u_0 \\
&= \frac{1}{2}\left(1 + \frac{\Delta}{2}\right)^{-1} u_0 \\
&= \frac{1}{2}\left[1 - \frac{\Delta}{2} + \frac{1}{2^2}\Delta^2 - \frac{1}{2^3}\Delta^3 + \cdots\right] u_0 \\
&= \frac{1}{2}u_0 - \frac{1}{4}\Delta u_0 + \frac{1}{8}\Delta^2 u_0 - \frac{1}{16}\Delta^3 u_0 + \cdots \\
&= \text{RHS}
\end{aligned}$$

(c) LHS $= u_1 x + x^2 E u_1 + x^3 E^2 u_1 + \cdots$

$$= x\left[1 + xE + x^2E^2 + \cdots\right] u_1$$

$$= x(1 - xE)^{-1} u_1$$

$$= x\{1 - x(1+\Delta)\}^{-1} u_1$$

$$= x\{(1 - x) - x\Delta\}^{-1} u_1$$

$$= \frac{x}{1-x}\left(1 - \frac{x}{1-x}\Delta\right)^{-1} u_1$$

$$= \frac{x}{1-x}\left[1 + \frac{x}{1-x}\Delta + \frac{x^2}{(1-x)^2}\Delta^2 + \cdots\right] u_1$$

$$= \frac{x}{1-x}u_1 + \frac{x^2}{(1-x)^2}\Delta u_1 + \frac{x^3}{(1-x)^3}\Delta^2 u_1 + \cdots$$

$$= \text{RHS}$$

Example 2.10.2.

(a) Show that

$$u_{2n} - \binom{n}{1} 2 \cdot u_{2n-1} + \binom{n}{2} 2^2 \cdot u_{2n-2} - \cdots + (-2)^n u_n = (-1)^n \cdot (c - 2an)$$

where $u_x = ax^2 + bx + c$;

(b) $u_x - u_{x+1} + u_{x+2} - u_{x+3} + \cdots = \frac{1}{2}\left[u_{x-\frac{1}{2}} - \frac{1}{8}\Delta^2 u_{x-\frac{3}{2}} + \frac{1.3}{2!8^2}\Delta^4 u_{x-\frac{5}{2}} - \cdots\right]$;

(c) $u_x = u_{x-1} + \Delta u_{x-2} + \Delta^2 u_{x-3} + \cdots + \Delta^{n-1} u_{x-n} + \Delta^n u_{x-n}$.

Solution:

(a) LHS $= \left[E^n - \binom{n}{1} \cdot 2 \cdot E^{n-1} + \binom{n}{2} \cdot 2^2 E^{n-2} - \cdots + (-1)^n \cdot 2^n\right] u_n$

$$= (E-2)^n u_n = (\Delta - 1)^n u_n \quad [\because E \equiv 1 + \Delta]$$

$$= (-1)^n \left\{1 - \binom{n}{1}\Delta + \binom{n}{2}\Delta^2\right\} u_n$$

[Since u_n is a polynomial of degree 2 in n, we have retained differences up to order 2 in the above binomial expansion]

$$= (-1)^n \left\{u_n - n\Delta u_n + \frac{n(n-1)}{2!}\Delta^2 u_n\right\}$$

Now,

$$u_n = an^2 + bn + c$$

$$\Delta u_n = a(n+1)^2 + b(n+1) + c - an^2 - bn - c \quad \text{[interval of differencing = 1]}$$

$$= 2an + (a + b)$$

$$\Delta^2 u_x = 2a(n+1) + (a+b) - 2ax - (a+b)$$

$$= 2a$$

$$\therefore \text{LHS} = (-1)^n \left\{an^2 + bn + c - n(2an + a + b) + \frac{n(n-1)}{2} \cdot 2a\right\}$$

$$= (-1)^n \left\{(an^2 - 2an^2 + an^2) + (bn - bn) + c - 2an\right\}$$

$$= (-1)^n \cdot (c - 2an)$$

$$= \text{RHS}$$

(b) $$\text{RHS} = \frac{1}{2}\left[E^{-\frac{1}{2}} - \frac{1}{8}\Delta^2 E^{-\frac{3}{2}} + \frac{1.3}{2!}\frac{1}{8^2}\Delta^4 E^{-\frac{5}{2}} - \frac{1.3.5}{3!}\frac{1}{8^3}\Delta^6 E^{-\frac{7}{2}} + \cdots\right] u_x$$

$$= \frac{1}{2}E^{-\frac{1}{2}}\left[1 + \left(-\frac{1}{2}\right)\left(\frac{1}{4}\frac{\Delta^2}{E}\right) + \frac{\left(-\frac{1}{2}\right)\left(-\frac{3}{2}\right)}{2!}\left(\frac{1}{4}\frac{\Delta^2}{E}\right)^2 + \frac{\left(-\frac{1}{2}\right)\left(-\frac{3}{2}\right)\left(-\frac{5}{2}\right)}{3!}\left(\frac{1}{4}\frac{\Delta^2}{E}\right)^3 + \cdots\right] u_x$$

$$= \frac{1}{2}E^{-\frac{1}{2}}\left(1 + \frac{1}{4}\frac{\Delta^2}{E}\right)^{-(1/2)} u_x = \frac{1}{2}E^{-\frac{1}{2}}\left(\frac{4E + \Delta^2}{4E}\right)^{-(1/2)} u_x$$

$$= (4E + \Delta^2)^{-(1/2)} u_x = \left\{4E + (E-1)^2\right\}^{-\frac{1}{2}} u_x$$

$$= (E^2 + 2E + 1)^{-(1/2)} u_x = (E+1)^{-1} u_x$$

$$= (1 - E + E^2 - E^3 + \cdots) u_x$$

$$= u_x - u_{x+1} + u_{x+2} - u_{x+3} + \cdots$$

$$= \text{LHS}$$

(c) $$\text{LHS} = u_x = u_x - \Delta^n u_{x-n} + \Delta^n u_{x-n}$$

$$= \left(1 - \frac{\Delta^n}{E^n}\right) u_x + \Delta^n u_{x-n} = (E^n - \Delta^n) u_{x-n} + \Delta^n u_{x-n}$$

$$= \left(\frac{E^n - \Delta^n}{E - \Delta}\right) u_{x-n} + \Delta^n u_{x-n} \; [\because E - 4 \equiv 1]$$

$$= \left(E^{n-1} + E^{n-2}\Delta + E^{n-3}\Delta^2 + \cdots\right) u_{x-n} + \Delta^n u_{x-n}$$

$$= u_{x-1} + \Delta u_{x-2} + \Delta^2 u_{x-3} + \cdots + \Delta^{n-1} u_{x-n} + \Delta^m u_{x-n}$$

$$= \text{RHS}$$

2.11 Factorial notation

The product of m consecutive factors, each at a constant difference, say h (*unless it is specified to be unity*), the first factor being x, is called a factorial function, and is denoted by $x^{(m)}$. Thus,

$$x^{(m)} = x(x-h)(x-2h)\cdots(x - \overline{m-1}h) \qquad (2.11.1)$$

In particular, if $h = 1$

$$x^{(m)} = x(x-1)(x-2)\cdots(x-m+1) \tag{2.11.2}$$

If interval of differencing be h

$$\begin{aligned}\Delta x^{(m)} &= (x+h)^{(m)} - x^{(m)} \\ &= (x+h)x(x-h)\cdots(x-\overline{m-2}h) - x(x-h)\cdots(x-\overline{m-1}h) \\ &= x(x-h)\cdots(x-\overline{m-2}h)\{(x+h)-(x-\overline{m-1}h)\} \\ &= mhx^{(m-1)}\end{aligned}$$

Similarly, we may show

$$\Delta^2\Delta^{(m)} = mh\Delta x^{(m-1)} = \{mh\}\cdot\{(m-1)h\}x^{(m-2)}$$

Proceeding in this way, and noting the symmetry in the result at each stage, we may write

$$\Delta^m x^{(m)} = (mh)(\overline{m-1}h)(\overline{m-2}h)\cdots(1h) = m!h^m$$

We note that the result of differencing is analogous to that of differentiating x^m. If the interval of differencing be unity

$$\Delta^m x^{(m)} = m!$$

Again from

$$x^{(m)} = x(x-h)(x-2h)\cdots(x-\overline{m-1}h) = (x-\overline{m-1}h)x^{(m-1)} \tag{2.11.3}$$

We have for $m = 0$

$$x^{(0)} = (x+h)x^{(-1)}$$

By convention, we take $x^{(0)} = 1$.

$$\therefore x^{(-1)} = \frac{1}{x+h} = \frac{1}{(x+h)^{(1)}}$$

Again by putting $m = -1$ in (2.11.3), we get

$$x^{(-1)} = (x+2h)\cdot x^{(2)} \Rightarrow \frac{1}{x+h} = (x+2h)\cdot x^{(-2)}$$

$$\therefore x^{(-2)} = \frac{1}{(x+h)(x+2h)} = \frac{1}{(x+2h)^{(2)}}$$

In general, we get

$$x^{(-m)} = \frac{1}{(x+h)(x+2h)\cdots(x+mh)} = \frac{1}{(x+mh)^{(m)}} \tag{2.11.4}$$

If the interval of differencing be unity

$$x^{(-m)} = \frac{1}{(x+1)(x+2)\cdots(x+m)} = \frac{1}{(x+m)^{(m)}} \tag{2.11.5}$$

Also

$$\Delta x^{(-m)} = (x+h)^{(-m)} - x^{(-m)}$$

$$= \frac{1}{(x+2h)(x+3h)\cdots(x+\overline{m+1}h)} - \frac{1}{(x+h)\cdots(x+mh)}$$

$$= \frac{(x+h)-(x+\overline{m+1}h)}{(x+h)(x+2h)\cdots(x+mh)(x+\overline{m+1}h)}$$

$$= (-1)\cdot mh\cdot x^{(-m-1)}$$

$$\Delta^2 x^{(-m)} = (-1)mh\Delta x^{(-m-1)}$$

$$= (-1)^2\cdot(mh)(\overline{m+1}h)x^{(-m-2)}$$

Proceeding in this way, we may show that in general

$$\Delta^m x^{(-m)} = (-1)^m\cdot m(m+1)\cdots(m+m-1)h^m x^{(-m-m)}$$

Thus, for negative index for factorial function also, the operator $\Delta \equiv D$.

2.12 To express any polynomial $f(x)$, as sum of several factorial functions

Example 2.12.1. Express $f(x) = 3x^3 - 4x^2 + 3x - 11$ and its differences as sums of several factorial functions.

Solution: Let

$$f(x) = 3x^3 - 4x^2 + 3x - 11$$

$$= Ax(x-1)(x-2) + Bx(x-1) + Cx + D$$

Putting $x = 0, 1, 2$, and 3, we obtain

$$f(0) = -11 = D$$

$$f(1) = -9 = C + D \Rightarrow C = -9 - D = -9 + 11 = 2$$

$$f(2) = 3 = 2B + 2C + D \Rightarrow 2B = 3 - D - 2C = 3 + 11 - 4 = 10$$

$$\therefore B = 5$$

$$f(3) = 43 = 6A + 6B + 3C + D \Rightarrow 6A = 43 - 30 - 6 + 11 = 18$$

$$\therefore A = 3$$

$$\therefore f(x) = 3x(x-1)(x-2) + 5x(x-1) + 2x - 11$$

$$= 3x^{(3)} + 5x^{(2)} + 2x^{(1)} - 11$$

$$\Delta f(x) = 3\Delta x^{(3)} + 5\Delta x^{(2)} + 2\Delta x^{(1)}$$

$$= 3.3x^{(2)} + 5.2x^{(1)} + 2 \quad \text{[interval of differencing 1]}$$

$$= 9x(x-1) + 10x + 2 = 9x^2 + x$$

$$\Delta^2 f(x) = 9\Delta x^{(2)} + 10\Delta x^{(1)} = 18x + 10$$

$$\Delta^3 f(x) = 18 (= 3 \times 3!)$$

The usefulness of factorial function is very clear from the ease of computations of differences of a polynomial function.

An alternative method for expressing $\phi_n(x)$ in factorial notation is by the use of detached coefficients. By this method any such function can be written down in the form:

$A_0x^{(n)} + A_1x^{(n-1)} + \dots + A_{n-1}x^{(1)} + A_n$

The principal of the method can best be illustrated by an example.

Example 2.12.2. Write down $3x^3 - 4x^2 + 3x - 11$ in factorial notation.

Let

$$f(x) = A_0x^{(3)} + A_1x^{(2)} + A_2x^{(1)} + A_3$$
$$= A_0x(x-1)(x-2) + A_1x(x-1) + A_2x + A_3$$

If we divide $f(x)$ by x, the quotient will be $3x^2 - 4x + 3$ and the remainder -11 $(= A_3)$.

Divide $3x^2 - 4x + 3$ by $x - 1$:

$$\begin{array}{r|l|l} x-1 & 3x^2 - 4x + 3 & 3x-1 \\ & 3x^2 - 3x & \\ \hline & \quad -x + 3 & \\ & \quad -x + 1 & \\ \hline & \qquad\quad 2 & \end{array} \quad \text{(remainder)} \quad (= A_2)$$

Divide $3x - 1$ by $x - 2$:

$$\begin{array}{r|l|l} x-2 & 3x - 1 & 3 \quad (= A_0) \\ & 3x - 6 & \\ \hline & \quad\ \ 5 & \end{array} \quad \text{(remainder)} \quad (= A_1)$$

The above process may be appreciably shortened by adopting the following procedure:

1. omit x^3, x^2, x and work on coefficients alone;
2. change the sign of the constant term in $x - 1$, $x - 2$, $x - 3$, so that addition takes the place of subtraction.

The remainder can then be easily obtained.

$$\begin{array}{lcccl} \text{Thus, 1.} & 3 & -4 & 3 & \big|\ -11 \\ & 0 & 3 & -1 & \big|\ (= A_3) \\ \hline 2. & 3 & -1 & 2 & \\ & 0 & 6 & (= A_2) & \\ \hline & 3 & 5 & & \\ & (= A_0) & (A_1) & & \end{array}$$

gives $f(x) = 3x^{(3)} + 5x^{(2)} + 2x^{(1)} - 11$.

Example 2.12.3. Find the function $f(x)$, whose first difference is $x^3 + 3x^2 + 5x + 12$.

Solution: Let $f(x)$ be the function, whose

$\Delta f(x) = x^3 + 3x^2 + 5x + 12$

We shall first express $\Delta f(x)$ in factorial notation.

$$\begin{array}{lcccl} 1. & 1 & 3 & 5 & \big|\ 12 \\ & 0 & 1 & 4 & \\ \hline 2. & 1 & 4 & 9 & \\ & 0 & 2 & & \\ \hline & 1 & 6 & & \end{array}$$

$$\therefore \Delta f(x) = x^{(3)} + 6x^{(2)} + 9x^{(1)} + 12$$

$$\therefore f(x) = \frac{1}{\Delta}x^{(3)} + 6\cdot\frac{1}{\Delta}x^{(2)} + 9\cdot\Delta x^{(1)} + \Delta\cdot 12$$

$$= \frac{1}{4}x^{(4)} + \frac{6}{3}x^{(3)} + \frac{9}{2}x^{(2)} + 12x^{(1)} + k \quad [k \text{ is an arbitrary constant}]$$

$$= \frac{1}{4}x(x-1)(x-2)(x-3) + 2x(x-1)(x-2) + \frac{9}{2}x(x-1) + 12x + k$$

[NOTE: $\Delta x^{(m)} = mx^{(m-1)} \Rightarrow x^{(m)} = m\cdot\frac{1}{\Delta}x^{(m-1)} \Rightarrow \frac{1}{m}x^{(m)} = \frac{1}{\Delta}x^{(m-1)}$.]

2.13 Differences of zero

The name *differences of zero* is given to the quantities $\Delta^n x^m\big]_{x=0}$, and is generally written as $\Delta^n O^m$. We may note that n and m are both positive integers.

THEOREM 2.13.1. Assume n and m are both positive integers. Then

$$\Delta^n O^m = n^m - \binom{n}{1}(n-1)^m + \binom{n}{2}(n-2)^m - \cdots + (-1)^{m+n-1}\binom{n}{n-1} + (-1)^n O^m$$

Proof. We know

$$\Delta^n x^m = (E-1)^n x^m = \left[E^n - \binom{n}{1}E^{n-1} + \binom{n}{2}E^{n-2} - \cdots + (-1)^n\right]x^m$$

$$= (x+n)^m - \binom{n}{1}(x+n-1)^m + \binom{n}{2}(x+n-2)^m - \cdots + (-1)^n x^m$$

Putting $x = 0$

$$\Delta^n O^m = n^m - \binom{n}{1}(n-1)^m + \binom{n}{2}(n-2)^m - \binom{n}{3}(n-3)^m + \cdots + (n-1)^n O^m$$

THEOREM 2.13.2. If $f(x)$ is a polynomial in x of degree m, then

$$x^m = \sum_{r=0}^{r=m} \frac{x^{(r)}}{r!}\Delta^r O^m$$

Proof.

$$f(x) = E^x f(0) = (1+\Delta)^x f(0)$$

$$= \left\{1 + \binom{x}{1}\Delta + \binom{x}{2}\Delta^2 + \cdots + \binom{x}{x}\Delta^x\right\} f(0)$$

$$= f(0) + \binom{x}{1}\Delta f(0) + \binom{x}{2}\Delta^2 f(0) + \cdots + \binom{x}{x}\Delta^x f(0)$$

Writing $f(x) = x^m$, we have

$$x^m = O^m + \binom{x}{1}\Delta O^m + \binom{x}{2}\Delta^2 O^m + \cdots + \binom{x}{x}\Delta^r O^m$$

$$= \sum_{r=0}^{r=m}\binom{x}{r}\Delta^r O^m$$

EXERCISES

1. If $f(x) = x^3 - 4x^2 + 5x - 13$. Compute the entry $y = f(x)$, for $x = 1, 3, 5, 7, 9, 11$ and prepare a diagonal difference table. Explain why third order differences $\Delta^3 y_x$ are constant throughout. Verify that this constant is 48.
2. Given $y_x = 2x^3 - x^2 + 5x + 1$. Prepare horizontal difference table by taking the values of argument $x = 0, 1, 2, 3, 4, 5$. Show that the third order differences are constant and equal to 12.
3. The following table gives the values of entry y_x for some equispaced values of the argument x.
   y_x : 0.09983, 0.14944, 0.19867, 0.24740, 0.29552, 0.34290, 0.38945, 0.43497, 0.47943, 0.52269, 0.56464, 0.60519, 0.64422.
   Excepting one erroneous entry, which lies somewhere near the middle, all other entries are free from error. By preparing a horizontal difference table, locate the erroneous entry and rectify the error.
   [*Ans:* Erroneous entry 0.38945; correct entry 0.38942.]
4. Find:
   (a) Δc^x. [*Ans:* $(c^h - 1)c^x$.]
   (b) $\Delta(ka^{bx})$. [*Ans:* $(a^{bh} - 1)ka^{bx}$.]
   (c) $\Delta \sin 2x$. [*Ans:* $2\cos(2x + h) \cdot \sin h$.]
   (d) $\Delta \cosh(cx + d)$. [*Ans:* $2\sinh\left(cx + d + \frac{ch}{2}\right)\sinh\frac{ch}{2}$.]
   (e) $\Delta \log cx$. [*Ans:* $\log\frac{u + h}{x}$.]
   (f) $\Delta^2(ka^{bx})$. [*Ans:* $(a^{bh} - 1)^2 ke^{bx}$ (interval of differencing being h in all these cases).
5. (a) Evaluate
   i. $\frac{\Delta^2}{E}x^4$. [*Ans:* $(x + h)^4 + (x + h)64 - 2x^4$.]
   ii. $\frac{\Delta^2 x^3}{Ex}$. [*Ans:* $\frac{(x + 2h)^{-3} - 2(x + h)^3 + x^3}{(x + h)}$.]
   iii. $\frac{\Delta^2}{E}c^x \cdot \frac{Ec^x}{\Delta^2 c^x}$. [*Ans:* c^x.]
   iv. $\frac{\Delta}{E}\sin 2x$. [*Ans:* $2\sinh\cos(2x - h)$.]
   v. $\frac{\Delta^2}{E}\sin(x - h)$.
   [*Ans:* $\sin(x + 2h) - 2\sin(x + h) + \sin x$ (interval of differencing is h).]
   (b) Show that, if $f(E)$ is a polynomial in E, $f(E)a^x = a^x f(a)$.
6. (a) Find $\Delta f(x)$, where $f(x) = (2x + 1)(2 - 3x)$. [*Ans:* $-5 - 12x$.]
   (b) Find $\Delta^n f(x)$, where $f(x)$ is
   i. $2x^n + 3x^{n-1}$, [*Ans:* $2(n!)$.]
   ii. e^{kx+c}. [*Ans:* $(e^{k+c} - 1)^n e^{kn+c}$.]
   (c) Find $\Delta f(x)$, where $f(x) = \frac{2x - 1}{3 - x}$.
   [*Ans:* $\frac{2x + 1}{2 - x} - \frac{2x - 1}{3 - x}$ [unit of differencing 1 in the above cases (a), (b), (c)].
7. Express the following functions in factorial notation:
   (a) $3x^4 - 5x^3 + 17x^2 + 9x - 12$. [*Ans:* $3x^{(4)} + 13x^{(3)} + 23x^{(2)} + 24x^{(1)} - 12$.]
   (b) $2x^4 - 7x^2 + 5x - 13$. [*Ans:* $2x^{(4)} + 12x^{(3)} + 7x^{(2)} - 13$.]
   (c) $x^4 - 12x^3 + 42x^2 - 30x + 9$. [*Ans:* $x^{(4)} - 6x^{(3)} + 13x^{(2)} + x^{(1)} + 9$.]

8. Obtain the function whose first difference is
   (a) $2x^3 + 5x^2 - 6x + 13$, [*Ans:* $\frac{1}{2}x^{(4)} + \frac{11}{3}x^{(3)} + \frac{1}{2}x^{(2)} + 13x^{(1)} + k$.]
   (b) e^x, [*Ans:* $\frac{e^x}{e-1}$.]
   (c) c^x. [*Ans:* $\frac{c^x}{c-1}$.]
9. Prove that

$$\Delta(\tan^{-1} x) = \tan^{-1}\left\{\frac{h}{1 + xh + x^2}\right\},$$

where h is the interval of differencing.

10. Show that, if Δ operates on n, then $\Delta\binom{n}{x+1} = \binom{n}{x}$ and hence

$$\sum_{n=1}^{N}\binom{n}{x} = \binom{N+1}{x+1} - \binom{1}{x+1}$$

11. Show that the rth difference of a polynomial of nth degree in x, is a polynomial of $(n - r)$th degree, if $r < n$. What happens when: (a) $r = n$, (b) $r > n$?
12. Define the function $x^{(m)}$ and $x^{(-m)}$. Obtain their nth differences, distinguishing between the cases when $n < m$, $n = m$, $n > m$.
13. By constructing the table of differences, find:
    (a) sixth term of the series: 8, 12, 19, 29, 42, . . . , [*Ans:* 58.]
    (b) seventh term of the series 0, 0, 2, 6, 12, 20, · · · . [*Ans:* 30.]
14. Given

$$u_0 + u_8 = 1.9243,$$
$$u_1 + u_7 = 1.9590,$$
$$u_2 + u_6 = 1.9823,$$
$$u_3 + u_5 = 1.9956.$$

Find u_4, stating the assumptions made. [*Ans:* $u_4 = 0.99996 \simeq 1$.]
[*Hints:* Assume u_x to be a polynomial of degree 7 in x.]

15. If u_x be a function whose differences, when the increment of x is unity, are denoted by $\delta u_x, \delta^2 u_x, \ldots$ and by $\Delta u_x, \Delta^2 u_x, \ldots$ when the increment of x is n, then if

$$\delta^2 u_x, \delta^2 u_{x+1}, \delta^2 u_{x+2}, \ldots$$

are the GP with common ratio q, show that

$$\frac{\Delta u_x - n\delta u_x}{(q^n - 1) - n(q - 1)} = \frac{\delta^2 u_x}{(q - 1)^2}$$

16. Find the relation among α, β and γ in order that $\alpha + \beta x + \gamma x^2$ may be expressible in one term in the factorial notation. [*Ans:* $\beta^2 - \gamma^2 = 4\alpha\gamma$.]
[*Hints:* Let $\alpha + \beta x + \gamma x^2 \equiv \gamma(x + a_0)^{(2)} \equiv \gamma(x + a_0)(x + a_0 - 1) \equiv \gamma\left\{x^2 + x(2a_0 - 1) + a_0(a_0 - 1)\right\}$. Equating coefficients, etc.]
17. Use the method of separation of symbols to prove the following identities:
    (a) $\Delta^n u_{x-n} = u_x - \binom{n}{1}u_{x-1} + \binom{n}{2}u_{x-2} - \binom{n}{3}u_{x-3} + \cdots$.
    (b) $u_{x+n} = u_n + \binom{x}{1}\Delta u_{n-1} + \binom{x+1}{2}\Delta^2 u_{n-2} + \binom{x+2}{3}\Delta^3 u_{n-3} + \cdots$.

(c) $u_0 + \binom{x}{1}\Delta u_1 + \binom{x}{2}\Delta^2 u_2 + \binom{x}{3}\Delta^3 u_3 + \cdots$

$= u_x + \binom{x}{1}\Delta^2 u_{x-1} + \binom{x}{2}\Delta^4 u_{x-2} + \cdots.$

(d) $\Delta^n y_x = y_{x+n} - \binom{n}{1} y_{x+n-1} + \binom{n}{2} y_{x+n-2} - \cdots + (-1)^n y_x.$

(e) $u_x - \frac{1}{8}\Delta^2 u_{x-1} + \frac{1.3}{8.16}\Delta^4 u_{x-2} - \frac{1.3.5}{8.16.24}\Delta^6 u_{x-3} + \cdots$

$= u_{x+\frac{1}{2}} - \frac{1}{2}\Delta u_{x+\frac{1}{2}} + \frac{1}{4}\Delta^2 u_{x+\frac{1}{2}} - \frac{1}{8}\Delta^3 u_{x+\frac{1}{2}} + \cdots.$

(f) Sum to n terms $1.2\Delta x^n - 2.3\Delta^2 x^n + 3.4 \cdot \Delta^3 x^n - \cdots.$

(g) If p, q, r, s be the successive entries corresponding to equidistant arguments in a table, show that when the third differences are taken into account, the entry corresponding to the argument half-way between the arguments of q and r is $A + \frac{1}{24}B$, where A is the arithmetic mean of q and r and B is the arithmetic mean of $3q - 2p - s$ and $3r - 2s - p$.

[*Hints:* Let the equidistant arguments be x_0, $x_0 + h$, $x_0 + 2h$, and $x_0 + 3h$, h being the interval of dfferencing. The corresponding entries are p, q, r and s.
Then prepare the difference table as under:

x	$f(x)$	$\Delta f(x)$	$\Delta^2 f(x)$	$\Delta^3 f(x)$
x_0	p			
		$q - p$		
$x_0 + h$	q		$r - 2q + p$	
		$r - q$		$s - 3r + 3q - p$
$x_0 + 2h$	r		$s - 2r + q$	
		$s - r$		
$x_0 + 3h$	s			

The argument half-way between the arguments of q and r is

$\frac{1}{2}\{(x_0 + h) + (x_0 + 2h)\} = x_0 + \frac{3h}{2}$

Hence the required entry is $f\left(x_0 + \frac{3h}{2}\right)$.

Thus,

$$f\left(x_0 + \frac{3h}{2}\right) = E^{3/2} f(x_0) = (1 + \Delta)^{3/2} f(x_0)$$

$$= \left[1 + \frac{3}{2}\Delta + \frac{\frac{3}{2}\left(\frac{3}{2} - 1\right)}{2}\Delta^2 + \frac{\frac{3}{2}\left(\frac{3}{2} - 1\right)\left(\frac{3}{2} - 2\right)}{6}\Delta^3\right] f(x_0)$$

$$= f(x_0) + \frac{3}{2}\Delta f(x_0) + \frac{3}{8}\Delta^2 f(x_0) - \frac{1}{16}\Delta^3 f(x_0)$$

$$= p + \frac{3}{2}(q - p) + \frac{3}{8}(r - 2q + p) - \frac{1}{16}(s - 3r + 3q - p)$$

$$= \left(1 - \frac{3}{2} + \frac{3}{8} + \frac{1}{16}\right)p + \left(\frac{3}{2} - \frac{3}{4} - \frac{3}{16}\right)q + \left(\frac{3}{8} + \frac{3}{16}\right)r - \frac{1}{16}s$$

$$= -\frac{1}{16}p + \frac{9}{16}q + \frac{9}{16}r - \frac{1}{16}s = \frac{1}{2}(q + r) + \frac{1}{16}(q + r - p - s)$$

Now,

$$A = \text{A.M. of } q \text{ and } r = \frac{1}{2}(q + r)$$

$$\text{and } B = \text{A.M. of } 3q - 2p - s \text{ and } 3r - 2s - p$$

$$= \frac{1}{2}(3q - 2p - s + 3r - 2r - p) = \frac{3}{2}(q + r - p - s)$$

$$\text{or, } \frac{1}{24}B = \frac{1}{16}(q + r - p - s)$$

$$\therefore A + \frac{1}{24}B = \frac{1}{2}(q + r) + \frac{1}{16}(q + r - p - s) = f\left(x_0 + \frac{3}{2}h\right).]$$

18. Prove that if m is a positive integer,
 (a) $\Delta^2 x^{(m)} = m(m - 1)x^{(m-2)}$,
 (b) $\Delta^2 x^{(-m)} = m(m + 1)x^{(-m-2)}$,
 (c) $\dfrac{(x + 1)^{(m)}}{m!} = \dfrac{x^{(m)}}{m!} + \dfrac{x^{(m-2)}}{(m - 1)!}$.

19. Defining $\nabla f(x + 1) = f(x + 1) - f(x)$ show that the operators as given under are equivalent $\Delta E \equiv \Delta$, where $Ef(x) = f(x + 1)$ and $\Delta f(x) = f(x + 1) - f(x)$.
 [*Hints:*In some texts, backward difference operator $\nabla f(x + 1)$ is written as $\Delta_1 f(x + 1)$, the notation followed in this book.]

20. (a) Show that $\Delta^n O^m = n\left(\Delta^{n-1} O^{m-1} + \Delta^n O^{m-1}\right)$.
 [*Hints:*

$$\Delta^n x^m = \left[E^n - \binom{n}{1}E^{n-1} + \binom{n}{2}E^{n-2} - \cdots + (-)^n\right]x^m$$

$$= (x + n)^m - \binom{n}{1}(x + n - 1)^m + \binom{n}{2}(x + n - 2)^m - \cdots + (-1)^n x^m$$

Putting $x = 0$

$$\Delta^n O^m = n^m - \binom{n}{1}(n - 1)^m + \binom{n}{2}(n - 2)^m - \cdots + (-1)^n O^m$$

$$= n\left[(1 + n - 1)^{m-1} - \binom{n-1}{1}(1 + n - 2)^{m-1} + \binom{n-1}{2}(1 + n - 3)^n - \cdots\right]$$

$$= n\left[(E - 1)^{n-1} x^{m-1}\right]_{x=1} = n\Delta^{n-1} EO^{m-1} = n\Delta^{n-1}(1 + \Delta)O^{m-1}$$

$$= n\Delta^{n-1} O^{m-1} + n\Delta^n O^{m-1} = n(\Delta^{n-1} O^{m-1} + \Delta^n O^{m-1}.]$$

 (b) Hence show that

$$\frac{\Delta^n O^m}{n!} = \frac{n\Delta^n O^m}{n!} + \frac{\Delta^{n-1} O^{m-1}}{(n - 1)!}$$

21. Show that

$$\left(\frac{\Delta^2}{E}\right)e^x \times \frac{Ee^x}{\Delta^2 e^x} = e^x$$

22. Show that

$$f(x) - \binom{n}{1}f(x - y) + \binom{n}{2}f(x - 2y) - \cdots + (-1)^n\binom{x}{n}f(x - xy) = y^n \cdot n!$$

where $f(x)$ in a polynomial of degree n is x.
 [*Hints:* Unit of differencing is y. Expression

$$\left[1 - \binom{n}{1}E^{-1} + \binom{n}{2}E^{-2} - \cdots + (-1)^n\binom{x}{n}E^{-n}\right]f(x) = (1 - E^{-1})^n f(x), \text{ etc.}]$$

23. (a) $\Delta O^m - \frac{1}{2}\Delta^2 O^m + \frac{1}{3}\Delta^3 O^m - \cdots = 0$, if $m > 1$.
(b) $\Delta^n O^{n+1} = \frac{1}{2}n(n+1)\Delta^n O^n = \frac{1}{2}n(n+1)!$
(c) $(n+1)\Delta^n O^n = 2(\Delta^{n-1}O^n + \Delta^n O^n)$.

24. For all integral values of n,

$$f(a+nh) = f(a) + \binom{n}{1}\Delta f(a) + \binom{n}{2}\Delta^2 f(a) + \cdots + \binom{x}{n}\Delta^n f(a)$$

[*Hints:* $f(a+nh) = E^n f(a)$ [unit of differencing h] $= (1+\Delta)^n f(a)$, etc.]

25. Show that

$$u_x = u_0 + \binom{x}{1}\Delta u_0 + \binom{x}{2}\Delta^2 u_0 + \cdots$$

can be put in the form:

$$u_x = u_0 + x\Delta u_0 - xa\Delta^2 u_0 + xab\Delta^3 u_0 - xabc\Delta^4 u_0 + \cdots$$

where

$$a = 1 - \frac{1}{2}(1+x),\quad b = 1 - \frac{1}{3}(1+x),\quad c = 1 - \frac{1}{4}(1+x)$$

[*Hints:*

$$\binom{x}{2} = \frac{x(x-1)}{2!} = \frac{(-x)(1-x)}{2} = -x\left\{1 - \frac{1}{2}(1+x)\right\} = -ax$$

$$\binom{x}{3} = \frac{x(x-1)(x-2)}{3!} = \frac{(-1)^2 x(1-x)(2-x)}{3!} = x\left(\frac{1-x}{2}\right)\left(\frac{2-x}{3}\right)$$

$$= x\left\{1 - \frac{1}{2}(1+x)\right\}\left\{1 - \frac{1}{3}(1+x)\right\} = abx$$

$$\binom{x}{4} = \frac{x(x-1)(x-2)(x-3)}{4!} = \frac{(-x)(1-x)(2-x)(3-x)}{2.3.4}$$

$$= -x\left(\frac{1-x}{2}\right)\left(\frac{2-x}{3}\right)\left(\frac{3-x}{4}\right)$$

$$= -x\left\{1 - \frac{1}{2}(1+x)\right\}\left\{1 - \frac{1}{3}(1+x)\right\}\left\{1 - \frac{1}{4}(1+x)\right\} = -abc\,x;$$ etc.]

26. Establish

$$\Delta x^m - \frac{1}{2}\Delta^2 x^m + \frac{1.3}{2.4}\Delta^3 x^m - \frac{1.3.5}{2.4.6}\Delta^4 x^m + \cdots \text{ to } m \text{ terms}$$

$$= \left(x + \frac{1}{2}\right)^m - \left(x + \frac{1}{2}\right)^m.$$

27. (a) $\Delta O^m - \frac{1}{2}\Delta^2 O^m + \frac{1}{3}\Delta^3 O^m - \cdots = 0$, if $m > 1$.
(b) $\Delta^n O^{n+1} = \frac{1}{2}n(n+1)\Delta^n O^n = \frac{1}{2}n(n+1)!$.
(c) $(n+1)\Delta^n O^n = 2(\Delta^{n-1}O^n + \Delta^n O^n)$.

Chapter 3

Interpolation with Equal Intervals

Summary

3.1 Introduction

Let a function $f(x)$ be given for some values of argument

$x_0, x_1, x_2, \ldots, x_n$

The corresponding entries

$y_0, y_1, y_2, \ldots, y_n$

are also available, where

$y_i = f(x_i), \quad i = 0, 1, 2, \ldots, n$

By interpolation, we mean *the art of reading between the lines of the tabulated values of the entries.* In elementary mathematics, the term usually denotes the process of computing intermediate values of $f(x)$ from a set of given values of that function. Generally, however, the form of the function $f(x)$ is either not known, or if known, it is so complicated that it cannot be subjected to such operations as we are required to perform. In either of the cases, we may replace the given function $f(x)$ (unknown or otherwise) by another which can be more easily handled. Using a theorem due to Weierstrass, we may state: Every function which is continuous in an interval (x_0, x_n) can be represented in that interval, to any desired degree of accuracy, by a polynomial $\phi(x)$, such that

$|f(x) - \phi(x)| < \varepsilon$

for every value of x in (x_0, x_n) where ε is a preassigned positive quantity. The technique of interpolation is actually applied to the approximating polynomial $\phi(x)$. When $(n + 1)$

independent values of the argument

$x_0, x_1, x_2, \ldots, x_n$

are given with the corresponding entries

$y_0, y_1, y_2, \ldots, y_n$

we may propose to replace the known function $f(x)$ by the interpolating polynomial $\phi_n(x)$ of degree n in x; because in that case, we may determine $(n + 1)$ unknown constants of $\phi_n(x)$ from the $(n + 1)$ independent entries

$y_0, y_1, \ldots, y_n$

We only assume that the unknown function $f(x)$ is continuous in the given range (x_0, x_n).

In our present discussion, we shall confine our attention to two types of problems:

Type I

In this case, the values of the argument

$x_0, x_1, x_2, \ldots, x_n$

are equispaced at an interval of h, say. Thus,

$x_i = x_0 + ih, \quad i = 0, 1, 2, \ldots, n$

The corresponding entries

$y_i = f(x_i), \quad i = 0, 1, 2, \ldots, n$

are also given.

1. We are required to interpolate y for some x in $x_0 < x < x_0 + h$, or
2. We are required to interpolate y for some x in $x_{n-1}, x < x_n$.

If x is located somewhere near the middle of the tabulated values of the argument, we reserve our discussion for the present.

Type II

In this case, the values of the argument, for which the corresponding entries are available, are not equispaced, but when the value(s) of x, for which the entries are required to be interpolated, are inserted, the argument values become equispaced.

NOTE 3.1.1. The process of computing the value of the function outside the range of the given values of the variable is called *extrapolation*. This is outside our present area of study.

3.2 Newton's forward interpolation formula

Let $f(x)$ be given only for the $(n + 1)$ equispaced values of the argument,

$x_0, x_1, x_2, \ldots, x_n$

where

$x_i = x_0 + ih, \quad i = 0, 1, 2, \ldots, n$

The corresponding entries

$y_0, y_1, \ldots, y_n$

are also available. We are required to interpolate y for some x in $x_0 < x, x_0 + h$. Assuming $f(x)$ to be continuous in (x_0, x_n), we may replace $f(x)$ by a polynomial $\phi_n(x)$ of degree n

in x, such that

$f(x_i) = \phi_n(x_i), \quad i = 0, 1, 2, \ldots, n$

But at all other points

$f(x) = \phi_n(x) + R_n(x)$

where $R_n(x)$ is the error term or the remainder term in the interpolation formula. Ignoring $R_n(x)$ for the present we may write

$$f(x) \simeq \phi_n(x) = a_0 + a_1(x - x_0) + a_2(x - x_0)(x - x_1) + \cdots + a_n(x - x_0)(x - x_1)\cdots(x - x_{n-1}) \tag{3.2.1}$$

Putting $x = x_0$,

$f(x_0) = \phi_n(x_0) = a_0 \equiv y_0$

Putting $x = x_1$,

$f(x_1) = \phi_n(x_1) = a_0 + a_1(x_1 - x_0) = y_1$

$$\Rightarrow a_1 = \frac{y_1 - a_0}{x_1 - x_0} = \frac{y_1 - y_0}{h} = \frac{\Delta y_0}{h}$$

(Using forward difference notation in diagonal table)

Putting $x = x_2$

$f(x_2) = \phi_n(x_2) = a_0 + a_1(x_2 - x_0) + a_2(x_2 - x_0)(x_2 - x_1) = y_2$

$$\Rightarrow a_2 = \frac{y_2 - y_0 - \frac{\Delta y_0}{h}(2h)}{(x_2 - x_0)(x_2 - x_1)} = \frac{(y_2 - y_1) + (y_1 - y_0) - 2\Delta y_0}{2h \cdot h}$$

$$= \frac{\Delta y_1 + \Delta y_0 - 2\Delta y_0}{2h^2} = \frac{\Delta y_1 - \Delta y_0}{2h^2} = \frac{\Delta^2 y_0}{2!h^2}$$

Putting $x = x_3$.

$$f(x_3) = \phi_n(x_3)$$

$$= y_0 + \frac{\Delta y_0}{h}(x_3 - x_0) + \frac{\Delta^2 y_0}{2!h^2}(x_3 - x_0)(x_3 - x_1) + a_3(x_3 - x_0)(x_3 - x_1)(x_3 - x_2)$$

$$\equiv y_3$$

$$\Rightarrow a_3 = \frac{y_3 - y_0 - \frac{\Delta y_0}{h}(3h) - \frac{\Delta^2 y_0}{2!h^2} \cdot 3h \cdot 2h}{3h \cdot 2h \cdot h}$$

$$= \frac{(y_3 - y_2) + (y_2 - y_1) + (y_1 - y_0) - 3\Delta y_0 - 3\Delta^2 y_0}{3!h^3}$$

$$= \frac{\Delta y_2 + \Delta y_1 + \Delta y_0 - 3\Delta y_0 - 3\Delta^2 y_0}{3!h^3}$$

$$= \frac{\Delta y_2 - \Delta y_1 + 2(\Delta y_1 - \Delta y_0) - 3\Delta^2 y_0}{3!h^3}$$

$$= \frac{\Delta^2 y_1 + 2\Delta^2 y_0 - 3\Delta^2 y_0}{3!h^3} = \frac{(\Delta^2 y_1 - \Delta^2 y_0)}{3!h^3}$$

$$= \frac{\Delta^3 y_0}{3!h^3}$$

Proceeding in this way, and noting the symmetry, we write putting $x = x_u$.

$$a_n = \frac{\Delta^n y_0}{n!h^n}$$

Therefore, the interpolating polynomial $\phi_n(x)$, may be written as

$$f(x) \simeq \phi_n(x) = y_0 + \frac{\Delta y_0}{h}(x - x_0) + \frac{\Delta^2 y_0}{2!h^2}(x - x_0)(x - x_1) + \cdots$$
$$+ \frac{\Delta^n y_0}{n!h^n}(x - x_0)(x - x_1)\cdots(x - x_{n-1})$$

Writing $\frac{x - x_0}{h} = u$, we note

$$x - x_1 = x - x_0 + x_0 - x_1 = uh - h = (u - 1)h$$
$$x - x_2 = (u - 2)h$$
$$\dots\dots\dots\dots\dots\dots\dots\dots\dots\dots\dots\dots\dots$$
$$x - x_{n-1} = (u - n + 1)h$$

and

$$f(x) \simeq \phi_n(x) = y_0 + \binom{u}{1}\Delta y_0 + \binom{u}{2}\Delta^2 y_0 + \cdots + \binom{u}{n}\Delta^n y_0 \tag{3.2.2}$$

Here the leading or the pivotal entry is y_0 and all the leading differences lie along the diagonal through y_0.

The formula (3.2.2) is known as Newton's Forward Interpolation formula. The expression for $R_n(x)$ will be obtained in a separate chapter (Accuracy in Interpolation Formulae) together with other polynomial interpolation formula.

Newton's Forward formula is used to interpolate the value of y near the *beginning* of a set of tabular values, and for extrapolating values of y a short distance backward from y_0 (i.e. to the left of y_0).

NOTE 3.2.1. The name *forward* interpolation formula lies in the fact that the formula contains values of the tabulated function from y_0 onward to the right (i.e. forward from y_0) and none to the left of y_0

3.3 Newton's backward formula

The function $f(x)$ is given for $(n + 1)$ equispaced values of the argument

$x_0, x_1, x_2, \dots, x_n$

where

$x_i = x_0 + ih, \quad i = 0, 1, 2, \dots, n$

The corresponding entries

$y_0, y_1, \dots, y_n$

are given as in Section 3.2. Here we are required to interpolate y, for some x in

$x_0 + \overline{x - 1}h < x < x_0 + nh$

We may write the interpolating polynomial $\phi_n(x)$, of degree n in x, in a slightly different manner as

$$\phi_n(x) = a_n + a_{n-1}(x - x_n) + a_{n-2}(x - x_n)(x - x_{n-1}) + \cdots$$
$$+ a_0(x - x_n)(x - x_{n-1}) \cdots (x - x_1) \quad (3.3.1)$$

Putting $x = x_n$,

$$\phi_n(x_n) = f(x_n) = a_n \Rightarrow a_n = y_n \quad (3.3.2)$$

Putting $x = x_{n-1}$

$$\phi_n(x_{n-1}) = y_{n-1} = a_n + a_{n-1}(x_{n-1} - x_n)$$

$$\Rightarrow a_{n-1} = \frac{y_{n-1} - y_n}{x_{n-1} - x_n} = \frac{y_n - y_{n-1}}{x_n - x_{n-1}} = \frac{\Delta y_{n-1}}{h} \quad (3.3.3)$$

Putting $x = x_{n-2}$

$$\phi_n(x_{n-2}) = f(x_{n-2}) = y_{n-2} = y_n + \frac{\Delta y_{n-1}}{h}(x_{n-2} - x_n) + a_{n-2}(x_{n-2} - x_n)(x_{n-2} - x_{n-1})$$

$$\Rightarrow a_{n-2} = \frac{y_{n-2} - y_n - \dfrac{\Delta y_{n-1}}{h}(-2h)}{(-2h)(-h)} = \frac{2\Delta y_{n-1} - (y_n - y_{n-1} + y_{n-1} - y_{n-2})}{2h^2}$$

$$= \frac{2\Delta y_{n-1} - \Delta y_{n-1} - \Delta y_{n-2}}{2h^2} = \frac{\Delta y_{n-1} - \Delta y_{n-2}}{2h^2} = \frac{\Delta^2 y_{n-2}}{2!h^2} \quad (3.3.4)$$

Putting $x = x_{n-3}$

$$\phi_n(x_{n-3}) = y_{n-3} = y_n + \frac{\Delta y_{n-1}}{h}(x_{n-3} - x_n) + \frac{\Delta^2 y_{n-2}}{2!h^2}(x_{n-3} - x_n)(x_{n-3} - x_{n-1})$$
$$+ a_{n-3}(x_{n-3} - x_n)(x_{n-3} - x_{n-1})(x_{n-3} - x_{n-2})$$

$$\Rightarrow a_{n-3}(3h)(2h)(h) = (y_n - y_{n-3}) - 3\Delta y_{n-1} + 3\Delta^2 y_{n-2}$$

$$\therefore a_{n-3} = \frac{(y_n - y_{n-1} + y_{n-1} - y_{n-2} + y_{n-2} - y_{n-3}) - 3\Delta y_{n-1} + 3\Delta^2 y_{n-2}}{3!h^3}$$

$$= \frac{(\Delta y_{n-1} + \Delta y_{n-2} + \Delta y_{n-3}) - 3\Delta y_{n-1} + 3\Delta^2 y_{n-2}}{3!h^3}$$

$$= \frac{-2(\Delta y_{n-1} - \Delta y_{n-2}) - (\Delta y_{n-2} - \Delta y_{n-3}) + 3\Delta^2 y_{n-2}}{3!h^3}$$

$$= \frac{-2\Delta^2 y_{n-2} + 3\Delta^2 y_{n-2} - \Delta^2 y_{n-3}}{3!h^3}$$

$$= \frac{\Delta^2 y_{n-2} - \Delta^2 y_{n-3}}{3!h^3} = \frac{\Delta^3 y_{n-3}}{3!h^3} \quad (3.3.5)$$

Proceeding in this way and noting the symmetry at each stage, we may write, in general,

$$a_{n-r} = \frac{\Delta^r y_{n-r}}{r!h^r} \quad (3.3.6)$$

and finally

$$a_0 = \frac{\Delta^n y_0}{n!h^n} \quad (3.3.7)$$

Hence

$$\phi_n(x) = y_n + \frac{\Delta y_{n-1}}{h}(x - x_n) + \frac{\Delta^2 y_{n-2}}{2!h^2}(x - x_n)(x - x_{n-1}) + \cdots$$
$$+ \frac{\Delta^r y_{n-r}}{r!h^r}(x - x_n)(x - x_{n-1}) \cdots (x - x_{n-r+1}) + \cdots$$
$$+ \frac{\Delta^n y_0}{n!h^n}(x - x_n)(x - x_{n-1}) \cdots (x - x_1)$$

We write

$$\frac{x - x_n}{h} = u \Rightarrow x - x_{n-1} = uh + h, \text{ etc.}$$

$$\phi_n(x) = y_n + u\Delta y_{n-1} + \binom{u+1}{2}\Delta^2 y_{n-2} + \binom{u+2}{3}\Delta^3 y_{n-3} + \cdots + \binom{u+n-1}{n}\Delta^n y_0 \quad (3.3.8)$$

This is the form in which *Newton's backward interpolation formula* is usually written. This formula is used mainly for interpolating the values of $f(x)$ near the end of a set of tabular values of entry.

NOTE 3.3.1. Using notations of horizontal difference table, we may note that

$\Delta y_{n-1} \equiv \Delta_1 y_n$

$\Delta^2 y_{n-2} \equiv \Delta_2 y_n$

....

$\Delta^n y_0 \equiv \Delta_n y_n$

and Newton's backward interpolation formula (3.3.9) may be rewritten as

$$f(x) \simeq \phi_n(x) = y_n + \binom{u}{1}\Delta_1 y_n + \binom{u+1}{2}\Delta_2 y_n + \cdots + \binom{u+n-1}{n}\Delta_n y_n \quad (3.3.9)$$

We show below the construction of Horizontal difference (Table 3.1)

Table 3.1: *Horizontal difference table*

Argument x	Entries y	$\Delta_1 y$	$\Delta_2 y$	$\Delta_3 y$
x_0	y_0			
$x_0 + h$	y_1	$y_1 - y_0 = \Delta_1 y_1$		
$x_0 + 2h$	y_2	$y_2 - y_1 = \Delta_1 y_2$	$\Delta_1 y_2 - \Delta_1 y_1 = \Delta_2 y_2$	
$x_0 + 3h$	y_3	$y_3 - y_2 = \Delta_1 y_3$	$\Delta_1 y_3 - \Delta_1 y_2 = \Delta_2 y_3$	$\Delta_2 y_3 - \Delta_2 y_2 = \Delta_3 y_3$

NOTE 3.3.2. In Newton's backward interpolation formula in form (3.3.9) the pivotal entry is y_n and all the leading differences lie along the horizontal line through y_n. (here y_3 in Table 3.1). These backward difference notations are used in this manner in order to maintain symmetry in the formula

NOTE 3.3.3. In some test books, the horizontal difference table have been given as $\nabla y_n, \nabla^2 y_n, \nabla^3 y_n, \ldots, \nabla^n y_n$

Using these notations, Newton's backward interpolation formula has been rewritten as:

$$f(x) \simeq \phi_n(x) = y_n + u\nabla y_n + \binom{u+1}{2}\nabla^2 y_n + \cdots + \binom{u+n-1}{n}\nabla^n y_n \quad (3.3.10)$$

3.4 Use of operators Δ and *E* in polynomial interpolation

We now take up the case of interpolation in Type II of Section 3.1.

Here the values of the argument (x) for which the corresponding entries are made available, are not initially given at equispaced interval. But the series of argument will become equispaced, if the values of the argument for which the corresponding entries are to be interpolated, are inserted in the series of argument.

We assume that the unknown function $f(x)$ can be approximated by a polynomial. The degree of polynomial will be one less than the number of entries initially supplied. For example, if $n - r$ values of argument are such that the corresponding entries are available, and we are further given r values of the argument whose corresponding entries are to be interpolated such that the series of n values of x are *equispaced*, then the degree of polynomial to be used will be one less than $n - r$ (i.e. $n - r - 1$). Thus, we assume $f(x)$ to be a polynomial of degree $n - r - 1$ in x, then

$\Delta^{n-r-1} f(x)$ will be a constant

and $\Delta^{n-r} f(x)$ will be zero. (3.4.1)

Since r entries are to be interpolated, we must have r equations by putting r independent values of x in order from the beginning in the result (3.4.1). We shall be able to solve the problem, if we can solve r independent equations

$$\Delta^{n-r} f(x_j) = 0, \quad j = 0, 1, 2, \ldots, r - 1$$

The method can be better explained by considering an example:

EXAMPLE 3.4.1. Find the missing entries in the following:

Argument (x)	16	18	20	22	24	26
Entries (y)	39	85	—	151	—	388

Solution: Here the series of argument, for which the entries are given, are not equispaced initially. However, after insertions of the values of x, namely, 20 and 24 (corresponding to which the entries are to be obtained); the argument becomes equispace; and we may therefore use the method suggested in this section. The degree of polynomial will be one less than the number of available entries (here 4).

For a polynomial of degree 4 in x, we have

$$\Delta^4 y_x = 0, \quad \text{for } x = 16, \text{ and } x = 18$$

We consider only two values of the argument since only two entries are to be interpolated.

Now

$$\Delta^4 y_{16} = 0 \qquad (3.4.2)$$

and $\Delta^4 y_{18} = 0$ (interval of differencing 2) (3.4.3)

will be two equations involving two unknowns y_{20} and y_{24}.

From (3.4.2),

$$(E-1)^4 y_{16} = 0 \Rightarrow (E^4 - 4E^3 + 6E^2 - 4E + 1)y_{16} = 0$$
$$\Rightarrow y_{24} - 4y_{22} + 6y_{20} - 4y_{18} + y_{16} = 0$$
$$\Rightarrow y_{24} + 6y_{20} = 4(y_{22} + y_{18}) - y_{16} = 905 \tag{3.4.4}$$

Again (3.4.3) will give

$$y_{26} - 4y_{24} + 6y_{22} - 4y_{20} + y_{18} = 0$$
$$\Rightarrow 4(y_{24} + y_{20}) = y_{26} + 6y_{22} + y_{18} = 1379 \tag{3.4.5}$$

Solving (3.4.4) and (3.4.5), we may obtain

$y_{20} = 112$ (approx)

and $y_{24} = 233$ (approx).

Second method

We assume $f(x)$ to be a polynomial of degree 3 (since 4 independent values of entry are given). Hence third order finite difference will be a constant and fourth order difference will be zero.

Denoting the unknown entries by y_{20} and y_{24}, let us prepare the difference table with the given data with two unknown entries.

Table 3.2: *Diagonal difference table*

x	y_x	Δy	$\Delta^2 y$	$\Delta^3 y$	Δ^4
16	39				
		46			
18	85		$y_{20} - 131$		
		$y_{20} - 85$		$367 - 3y_{20}$	
20	y_{20}		$236 - 2y_{20}$		$6y_{20} + y_{24} - 905$
		$151 - y_{20}$		$3y_{20} + y_{24} - 538$	
22	151		$y_{24} + y_{20} - 302$		$1379 - 4y_{24} - 4y_{20}$
		$y_{24} - 151$		$841 - 3y_{24} - y_{20}$	
24	y_{24}		$539 - 2y_{24}$		
		$388 - y_{24}$			
26	388				

Since fourth order differences are zero, we have

$6y_{20} + y_{24} = 905$

$4(y_{24} + y_{20}) = 1379$

The same two equations, we obtained in (3.4.4) and (3.4.5), so we obtain the same answer as before:

$y_{20} = 112$

$y_{24} = 233$

EXAMPLE 3.4.2. Complete the following table by interpolation:

x	2	3	4	5	6	7	8
$f(x)$	0.135	—	0.111	0.100	—	0.082	0.074

Solution: We may assume $f(x)$ to be a polynomial of degree 4 in x.

$\therefore \Delta^5 f(x) = 0$, (for $x = 2$ and 3)

$\Rightarrow (E-1)^5 f(2) = 0$

and $(E-1)^5 f(3) = 0$ (interval of differencing is 1)

$$\Rightarrow (E^5 - 5E^4 + 10E^3 - 10E^2 + 5E - 1)f(2) = 0 \quad (3.4.6)$$

$$\text{and } (E^5 - 5E^4 + 10E^3 - 10E^4 + 5E - 1)f(3) = 0 \quad (3.4.7)$$

$\Rightarrow f(7) - 5f(6) + 10f(5) - 10f(4) + 5f(3) - f(2)$

and $f(8) - 5f(7) + 10f(6) - 10f(5) + 5f(4) - f(3)$

$$\Rightarrow f(3) - f(6) = 0.0326 \quad (3.4.8)$$

$$\text{and } 10f(6) - f(3) = 0.781 \quad (3.4.9)$$

Adding (3.4.8) and (3.4.9)

$9f(6) = 0.8136 \Rightarrow f(6) = 0.0904 \simeq 0.090$ (correct to 3 decimal places)

where from (3.4.8)

$f(3) = 0.123$

Therefore, by interpolation, we obtain

$f(3) = 0.123$, and $f(6) = 0.090$

NOTE 3.4.1. It is to be noted in this connection that the interpolated values of y_x will never be the values of the unknown function at $x = 3$ and $x = 6$ respectively. The unknown function is not always a polynomial exactly. It is approximated by an interpolating polynomial $\phi(x)$ of suitable degree such that $f(x_i) = \phi(x_i)$ only for those values of x for which the entries are available. For other values of x (for which entries are not supplied), $f(x) \neq \phi(x)$; $f(x) - \phi(x)$ will be the error of interpolation.

EXAMPLE 3.4.3. Let $y_x = 2^x + 7$ be given for $x = 1, 3, 4$ and 5 as under:

x	1	2	3	4	5
$y = f(x)$	9	—	15	23	39

We are required to interpolate y_n for $x = 2$

Solution: Since four independent values of the entry are given, we may assume y_x to be a polynomial of degree 3 in x

$\therefore \Delta^4 y_x = 0$, for $x = 1$

$\Rightarrow (E-1)^4 y_1 = 0$ (interval of differencing 1)

$\Rightarrow y_5 - 4y_4 + 6y_3 - 4y_2 + y_1 = 0$

$\Rightarrow 4y_2 = y_5 - 4y_4 + 6y_3 + y_1 = 46$

$\therefore y_2 = 11.5$

However, the actual value of $y_2 = 11$.

The discrepancy between the actual (= 11) and the interpolated value (= 11.5) of y_2 is due to the fact that y_x is not strictly a polynomial. The interpolating polynomial $\phi(x)$ and the function $f(x)$ will have the same value only at those values of x for which the corresponding entries are already available.

3.5 Subdivision of intervals

A frequent problem in actual mathematics is the interpolation for values of $f(x)$ at finer subintervals of the argument x. For example: $f(x)$ is given for equispaced values of argument at an interval of h, say,

$f(x_0), f(x_0 + h), f(x_0 + 2h)$, etc.

We are required to interpolate $f(x)$, for

$x = x_0 + k, x_0 + 2k, x_0 + 3k, \ldots, x_0 + tk$, (where $tk = h$)

We define

$$Ef(x) = f(x + h) = f(x + tk)$$
$$\text{and } ef(x) = f(x + k)$$
$$\Rightarrow e^t f(x) = f(x + tk) = f(x + h) = Ef(x)$$

We therefore write symbolically

$$e^t \equiv E, \quad \text{or,} \quad e \equiv E^{1/t}$$

$$\text{or, } 1 + \delta \equiv (1 + \Delta)^{1/t} \Rightarrow \delta = (1 + \Delta)^{1/t} - 1 \tag{3.5.1}$$

where (3.5.1) gives the relationship between operators δ and Δ.

Thus,

$$\delta = \left\{1 + \left(\frac{1}{t}\right)\Delta + \frac{\frac{1}{t}\left(\frac{1}{t} - 1\right)}{2!}\Delta^2 + \frac{\frac{1}{t}\left(\frac{1}{t} - 1\right)\left(\frac{1}{t} - 2\right)}{3!}\Delta^3 + \frac{\frac{1}{t}\left(\frac{1}{t} - 1\right)\left(\frac{1}{t} - 2\right)\left(\frac{1}{t} - 3\right)}{4!}\Delta^4 + \cdots\right\} - 1$$

$$= \frac{1}{t}\Delta + \frac{\frac{1}{t}\left(\frac{1}{t} - 1\right)}{2!}\Delta^2 + \frac{\frac{1}{t}\left(\frac{1}{t} - 1\right)\left(\frac{1}{t} - 2\right)}{3!}\Delta^3 + \frac{\frac{1}{t}\left(\frac{1}{t} - 1\right)\left(\frac{1}{t} - 2\right)\left(\frac{1}{t} - 3\right)}{4!}\Delta^4 + \cdots \tag{3.5.2}$$

$$\delta^2 = \left\{\frac{1}{t}\Delta + \frac{\frac{1}{t}\left(\frac{1}{t} - 1\right)}{2!}\Delta^2 + \frac{\frac{1}{t}\left(\frac{1}{t} - 1\right)\left(\frac{1}{t} - 2\right)}{3!}\Delta^3 + \frac{\frac{1}{t}\left(\frac{1}{t} - 1\right)\left(\frac{1}{t} - 2\right)\left(\frac{1}{t} - 3\right)}{4!}\Delta^4 + \cdots\right\}^2 \tag{3.5.3}$$

$$\delta^3 = \left\{\frac{1}{t}\Delta + \frac{\frac{1}{t}\left(\frac{1}{t}-1\right)}{2!}\Delta^2 + \frac{\frac{1}{t}\left(\frac{1}{t}-1\right)\left(\frac{1}{t}-2\right)}{3!}\Delta^3 \right.$$

$$\left. + \frac{\frac{1}{t}\left(\frac{1}{t}-1\right)\left(\frac{1}{t}-2\right)\left(\frac{1}{t}-3\right)}{4!}\Delta^4 + \cdots\right\}^3 \qquad (3.5.4)$$

$$\delta^4 = \left\{\frac{1}{t}\Delta + \frac{\frac{1}{t}\left(\frac{1}{t}-1\right)}{2!}\Delta^2 + \frac{\frac{1}{t}\left(\frac{1}{t}-1\right)\left(\frac{1}{t}-2\right)}{3!}\Delta^3 \right.$$

$$\left. + \frac{\frac{1}{t}\left(\frac{1}{t}-1\right)\left(\frac{1}{t}-2\right)\left(\frac{1}{t}-3\right)}{4!}\Delta^4 + \cdots\right\}^4, \text{ etc.} \qquad (3.5.5)$$

We shall illustrate this method for obtaining values where quinquennial values are known. Here $\delta f(x)$ denotes the difference for unit intervals of x, and $\Delta f(x)$ denotes the difference for quinquennial intervals.

Then

$$f(x+5) = Ef(x) = (1+\Delta)f(x) = (1+\delta)^5 f(x)$$

$$[\because e = 1 + \delta \text{ and } e^5 f(x) = Ef(x)]$$

Symbolically,

$$(1+\delta)^5 \equiv (1+4)$$

$$\Rightarrow \delta \equiv (1+\Delta)^{1/5} - 1$$

$$= \frac{1}{5}\Delta + \frac{\frac{1}{5}\left(\frac{1}{5}-1\right)}{2!}\Delta^2 + \frac{\frac{1}{5}\left(\frac{1}{5}-1\right)\left(\frac{1}{5}-2\right)}{3!}\Delta^3$$

$$+ \frac{\frac{1}{5}\left(\frac{1}{5}-1\right)\left(\frac{1}{5}-2\right)\left(\frac{1}{5}-3\right)}{4!}\Delta^4 + \cdots$$

$$= 0.24 - 0.08\Delta^2 + 0.048\Delta^3 - 0.0336\Delta^4 + \cdots$$

$$\delta f(x) = \left\{0.24 - 0.08\Delta^2 + 0.048\Delta^3 - 0.0336\Delta^4 + \cdots\right\} f(x)$$

$$\delta^2 f(x) = \left\{0.2\Delta - 0.08\Delta^2 + 0.048\Delta^3 - 0.0336\Delta^4 + \cdots\right\}^2 f(x)$$

$$= \left\{0.04\Delta^2 - 0.032\Delta^3 + 0.0256\Delta^4 + \cdots\right\} f(x)$$

$$\delta^3 f(x) = \left\{0.24 - 0.08\Delta^2 + 0.048\Delta^3 - 0.0336\Delta^4 + \cdots\right\}^3 f(x)$$

$$= \left(0.008\Delta^3 - 0.0096\Delta^4 + \cdots\right) f(x)$$

$$\delta^4 f(x) = 0.0016\Delta^4 f(x)$$

An example will show the application of the above technique.

EXAMPLE 3.5.1. The following table gives yearly premiums for policies maturing at quinquennial ages, estimate the net premium for all ages from 45 to 50 inclusive:

Age(x)	45	50	55	60	65
Annual Premiums $f(x)$	14.50	16.80	20.25	23.50	28.75

Solution: The difference table is as under:

x	$f(x)$	$\Delta f(x)$	$\Delta^2 f(x)$	$\Delta^3 f(x)$	$\Delta^4 f(x)$
45	14.50				
		2.30			
50	16.80		1.15		
		3.45		−1.35	
55	20.25		−0.20		3.55
		3.25		2.20	
60	23.50		2.00		
		5.25			
65	28.75				

$$\begin{aligned}\delta f(45) &= \left\{0.2\Delta f(45) - 0.08\Delta^2 f(45) + 0.048\Delta^3 f(45) - 0.0336\Delta^4 f(45)\right\}\\ &= 0.2 \times 2.30 - 0.08 \times 1.15 + 0.048 \times (-1.35) - 0.0336 \times 3.55\\ &= 0.18392\end{aligned}$$

$$\begin{aligned}\delta^2 f(45) &= \left\{0.04\Delta^2 f(45) - 0.032\Delta^3 f(45) + 0.0256\Delta^4 f(45)\right\}\\ &= 0.04 \times 1.15 - 0.032 \times (-1.35) + 0.0256 \times 3.55\\ &= 0.18008\end{aligned}$$

$$\begin{aligned}\delta^3 f(45) &= 0.008\Delta^3 f(45) - 0.0096\Delta^4 f(45)\\ &= 0.008 \times (-1.35) - 0.0096 \times 3.55\\ &= -0.04488\end{aligned}$$

$$\begin{aligned}\delta^4 f(45) &= 0.0016\Delta^4 f(45)\\ &= 0.00568\end{aligned}$$

assuming fourth differences constant.

We have therefore by completing the table of differences:

Age x	$f(x)$	$\delta f(x)$	$\delta^2 f(x)$	$\delta^3 f(x)$	$\delta^4 f(x)$
45	14.50				
		0.18392			
46	14.68		0.18008		
		0.36400		−0.04488	
47	15.05		0.1352		0.00568
		0.49920		−0.0392	
48	15.55		0.0960		0.00568
		0.59520		−0.03352	
49	16.14		0.06248		
		0.65768			
50	16.80				

NOTE 3.5.1. Since we require the value of the premium to the nearest paisa, two decimal places will be required in $f(x)$ column.

The same principle can be adopted if decennial values are given. In that case, $\Delta f(x)$, $\Delta^2 f(x)$, $\Delta^3 f(x)$, $\Delta^4 f(x), \ldots$ will represent differences for decennial intervals, and the in-dividual differences will be found from the identity

$$\equiv (1+\Delta)^{1/10} - 1$$

Decennial and quinquennial points are very often given in acturial work and we are re-quired to interpolate at intervening points

3.6 Newton-Gregory formula

he formula

$$f(x+nh) = E^n f(x) = (1+\Delta)^n f(x)$$

$$= \left\{1 + \binom{n}{1}\Delta + \binom{n}{2}\Delta^2 + \cdots + \binom{n}{n}\Delta^n\right\} f(x)$$

$$= f(x) + \binom{n}{1}\Delta f(x) + \binom{n}{2}\Delta^2 f(x) + \cdots + \binom{n}{n}\Delta^n f(x) \qquad (3.6.1)$$

3.6.1) is known as *Newton-Gregory formula* and is the fundamental formula used in in-erpolation, when it is given that the values of the argument are at equispaced interval of , (say).

There are many variations of the problem, examples of such variations are given below:

XAMPLE 3.6.1.

Age (x)	25	26	27	28	29
Annual Premium $f(x)$	40	42	45	50	51

etermine $f\left(27\frac{1}{2}\right)$

olution: Here the argument is given for the equispaced interval of 1. But we are required to interpolate $f(x)$ for some $x = 27\frac{1}{2}$ somewhere in the middle of the interval. Newton's orward and backward formula cannot be advocated in this case.

Taking the pivotal entry as $f(25)$, we may take 25 as origin for the independent variable and write

$$\left(27\frac{1}{2}\right) = E^{5/2} f(25)$$

$$= \left\{1 + \frac{5}{2}\Delta + \frac{\frac{5}{2}\left(\frac{5}{2}-1\right)}{2!}\Delta^2 + \frac{\frac{5}{2}\left(\frac{5}{2}-1\right)\left(\frac{5}{2}-2\right)}{3!}\Delta^3 + \frac{\frac{5}{2}\left(\frac{5}{2}-1\right)\left(\frac{5}{2}-2\right)\left(\frac{5}{2}-3\right)}{4!}\Delta^4 + \cdots\right\} f(25)$$

$$= 1 + 2.5\Delta f(25) + \frac{2.5\times 1.5}{2}\Delta^2 f(25) + \frac{2.5\times 1.5\times 0.5}{3!}\Delta^3 f(25) + \frac{2.5\times 1.5\times 0.5\times(-0.5)}{4!}\Delta^4 f(25) + \cdots$$

We next construct the difference table:

x	$f(x)$	$\Delta f(x)$	$\Delta^2 f(x)$	$\Delta^3 f(x)$	$\Delta^4 f(x)$
25	40				
		2			
26	42		1		
		3		1	
27	45		2		–7
		5		–6	
28	50		–4		
		1			
29	51				

$$f\left(27\frac{1}{2}\right) = 40 + 2.5 \times 2 + 1.875 \times 1 + 0.3125 \times 1 - 0.0391 \times (-7)$$

$$= 40 + 5 + 1.875 + 0.3125 + 0.2737 = 47.46$$

EXERCISES

1. Compute the value of

$$\left(\frac{2}{\sqrt{\lambda}}\right) \int_0^x e^{-x^2} dx$$

when $x = 0.6238$ and when $x = 0.6748$, stating clearly the assumptions made.

x	$\frac{2}{\sqrt{\lambda}} \int_0^x e^{-x^2} dx$
0.62	0.6194114
0.63	0.6270463
0.64	0.6345857
0.65	0.6420292
0.66	0.6493765
0.67	0.6566275
0.68	0.6637820

[*Ans:* 0.6223229; 0.6600737.]

2. (a) Given $y_0 = 580$, $y_1 = 556$, $y_2 = 520$, $y_4 = 385$; find y_3.
   (b) Given $y_{-1} = 202$, $y_0 = 175$, $y_1 = 82$, $y_2 = 55$; find $y_{1/2}$.
   (c) Given $y_0 = 6021$, $y_1 = 5229$, $y_2 = 4559$, $y_3 = 3979$, find $y_{3/4}$

[*Ans:* (a) 465, (b) 128, (c) 5414.]

3. Given $\log_x$ for $x = 210, 220, 230, 240, 250$ and 260 in the following table. Find the value of: (a) log 2585, (b) log 2178.

x	210	220	230	240	250	260
$\log x$	2.3222193	2.3424227	2.3617278	2.3802112	2.3979400	2.4149733

[*Ans:* (a) 3.3380577, (b) 3.4124605.]

[*Hints:* Obtain by interpolation, log 258.5 and log 217.8.]

4. Given the following score distribution of statistics:

Marks	30–40	40–50	50–60	60–70
No. of Students	52	36	21	14

Find:
(a) the number of students scoring below 35;
(b) the number of students scoring above 65;
(c) the number of students scoring between 35–45. [*Ans:* (a) 26, (b) 7, (c) 46.]

5. (a) Interpolate the missing entry in the following data:

x	1981	1982	1983	1984	1985	1986	1987
y	33.1	172.8	219.7	*	375.2	409.6	491.3

(b) Given

x	2.0	2.1	2.2	2.3	2.4	2.5	2.6
y	0.135	—	0.111	0.100	—	0.082	0.074

[*Ans:* (a) 299.1, (b) $y_{2.1} = 0.123$; $y_{2.4} = 0.090$.]

[*Hints:* Assume y_x to be a polynomial of degree 4 in x such that $\Delta^4 y_x$ is constant and $\Delta^5 y_x = 0$, i.e. $(E-1)^5 y_x$, unit of differencing being 0.1 we take $(E^5 - 5E^4 + 10E^3 - 10E^2 + 5E - 1)y_x = 0$ for $x = 2.1$ and $x = 2.4$. Solve for the two unknowns $y_{2.1}$ and $y_{2.4}$.]

6. If $u_0, u_1, u_2, \ldots, u_6$ be consecutive terms of the series, prove that, if fifth differences are constant,

$$u_3 = 0.05u_0 - 0.30u_1 + 0.75u_2 + 0.75u_4 - 0.30u_5 + 0.05u_6$$

Supply the missing term: $u_0 = 72795$, $u_1 = 71651$, $u_2 = 70458$, $u_4 = 67919$, $u_5 = 66566$, $u_6 = 65152$, (Freeman). [*Ans:* $u_3 = 69215$.]

[*Hints:* Assume u_3 missing and u_x is a polynomial of degree 5 in x. Unit of differencing 1. Use shift operator E, take $(E-1)^5 u_0 = 0$ and solve for u_3.]

7. Given $f(0) = 1$, $f(1) + f(2) = 10$, $f(3) + f(4) = f(5) = 65$; find $f(4)$.

[*Ans:* $f(4) = 21$.]

[*Hints:* Assume $f(x)$ a polynomial of degree 2 in x.
Then $f(x) = a + bx + cx^2$. Obtain three equations by taking $f(0) = 1$, $f(1) + f(2) = 10$ and $f(3) + f(4) + f(5) = 65$ and solve for a, b and c. Hence obtain $f(4)$.]

8. The table below gives the average length of life remaining to l_x persons who attain exact age x, for male population in a certain region.

x	e_x^0
20	57.64
25	51.35
30	48.15
35	44.32
40	40.62

Obtain e_x^0 for $x = 21, 22, 23$ and 24.

[*Ans:* $e_{21}^0 = 55.81$, $e_{22}^0 = 54.33$, $e_{23}^0 = 53.14$, $e_{24}^0 = 52.16$.]

9. Show that if $y_1, y_2, \ldots, y_4$ are the first nine consecutive terms of a series, then

$$y_5 = 0.8(y_4 + y_6) - 0.4(y_3 + y_7) + 0.114(y_2 + y_8) - 0.014(y_1 + y_9)$$

[*Hints:* We assume y_5 is missing. Therefore, we are given 9 entries. Therefore, $\Delta^8 y_1 = 0$; because we may assume y_x to be a polynomial of degree 7 in x.

$$\therefore (E-1)^8 y_1 = 0 \Rightarrow y_9 - 8y_8 + 28y_7 - 56y_6 + 70y_5 - 56y_4 + 28y_3 - 8/2 + y_1 = 0$$
$$\Rightarrow 70y_5 = 8(y_2 + y_8) - 28(y_3 + y_7) + 56(y_4 + y_6) - (y_1 + y_9)$$
$$\Rightarrow y_5 = 0.114(y_2 + y_8) - 0.4(y_3 + y_7) + 0.8(y_4 + y_6) - 0.014(y_1 + y_9).]$$

10. Given:

$$y_0 + y_8 = 21.9243, \quad y_1 + y_7 = 21.9590$$
$$y_2 + y_6 = 21.9823, \quad y_3 + y_5 = 21.9956$$

find y_4, stating clearly all the assumptions. [*Ans:* 11.0000.]
[*Hints:* Assume y_4 missing. On the basis of other 8 entries, we may assume y_x to be a polynomial of degree seven in x.

$\Delta^8 y_0 = 0$, or, $(E-1)^8 y_0 = 0$, etc.]

11. Following are the population (in million) of a district at decennial intervals, find the population of all years from 1951 to 1961.

Years (x)	1951	1961	1971	1981
Population $f(x)$ (in million)	48	56	63	69

[*Ans:* 48.755, 49.520, 50.295, 51.08, 51.875, 52.68, 53.495, 54.32, 55.155. (unit: million in each case).]
[*Hints:* Third order differences are assumed to be constant in subtabulation.]

12. If $y_0, y_1, y_2, y_3, y_4, y_5$ be the consecutive terms of a series, prove that if fourth order differences are constant:

$$10y_3 = y_0 - 5y_1 + 10y_2 + 5y_4 - y_5$$

Also obtain y_0, if $y_1 = 2$, $y_2 = 4$, $y_3 = 8$, $y_4 = 16$ and $y_5 = 32$. Bring out the following in this result. [Check $y_x = 2^x$.]

13. The function $f(x) = 3^x$ takes, as it should be, the values 1, 3, 9, and 81 for $x = 0, 1, 2$ and 4 respectively. Use any suitable method of finite difference to estimate $f(x)$ for $x = 3$. Explain why this result differs from the actual value 27.
[*Ans:* Assuming y_x to be a polynomial of degree 3 in x, $y = 31$ (Actually $y_x = 3^x$ is not exactly a polynomial, hence the discrepancy).]

14. Given:

$$\sum_{x=1}^{x=10} f(x) = 500426, \quad \sum_{x=4}^{x=10} f(x) = 329240, \quad \sum_{x=7}^{x=10} f(x) = 175212$$

and $f(10) = 40365$; find $f(1)$. [*Ans:* 58835.]
[*Hints:* Let

$$u_1 = \sum_{x=1}^{10} f(x), \quad u_4 = \sum_{x=4}^{x=10} f(x), \quad u_7 = \sum_{x=7}^{10} f(x) \text{ and } u_{10} = \sum_{x=10}^{x=10} f(x)$$

From the given u_1, u_4, u_7, u_{10} which are given for the equispaced values of argument 1, 4, 7 and 10, find

$$u_2 = \sum_{x=2}^{x=10} f(x)$$

by Newton's forward formula or subdivision of interval: $f(1) = u_1 - u_2$.]

15. Given: $u_1 = 1$; $u_2 + u_3 = 5.41$, $u_4 + u_5 + u_6 = 18.47$, $u_7 + u_8 + u_9 + u_{10} + u_{11} + u_{12} = 90.36$. Find the value of u_x, for all values of x from 1 to 12 inclusive.

[*Ans:* 1; 2.10; 3.31; 4.64; 6.11; 7.73; 9.51; 11.47; 13.62; 15.97.]

16. If you are given u_0, u_1, u_2 and $\sum_{x=1}^{10} u_x$, how will you complete the table of u_x up to u_{10}?

Chapter 4

Interpolation with Unequal Intervals of the Argument

Summary

4.1 Introduction

The interpolation formulae derived in chapter 3 are applicable only when the values of the functions are given at equispaced intervals of the independent variable or argument. Sometimes, for some practical limitations or inconvenience, we are not in a position to obtain values of the function at equidistant values of the argument. We shall introduce in this chapter two such formulae when the values of the argument are not given at equidistant intervals. One is Newton's general divided difference formula and the other is Lagrange's formula. Divided difference is just the ordinary difference divided by certain differences of the argument. It is therefore so named.

Let

$$y_i = f(x_i), \quad i = 0, 1, 2, n$$

denote functional value corresponding to any values $x_0, x_1, x_2, \ldots, x_n$ of the argument. Then the divided differences of y in ascending order are defined as follows:

First order divided difference with argument x_0, x_1, to be denoted by the symbol $\delta[x_0, x_1]$ is given by

$$\delta[x_0, x_1] = \frac{f(x_0) - f(x_1)}{x_0 - x_1} = \frac{y_0 - y_1}{x_0 - x_1}$$
$$= \delta[x_1, x_0]$$

First order divided difference of argument x_1, x_2

$$\delta[x_1, x_2] = \frac{y_1 - y_2}{x_1 - x_2}$$
$$= \delta[x_2, x_1], \text{ etc.}$$

Second order differences are:

$$\delta[x_0, x_1, x_2] = \frac{\delta[x_0, x_1] - \delta[x_1, x_2]}{x_0 - x_2}$$
$$= \delta[x_2, x_1, x_0]$$

$$\delta[x_1, x_2, x_3] = \frac{\delta[x_1, x_2] - \delta[x_2, x_3]}{x_1 - x_3}$$
$$= \delta[x_3, x_2, x_1], \text{ etc.}$$

Third order differences are:

$$\delta[x_0, x_1, x_2, x_3] = \frac{\delta[x_0, x_1, x_2] - \delta[x_1, x_2, x_3]}{x_0 - x_3}$$
$$= \delta[x_3, x_2, x_1, x_0], \text{ etc.}$$

It may be noted that the order of any divided difference is one less than the number of values of the argument in it.

For the simplicity of presentation, we shall denote divided differences of different order only by the number of values of the argument in it inside a box bracket.

Thus,

$\delta[x_0, x_1] \Rightarrow [x_0, x_1]$

$\delta[x_1, x_2] \Rightarrow [x_1, x_2]$

$\delta[x_0, x_1, x_2] \Rightarrow [x_0, x_1, x_2]$, etc.

We may give the following geometrical interpretation of the first order divided differences:

If $(x_0, y_0), (x_1, y_1), (x_2, y_2)$, etc. are points on a curve, the first order divided difference may be taken as the the slope of the secant line through any *two* points.

Table 4.1: *Tables of divided differences*

x	y	1st order Divided Difference	2nd order Divided Difference	3rd order Divided Difference	4th order Divided Difference
x_0	y_0				
		$[x_0, x_1]$			
x_1	y_1		$[x_0, x_1, x_2]$		
		$[x_1, x_2]$		$[x_0, x_1, x_2, x_3]$	
x_2	y_2		$[x_1, x_2, x_3]$		$[x_0, x_1, x_2, x_3, x_4]$
		$[x_2, x_3]$		$[x_1, x_2, x_3, x_4]$	
x_3	y_3		$[x_2, x_3, x_4]$		
		$[x_3, x_4]$			
x_4	y_4				

Example 4.1.1. Construction of divided difference table.

x	y	1st order Divided Difference	2nd order Divided Difference	3rd order Divided Difference	4th order Divided Difference
1	4				
		$\frac{30-4}{3-1} = 13$			
3	30		$\frac{49-13}{5-1} = 9$		
		$\frac{128-30}{5-3} = 49$		$\frac{15-9}{7-1} = 1$	
5	128		$\frac{109-49}{7-3} = 15$		0
		$\frac{346-128}{7-5} = 109$		$\frac{23-15}{11-3} = 1$	
7	346		$\frac{247-109}{11-5} = 23$		
		$\frac{1334-346}{11-7} = 247$			
11	1334				

If may be noted that y is given by $y = x^3 + 3$ in the above table of values. In other words, y is a polynomial of degree 3 in x. Like ordinary finite difference, the third order divided differences also give constant (= 1). We shall give a general proof of this property of divided differences later in this chapter.

4.2 Symmetry of divided differences

Divided differences are symmetric functions of their arguments, as indicated below:

$$[x_0, x_1] = \frac{y_0 - y_1}{x_0 - x_1} = \frac{y_0}{x_0 - x_1} + \frac{y_1}{x_1 - x_0}$$

$$= \frac{f(x_0)}{x_0 - x_1} + \frac{f(x_1)}{x_1 - x_0} \equiv [x_1, x_0] \tag{4.2.1}$$

Also

$$[x_0, x_1, x_2] = \frac{[x_0, x_1] - [x_1, x_2]}{x_0 - x_2}$$

$$= \frac{\dfrac{f(x_0)}{(x_0 - x_1)} + \dfrac{f(x_1)}{(x_1 - x_0)} - \dfrac{f(x_1)}{(x_1 - x_2)} - \dfrac{f(x_2)}{(x_2 - x_1)}}{x_0 - x_2}$$

$$\Rightarrow [x_0, x_1, x_2] = \frac{1}{x_0 - x_2}\left[\frac{f(x_0)}{x_0 - x_1} + \left(\frac{x_1 - x_2 - x_1 + x_0}{(x_1 - x_0)(x_1 - x_2)}\right) f(x_1) - \frac{f(x_2)}{(x_2 - x_1)}\right]$$

$$= \frac{f(x_0)}{(x_0 - x_1)(x_0 - x_2)} + \frac{f(x_1)}{(x_1 - x_0)(x_1 - x_2)} + \frac{f(x_2)}{(x_2 - x_0)(x_2 - x_1)} \tag{4.2.2}$$

Similarly,

$$[x_0, x_1, x_2, x_3] = \frac{f(x_0)}{(x_0 - x_1)(x_0 - x_2)(x_0 - x_3)} + \frac{f(x_1)}{(x_1 - x_0)(x_1 - x_2)(x_1 - x_3)} + \frac{f(x_2)}{(x_2 - x_0)(x_2 - x_1)(x_2 - x_3)} + \frac{f(x_3)}{(x_3 - x_0)(x_3 - x_1)(x_3 - x_2)} \tag{4.2.3}$$

We may easily verify that the right-hand members of the equations (4.2.1), (4.2.2) and (4.2.3) remain unchanged when *any two values of x are interchanged* and the corresponding y's are also interchanged. This implies that a divided difference will remain unchanged regardless of how much its arguments are interchanged. Thus,

$[x_0, x_1, x_2] = [x_1, x_2, x_0] = [x_2, x_1, x_0]$, etc.

we may also prove by mathematical induction that

$$[x, x_0, x_1, x_2, \ldots, x_n] = \frac{f(x)}{(x - x_0)(x - x_1)\cdots(x - x_n)} + \frac{f(x_0)}{(x_0 - x)(x_0 - x_1)\cdots(x_0 - x_n)} + \cdots + \frac{f(x_n)}{(x_n - x)(x_n - x_0)\cdots(x_n - x_{n-1})} \tag{4.2.4}$$

4.3 Relation between divided differences and simple differences

The relation between the divided differences and the simple differences of the corresponding order can be easily established when the arguments

$x_0, x_1, x_2, \ldots, x_n$

are given at equispaced interval of h (say).

Let

$y_0, y_1, y_2, \ldots, y_n$

be the functional values corresponding to argument

$x_0, x_1, x_2, \ldots, x_n,$

where $x_i = x_0 + ih$, $i = 0, 1, 2, \ldots, n$.

First order divided difference of argument x_0, x_1 is

$$[x_0, x_1] = \frac{f(x_0) - f(x_1)}{x_0 - x_1} = \frac{y_0 - y_1}{-h} = \frac{y_1 - y_0}{h} = \frac{\Delta y_0}{h} \tag{4.3.1}$$

$$[x_0, x_1, x_2] = \frac{[x_0, x_1] - [x_1, x_2]}{x_0 - x_2} = \frac{\frac{\Delta y_0}{h} - \frac{\Delta y_1}{h}}{-2h} = \frac{\Delta y_1 - \Delta y_0}{2 \cdot h^2} = \frac{\Delta^2 y_0}{2! h^2} \tag{4.3.2}$$

$$[x_0, x_1, x_2, x_3] = \frac{[x_0, x_1, x_2] - [x_1, x_2, x_3]}{x_0 - x_3} = \frac{\frac{\Delta^2 y_0}{2h^2} - \frac{\Delta^2 y_1}{2h^2}}{-3h^2} = \frac{\Delta^2 y_1 - \Delta^2 y_0}{2 \cdot 3h^3} = \frac{\Delta^3 y_0}{3! h^3} \tag{4.3.3}$$

Proceeding in this way and noting the symmetry at each stage, we may see that the divided difference of the nth order of argument

$x_0, x_1, x_2, \ldots, x_n$,

where $x_i = x_0 + ih$, $i = 0, 1, 2, \ldots, n$,is

$$[x_0, x_1, \ldots, x_n] = \frac{\Delta^n y_0}{n! h^n} \tag{4.3.4}$$

We have already established the result that if $f(x)$ be a polynomial of the nth degree is x, then nth order finite difference $\Delta^n y_0$ will be a constant.

Since nth order divided difference is equivalent to the nth order ordinary (simple) difference divided by the constant $n! h^n$, it follows that nth order divided difference of a polynomial of nth degree will also be a constant. We now give a proof based on mathematical induction to establish this result.

Theorem 4.3.1. If $y = f(x)$ be a polynomial of degree n in x, then nth order divided difference of y will be a constant.

Proof. Let $y = f(x)$ be given for the values of the argument $x_0, x_1, x_2, \ldots, x_n$. The corresponding entries $y_0, y_1, y_2, \ldots, y_n$ are also given. We shall prove that

$[x, x_0, x_1, \ldots, x_{n-1}] =$ Constant.

$$[x, x_0] = \frac{f(x) - f(x_0)}{x - x_0}$$

Let

$f(x) = a_0 x^n + a_1 x^{n-1} + \cdots + a_{n-1} x + a_n$, $(a_0 \neq 0)$

be a polynomial of degree n in x. Then

$$\begin{aligned}
[x, x_0] &= a_0 \frac{x^n - x_0^n}{x - x_0} + a_1 \frac{x^{n-1} - x_0^{n-1}}{x - x_0} + \cdots + a_{n-1} \frac{x - x_0}{x - x_0} \\
&= a_0 \{x^{n-1} + x^{n-2} x_0 + \cdots + x_0^{n-1}\} + a_1 \{x^{n-2} + x^{n-3} x_0 + \cdots + x_0^{n-2}\} + \cdots + a_{n-1} \\
&= a_0 x^{n-1} + (a_1 + a_0 x_0) x^{n-2} + (a_2 + a_1 x_0 + a_0 x_0^2) x^{n-3} + \cdots + a_{n-1} \\
&\equiv b_0 x^{n-1} + b_1 x^{n-2} + b_2 x^{n-3} + \cdots + b_{n-1}
\end{aligned}$$

where $b_0 = a_0$, $b_1 = a_1 + a_0 x_0$, $b_2 = a_2 + a_1 x_0 + a_2 x_0^2$, etc.

Thus, the first order divided difference of argument x_0, x of a polynomial of nth degree will give a polynomial of degree $n - 1$ in x.

Again

$$[x, x_0, x_1] = [x_0, x, x_1] = \frac{[x_0 x] - [x, x_1]}{x - x_1}$$

$$= \frac{\{b_0 x^{n-1} + b_1 x^{n-2} + b_2 x^{n-3} + \cdots + b_{n-1}\} - \{b_0 x_1^{n-1} + \cdots + b_{n-1}\}}{x - x_1}$$

$$= b_0 \frac{(x^{n-1} - x_1^{n-1})}{x - x_1} + b_1 \frac{x^{n-2} - x_1^{n-2}}{x - x_1} + \cdots + b_{n-2} \frac{x - x_1}{x - x_1}$$

$$= b_0 \left\{x^{n-2} + x^{n-3} x_1 + x^{n-4} x_1^2 + \cdots + x_1^{n-2}\right\}$$

$$+ b_1 \left\{x^{n-3} + x^{n-4} x_1 + x^{n-5} x_1^2 + \cdots + x_1^{n-3}\right\} + \cdots + b_{n-2}$$

$$= b_0 x^{n-2} + (b_0 x_1 + b_1) x^{n-3} + (b_2 + b_1 x_1 + b_0 x_1^2) x^{n-4} + \cdots + b_{n-2}$$

$$\equiv c_0 x^{n-2} + 4x^{n-3} + c_2 x^{n-4} + \cdots + c_{n-2},$$

where $b_0 = c_0$, $c_1 = (b_1 + b_0 x_1)$, $c_2 = (b_2 + b_1 x_1 + b_0 x_1^2) \ldots$, etc. which is a polynomial of degree $n - 2$ in x. Thus, the second order divided difference of argument x, x_0, x_1 gives a polynomial of degree $n - 2$ in x.

Let us assume that the rth order ($r < n$), divided difference of argument $x, x_0, \ldots, x_{r-1}$, gives a polynomial of degree $n - r$ in x. In other words,

$$[x, x_0, x_1, \ldots, x_{r-1}] = d_0 x^{n-r} + d_1 x^{n-r-1} + \cdots + d_{n-r} = 0, \quad (d_0 \neq 0)$$

Now $(r + 1)$th order divided difference of argument $x, x_0, \ldots, x_r$ is

$$[x, x_0, \ldots, x_r] = \frac{[x_0, x_1, \ldots, x_{r-1}, x] - [x_0, x_1, \ldots, x_r]}{x - x_r}$$

$$= d_0 \frac{x^{n-r} - x_r^{n-r}}{x - x_r} + d_1 \frac{x^{n-r-1} - x_r^{n-r-1}}{x - x_r} + \cdots + d_{n-r-1} \frac{x - x_r}{x - x_r}$$

$$= d_0 \left\{x^{n-r-1} + x^{n-r-2} x_r + x^{n-r-3} x_r^2 + \cdots + x_r^{n-r-1}\right\}$$

$$+ d_1 \left\{x^{n-r-2} + x^{n-r-3} x_r + \cdots + x_r^{n-r-2}\right\} + \cdots + d_{n-r-1}$$

$$\equiv e_0 x^{n-r-1} + e_1 x^{n-r-2} + \cdots + e_{n-r-1}$$

where $e_0 = d_0$; $e_1 = d_1 + d_0 x_r$, etc.

Thus, the divided difference of order $(r + 1)$ will be a polynomial of degree $n - r - 1$; provided the divided difference of order r produces a polynomial of degree $n - r$ in x. But we have already shown that the first and the second order divided difference produced a polynomial of degree $n - 1$ and $n - 2$ respectively. Hence by induction, we may write that the divided difference of order n of a polynomial of degree n in x, will result in a constant, and obviously, the $(n + 1)$th order divided difference will be zero.

4.4 Newton's general divided difference formula

Let a function $f(x)$ be given for $(n + 1)$ values of argument $x_0, x_1, x_2, \ldots, x_n$ as $y_0, y_1, y_2, \ldots, y_n$. The argument $x_0, x_1, \ldots$ are not generally equispaced. Let it be required to interpolate $f(x)$ for some given x in $x_0 < x < x_n$. We assume that in the interval $[x_0, x_n]$ $f(x)$, can be replaced by a polynomial $\phi(x)$ of degree n in x, such that

$$f(x_i) = \phi(x_i), \quad i = 0, 1, 2, \ldots, n$$

At all other points

$f(x) = \phi(x) + R(x),$

where $R(x)$ is the remainder or error term. Ignoring the error term $R(x)$ for the present, we may write

$f(x) \simeq \phi(x)$

From the definition of divided difference

$$[x_0, x] = \frac{f(x) - f(x_0)}{x - x_0}$$

$$\Rightarrow f(x) = f(x_0) + (x - x_0)[x_0, x]$$
$$= f(x_0) + (x - x_0)[x, x_0] \qquad (4.4.1)$$

Also

$$[x, x_0, x_1] = \frac{[x, x_0] - [x_0, x_1]}{x - x_1}$$

$$\Rightarrow [x, x_0] = [x_0, x_1] + (x - x_1)[x, x_0, x_1]$$

$$\Rightarrow \frac{f(x) - f(x_0)}{x - x_0} = [x_0, x_1] + (x - x_1)[x, x_0, x_1]$$

$$\Rightarrow f(x) = f(x_0) + (x - x_0)[x_0, x_1] + (x - x_0)(x - x_1)[x, x_0, x_1] \qquad (4.4.2)$$

Proceeding in this way and noting the symmetry at each stage, we may note that $[x, x_0, x_1, \ldots, x_n]$, which is a divided difference of $(n + 1)$th order, will yield

$$f(x) = f(x_0) + (x - x_0)[x_0, x_1] + (x - x_0)(x - x_1)[x_0, x_1, x_2] + \cdots$$
$$+ (x - x_0)(x - x_1)\cdots(x - x_{n-1})(x - x_n)[x, x_0, x_1, \ldots, x_n] \qquad (4.4.3)$$

The last term of (4.4.3), $[x, x_0, x_1, \ldots, x_n]$, is the divided difference of order $(n + 1)$ of the polynomial, $\phi(x)$, of degree n in x, and is therefore equal to zero.

Hence we obtain from (4.4.3)

$$f(x) \simeq \phi(x) = f(x_0) + (x - x_0)[x_0, x_1) + (x - x_0)(x - x_1)[x_0, x_1, x_2] + \cdots$$
$$+ (x - x_0)(x - x_1)\cdots(x - x_{n-1})[x_0, x_1, \ldots, x_n] \qquad (4.4.4)$$

Formula (4.4.4) is known as *Newton's General Divided Difference formula* with values of argument $x_0, x_1, \ldots, x_n$.

NOTE 4.4.1. The error term is

$$R(x) = f(x) - \phi(x) = (x - x_0)(x - x_1)\cdots(x - x_n)[x, x_0, x_1, \ldots, x_n] \qquad (4.4.5)$$

EXAMPLE 4.4.1. Find $f(7)$ from the following:

x	1	3	4	6	10
$f(x)$	0	22	53	196	961

Solution: We assume that the interpolating polynomial $\phi(x)$ is of degree 4 in x. We prepare the table of divided differences as follows:

x	$y = f(x)$	First order divided differences	Second order divided differences	Third order divided differences	Fourth order divided difference
1	0				
		11			
3	22		6.6667		
		31		1.3667	
4	53		13.5		2.0360
		71.5		19.6905	
6	196		151.3333		
		191.25			
10	961				

$$\begin{aligned} f(7) &\simeq f(1) + (7-1)\times 11 + (7-1)(7-3)\times 6.6667 + (7-1)(7-3)(7-4) \\ &\quad \times 1.3667 + (7-1)(7-3)(7-4)(7-6)\times 2.0360 \\ &= 0 + 66 + 160.0008 + 98.4024 + 146.5920 \\ &= 470.9952 \\ &\simeq 471 \text{ (approx)} \end{aligned}$$

Example 4.4.2. Find the form of the function $f(x)$, under suitable assumption, from the following data. Hence find $f(10)$

x	0	2	5	9	11
$f(x)$	1	5	116	712	1310

Solution: We assume $f(x)$ to be a polynomial of degree 3 in x. Divided difference Table is given as under:

x	$y = f(x)$	First order divided differences	Second order divided differences	Third order divided differences	Fourth order divided difference
0	1				
		2			
2	5		7		
		37		1	
5	116		16		0
		149		1	
9	712		25		
		299			
11	1310				

Newton's General Divided Difference formula gives

$$f(x) \simeq f(0) + (x-0)\times 2 + (x-0)(x-2)\times 7 + (x-0)(x-2)(x-5)\times 1$$
$$+ (x-0)(x-2)(x-5)(x-9)\times 0$$
$$= 1 + 2x + 7x(x-2) + x(x-2)(x-5) + 0$$
$$= 1 + 2x + 7x^2 - 14x + x^3 - 7x^2 + 10x$$
$$= x^3 - 2x + 1$$

[*Note.* Since third order differences are constant, $f(x)$ is actually a polynomial of degree 3 in x].

$\therefore f(10) = 10^3 - 2.10 + 1 = 981$

4.5 Lagrange's interpolation

Let the functional values of $f(x)$ be $f(x_0), f(x_1), \ldots, f(x_n)$, for $(n+1)$ values of argument $x_0, x_1, \ldots, x_n$, which are not necessarily equispaced. We are required to interpolate $f(x)$ for some x in $x_0 < x < x_n$.

We assume that $f(x)$ can be replaced by a polynomial $\phi(x)$ of degree n, such that

$$f(x_i) = \phi(x_i), \quad i = 1, 2, \ldots, n$$

At all other points

$$f(x) = \phi(x) + R(x),$$

where $R(x)$ is the remainder term.

We start from the definition of divided difference and show as in (4.2.4) that the divided difference of order $(n+1)$ with argument $x, x_0, x_1, \ldots, x_n$ can be written as

$$[x, x_0, x_1, \ldots, x_n] = \frac{f(x)}{(x-x_0)(x-x_1)\cdots(x-x_n)} + \frac{f(x_0)}{(x_0-x)(x_0-x_1)\cdots(x_0-x_n)}$$
$$+ \frac{f(x_1)}{(x_1-x)(x_1-x_0)(x_1-x_2)\cdots(x_1-x_n)} + \cdots$$
$$+ \frac{f(x_i)}{(x_i-x)(x_i-x_0)\cdots(x_i-x_{i-1})(x_i-x_{i+1})\cdots(x_i-x_n)} + \cdots$$
$$+ \frac{f(x_n)}{(x_n-x)(x_n-x_0)\cdots(x_n-x_{n-1})} \qquad (4.5.1)$$

Since $f(x)$ is approximated by a polynomial of nth degree in x, therefore $(n+1)$th order divided difference of the function will be zero. Hence

$$[x, x_0, x_1, \ldots, x_n] = 0$$

Therefore, from (4.5.1), we obtain

$$-\frac{f(x)}{(x-x_0)(x-x_1)\cdots(x-x_n)} = \frac{f(x_0)}{(x_0-x)(x_0-x_1)\cdots(x_0-x_n)} + \frac{f(x_1)}{(x_1-x)(x_1-x_0)(x_1-x_2)\cdots(x_1-x_n)} + \cdots + \frac{f(x_i)}{(x_i-x)(x_i-x_0)(x_i-x_1)\cdots(x_i-x_n)} + \cdots + \frac{f(x_n)}{(x_n-x)(x_n-x_0)\cdots(x_n-x_{n-1})}$$

$$\Rightarrow f(x) = \frac{(x-x_1)(x-x_2)\cdots(x-x_n)}{x_0-x_1)(x_0-x_2)\cdots(x_0-x_n)}\cdot f(x_0) + \frac{(x-x_0)(x-x_2)\cdots(x-x_n)}{(x_1-x_0)(x_1-x_2)\cdots(x_1-x_n)}\cdot f(x_1) + \frac{(x-x_0)(x-x_1)\cdots(x-x_{n-1})}{(x_n-x_0)(x_n-x_1)\cdots(x_n-x_{n-1})}\cdot f(x_n) \tag{4.5.2}$$

Formula given in (4.5.2) is known as Lagrange's Interpolation formula.

Corollary 4.5.1. Lagrange's interpolation formula can be written in the following compact form:

$$f(x) = \sum_{r=0}^{r=n} \frac{F(x)}{(x-x_r)F'(x_r)}\cdot f(x_r) \tag{4.5.3}$$

where $F(x) = (x-x_0)(x-x_1)\cdots(x-x_n)$ and $F'(x_r) = \frac{d}{dx}F(x)$ at $x = x_r$.

Proof. We have

$$F(x) = (x-x_0)(x-x_1)\cdots(x-x_n)$$

$$F'(x) = (x-x_1)(x-x_2)\cdots(x-x_n) + (x-x_0)(x-x_2)\cdots(x-x_n) + (x-x_0)(x-x_1)(x-x_3)\cdots(x-x_n) + (x-x_0)(x-x_1)(x-x_2)(x-x_4)\cdots(x-x_n) + \cdots + (x-x_0)(x-x_1)\cdots(x-x_{n-1}) \tag{4.5.4}$$

$$\therefore F'(x_r) = (x_r-x_0)(x_r-x_1)\cdots(x_r-x_{r-1})(x_r-x_{r+1})\cdots(x_r-x_n)$$

[all other terms in (4.5.4) vanish]

From (4.5.2), we obtain

$$\therefore f(x) = \sum_{r=0}^{r-n} \frac{F(x)}{(x-x_r)F'(x_r)}\cdot f(x_r) \tag{4.5.5}$$

Corollary 4.5.2. Show that the sum of the coefficients of $y_r = f(x_r)$, $r = 0, 1, 2, \ldots, n$ in the Lagrange's interpolation formula is unity

Proof. Let

$$F(x) = (x - x_0)(x - x_1) \cdots (x - x_n)$$

$$\therefore \frac{1}{F(x)} \equiv \frac{A_0}{x - x_0} + \frac{A_1}{x - x_1} + \cdots + \frac{A_r}{x - x_r} + \cdots + \frac{A_n}{x - x_n}$$

$$\therefore \frac{1}{F(x)} \equiv \frac{A_0(x - x_1)\cdots(x - x_n) + A_1(x - x_0)(x - x_2)\cdots(x - x_n) + \cdots + A_n(x - x_0)\cdots(x - x_{n-1})}{F(x)}$$

$$\begin{aligned} \therefore I \equiv{} & A_0(x - x_1)\cdots(x - x_n) + A_1(x - x_0)(x - x_2)\cdots(x - x_n) + \cdots \\ & + A_r(x - x_0)(x - x_1)\cdots(x - x_{r-1})(x - x_{r+1})\cdots(x - x_n) + \cdots \\ & + A_n(x - x_0)(x - x_1)\cdots(x - x_{n-1}) \end{aligned} \tag{4.5.6}$$

The above being an identity in x, is true for all values of x. Therefore, putting $x = x_0, x_1, \ldots, x_r, \ldots, x_n$

$$A_0(x_0 - x_1)\cdots(x_0 - x_n) = 1 \Rightarrow A_0 = \frac{1}{F'(x_0)}$$

$$A_1(x_1 - x_0)\cdots(x_1 - x_n) = 1 \Rightarrow A_1 = \frac{1}{F'(x_1)}$$

....

$$A_r(x_r - x_0)(x_r - x_1)\cdots(x_r - x_{r-1})(x_r - x_{r+1})\cdots(x_r - x_n) \Rightarrow A_r = \frac{1}{F'(x_r)}, \text{ etc.}$$

Substituting the values of $A_0, A_1, \ldots, A_n$ in (4.5.6), we get

$$\begin{aligned} 1 &\equiv \sum_{r=0}^{r-n} \frac{(x - x_0)(x - x_1)\cdots(x - x_{r-1})(x - x_{r+1})\cdots(x - x_n)}{(x_r - x_0)(x_r - x_1)\cdots(x_r - x_{r-1})(x_r - x_{r+1})\cdots(x_r - x_n)} \\ &\equiv \sum_{r=0}^{r=n} \frac{F(x)}{(x - x_r)F'(x_r)} \end{aligned}$$

$= \text{sum of the coefficients of } f(x_r), \quad r = 0, 1, 3, \ldots, n$

Hence the result.

Example 4.5.1. Find by Lagrange's formula the value of $f(48)$ given $f(40) = 15.22, f(45) = 13.99, f(50) = 12.62, f(55) = 11.13$.

Solution: Assuming $f(x)$ to be a polynomial of degree 3 in x, we write

$$\begin{aligned} f(x) ={} & \frac{(48-45)(48-50)(48-55)}{(40-45)(40-50)(40-55)} \cdot f(40) + \frac{(48-40)(48-50)(48-55)}{(45-40)(45-45)(45-50)(45-55)} \cdot f(45) \\ & + \frac{(48-40)(48-45)(48-55)}{(50-40)(50-45)(50-55)} \cdot f(50) + \frac{(48-40)(48-45)(48-50)}{(55-40)(55-45)(55-50)} \cdot f(55) \\ ={} & \frac{3 \times (-2) \times (-7)}{(-5)(-10)(-15)} \times 15.22 + \frac{8 \times (-2) \times (-7)}{5 \times (-5)(-10)} \times 13.99 \\ & + \frac{8 \times 3 \times (-7)}{10 \times 5 \times (-5)} \times 12.62 + \frac{8 \times 3 \times (-2)}{15 \times 10 \times 5} \times 11.13 \\ ={} & -0.85232 + 6.26752 + 8.48064 - 0.71232 \\ ={} & 13.18352 = 13.18 \end{aligned}$$

EXAMPLE 4.5.2. **Find by Lagrange's formula the approximate form of the function $y = f(x)$, given by**

x	0	2	3	6
$f(x)$	659	705	729	804

Solution: Assuming $f(x)$ can be replaced by a polynomial of degree 3 in x, we may write

$$\begin{aligned}
f(x) &= \frac{(x-2)(x-3)(x-6)}{(0-2)(0-3)(0-6)} \times 659 + \frac{x(x-3)(x-6)}{(2-0)(2-3)(2-6)} \times 705 \\
&\quad + \frac{x(x-2)(x-6)}{(3-0)(3-2)(3-6)} \times 729 + \frac{x(x-2)(x-3)}{(6-0)(6-2)(6-3)} \times 804 \\
&= -\frac{1}{36}(x-2)(x-3)(x-6) \times 659 + \frac{1}{8}x(x-3)(x-6) \times 705 \\
&\quad - \frac{1}{9}x(x-2)(x-6) \times 729 + \frac{1}{72}x(x-2)(x-3) \times 804 \\
&= x^3\left(-\frac{659}{36} + \frac{705}{8} - 81 + \frac{804}{72}\right) + x^2\left(\frac{11 \times 659}{36} - \frac{9 \times 705}{8} + 648 - \frac{4020}{72}\right) \\
&\quad + x\left(-659 + \frac{6345}{4} - 972 + 67\right) + 659 \\
&= -\frac{1}{72}x^3 + \frac{29}{72}x^2 + \frac{89}{4}x + 659
\end{aligned}$$

Lagrange's formula can be used whether the argument is given for equispaced interval or not. But its computational work is usually laborious and also requires very careful attention to signs. It must also be used with care and caution, for if the values of the argument are not taken close together, the results are liable to be very inaccurate. Divided difference formula by Newton is preferable because of its computational ease. However, Lagrange's formula has one special merit. It can be used for inverse interpolation, by just interchanging the x's and y's. This will be discussed later in chapter 7.

4.6 Confluent divided difference

We may find difficulty in defining divided differences when two or more of the argument values coincide. In this section we shall try to interpret $[x_0, x_0]$, $[x_0, x_0, x_0]$, $[x_0, x_0, x_0, x_0]$, etc. We propose to take limit in such situation. *The limiting values of the divided difference, which arises in case where two or more arguments coincide, will be called confluent divided difference.*

We define

$$\begin{aligned}
[x_0, x_0] &= \lim_{\varepsilon \to 0}[x_0 + \varepsilon, x_0] = \lim_{\varepsilon \to 0} \frac{f(x_0+\varepsilon) - f(x_0)}{\varepsilon} \\
&= f'(x_0) \quad \text{(assuming } f(x) \text{ to be differentiable)}
\end{aligned}$$

Again

$$[x_0, x_0, x_0] = \lim_{\substack{\varepsilon_1 \to 0 \\ \varepsilon_2 \to 0}} [x_0, x_0 + \varepsilon_1, x_0 + \varepsilon_2]$$

$$= \lim_{\substack{\varepsilon_1 \to 0 \\ \varepsilon_2 \to 0}} \frac{[x_0, x_0 + \varepsilon_1] - [x_0 + \varepsilon_1, x_0 + \varepsilon_2]}{x_0 - (x_0 + \varepsilon_2)}$$

$$= \lim_{\varepsilon_2 \to 0} \frac{f'(x_0) - [f(x_0 + \varepsilon_2) - f(x_0)]/\varepsilon_2}{-\varepsilon_2}$$

$$= \lim_{\varepsilon_2 \to 0} \frac{\varepsilon_2 f'(x_0) - f(x_0 + \varepsilon_2) + f(x_0)}{-\varepsilon_2^2}$$

This is $\frac{0}{0}$ form and applying L'Hôspital's rule twice, we obtain

$$[x_0, x_0, x_0] = \lim_{\varepsilon_2 \to 0} \frac{-f''(x_0 + \varepsilon_2)}{-2} = \frac{1}{2} f''(x_0) \tag{4.6.1}$$

In general, we may show with $(r + 1)$ equal arguments

$$[x_0, x_0, x_0, \ldots, x_0] = \frac{1}{r!} f^r(x_0) \tag{4.6.2}$$

It can also be shown that

$$[x, x_0, x_1, \ldots, x_n] = \frac{d}{dx}[x, x_0, x_1, \ldots, x_n] \tag{4.6.3}$$

4.7 Uses of confluent divided difference

Whenever the unknown function can be exactly represented by a polynomial of degree n in x, as will be evident from the prepared divided difference table on the basis of the given data, we may express $f(x)$ as a polynomial of degree n in $(x - x_0)$ by using the technique given below:

EXAMPLE 4.7.1. Given the following data, find $f(x)$ as a polynomial in powers of $x - 5$ by extending the table to include arguments $x = 5$, repeated as many times as may be necessary. Also find $f'(5)$, $f''(5)$ and $f'''(5)$; $f(2) = 14$, $f(3) = 36$, $f(4) = 76$, $f(6) = 234$, $f(7) = 364$, $f(9) = 756$.

Solution: We prepare the divided difference table:

x	$y = f(x)$	First order divided difference	Second order divided difference	Third order divided difference	Fourth order divided difference
2	14				
		22			
3	36		9		
		40		1	
4	76		13		0
		79		1	
6	234		17		0
		130		1	
7	364		22		0
		196		1	
9	756		a		0
		$\frac{c-756}{5-9} = b$		1	
5	c		a'		0
		b'		1	
5	c		a''		
		b''			
5	c				

Since the third order divided differences of $f(x)$ are constant, $f(x)$ is a polynomial of degree 3 in x and hence the argument 5 should appear three times successively in the divided difference table.

Taking (or *assuming*) the third order divided differences constant, we extend the divided difference table as above. The values of a, a', a'' and b, b', b'' and c are obtained as under:

$$\frac{c-756}{5-9} = b, \quad \frac{b-196}{5-7} = a, \quad \frac{a-22}{5-6} = 1$$

$$\Rightarrow \quad a = -1 + 22 = 21$$

and $b = 196 - 2 \times 21 = 154$

and $c = 756 + 154(-4) = 756 - 616 = 140$

Again from the table

$$\frac{b'-b}{5-9} = a', \quad \frac{a'-a}{5-7} = 1$$

$$\Rightarrow \quad a' = a - 2 = 19$$

and $b' = b - 4 \times 19 = 154 - 76 = 78$

Also from table

$$\frac{a''-a'}{5-9} = 1, \text{ or, } a'' = a' - 4 = 15$$

Now Newton's divided difference formula with arguments x_0, x_1, x_2, x_3 is:

$$f(x) = f(x_0) + (x - x_0)[x_0, x_1] + (x - x_0)(x - x_1)[x_0, x_1, x_2] + (x - x_0)(x - x_1)(x - x_2)[x_0, x_1, x_2, x_3]$$

Putting $x_0 = x_1 = x_2 = x_3 = 5$.

$$\begin{aligned} f(x) &= f(5) + (x - 5)[5, 5] + (x - 5)^2[5, 5, 5] + (x - 5)^3[5, 5, 5, 5] \\ &= 140 + (x - 5)b' + (x - 5)^2 \cdot a'' + (x - 5)^3 \cdot 1 \\ &= 140 + 78(x - 5) + 15(x - 5)^2 + (x - 5)^3 \end{aligned}$$

We now write

$$[5, 5] = f'(5) = 78$$

$$[5, 5, 5] = \frac{1}{\angle 2} f''(5) \Rightarrow f''(5) = \angle 2 \times [5, 5, 5] = \angle 2 \times a'' = 2 \times 15 = 30$$

and $[5, 5, 5, 5] = \dfrac{1}{\angle 3} f'''(5) = \dfrac{1}{6} \times f'''(5) \Rightarrow f'''(5) = 6 \times [5, 5, 5, 5] = 6 \times 1 = 6$

Example 4.7.2. Using the following table, find $f(x)$ as a polynomial in powers of $(x - 3)$:

x	5	11	27	34	42
$f(x)$	23	899	17315	35606	68510

Hence find $f'(3)$, $f''(3)$ and $f'''(3)$.

Solution:

x	$f(x)$	First order divided difference	Second order divided difference	Third order divided difference	Fourth order divided difference
5	23				
		146			
11	899		40		
		1026		1	
27	17315		69		0
		2613		1	
34	35606		100		0
		4113		1	
42	68510		a		0
		b		1	
3	c		a'		0
		b'		1	
3	c		a''		
		b''			
3	c				

Here again, the third differences of $f(x)$ is constant. $f(x)$ is therefore a polynomial of degree 3 in x and hence the argument 3, should appear 3 times (at the most) successively in the divided difference table.

From columns (3), (4) and (5), we have

$$\frac{c-68510}{3-42}=b,\quad \frac{b-4113}{3-34}=a,\quad \frac{a-100}{3-27}=1$$

Solving a, b and c, we obtain

$$a=76,\quad b=1757,\quad c=-13$$

Again from columns (4) and (5)

$$\frac{a'-a}{3-34}=1\Rightarrow a'=a-31=45$$

$$\frac{b'-b}{3-42}=a'\Rightarrow b'=b-39a'=2$$

Again from column (5)

$$\frac{a''-a'}{3-42}=1\Rightarrow a''=6$$

Now Newton's General Divided difference formula with arguments x_0, x_1, x_2, x_3 is

$$f(x)=f(x_0)+(x-x_0)[x_0,x_1]+(x-x_0)(x-x_1)[x_0,x_1,x_2]$$
$$+(x-x_0)(x-x_1)(x-x_2)[x_0,x_1,x_2,x_3]$$

Putting $x_0=x_1=x_2=x_3=3$.

$$f(x)=f(3)+(x-3)[3,3]+(x-3)^2[3,3,3]+(x-3)^3[3,3,3,3]$$
$$f'(3)=2,\quad f''(3)=12,\quad f'''(3)=6$$

Check

$$[3,3]=b'=2$$

$$[3,3,3]=\frac{1}{2!}f''(3)=a''=6$$

$$[3,3,3,3]=\frac{1}{3!}f'''(3)=1=-13+2(x-3)+6(x-3)^2+(x-3)^3$$

$$\therefore\ f'(3)=2,\quad f''(3)=12,\quad f'''(3)=6$$

4.8 Hermite's interpolation formula

Given $y=f(x)$ for n values of the argument $x_1, x_2, \ldots, x_n$. The corresponding entries are $f(x_1), f(x_2), \ldots, f(x_n)$. We may find a polynomial $\phi(x)$ of degree $(2n-1)$ such that at the points $x_1, x_2, \ldots, x_n$,

$$f(x_i)=\phi(x_i);\quad i=1,2,\ldots,n$$

and also

$$f'(x)_{x=x_i}=\phi'(x)_{x=x_i}$$

Thus, we shall be in a position to find $2n$, unknown constants of the polynomial $\phi(x)$ of degree $2n-1$ only from the n available arguments with corresponding entries. In our present discussion, we shall give the form of the Hermite's Interpolations formula without actually deriving it.

Hermite's Interpolation formula is of the form:

$$f(x) \simeq \phi(x) = \sum_{k=1}^{n} \{1 - 2H_k'(x_k)(x - x_k)\}\,(H_k(x))^2 f(x_k) + \sum_{k=1}^{n} (x - x_k)\,\{H_k(x)\}^2\, f'(x_k),$$

where $H_k'(x_k) = (d/dx)H_k(x)_{x=x_k}$ and $f(x_k) = y_k$, corresponding to $x = x_k$.

This kind of interpolation is named in some taxts osculating interpolation.

NOTE 4.8.1. Readers may carefully note that the raw data will contain three columns: (a) x_k, (b) $f(x_k)$, (c) $f'(x_k)$. Some illustrative examples may explain the computational steps.

EXAMPLE 4.8.1. Using Hermite's Interpolation formula, find an appropriate polynomial which meets the following specifications:

	x_k	y_k	y_k'
x_1	0	0	0
x_2	1	1	1

Solution: Here $n = 1$. Hence the degree of Hermite's polynomial will be $2n - 1 = 3$.

$$H_1(x) = \frac{x - x_2}{x_1 - x_2} = \frac{x - 1}{0 - 1} = -x + 1$$

$$\Rightarrow H_1'(x) = -1. \qquad \therefore H_1'(x_1) = -1$$

$$H_2(x) = \frac{x - x_1}{x_2 - x_1} = \frac{x - 0}{1 - 0} = +x$$

$$\Rightarrow H_1'(x) = +1. \qquad \therefore H_2'(x_2) = -1$$

$$\begin{aligned}
\phi(x) &= \sum_{k=1}^{2} \{1 - 2H_k'(x_k)(x - x_k)\}\,(H_k(x))^2 f(x_k) + \sum_{k=1}^{2} (x - x_l)\,\{H_k(x)\}^2\, f'(x_k) \\
&= \{1 - 2(-1)(x - 0)\}\,(-x + 1)^2 \times 0 + (x - 0)(-x + 1)^2 \times 0 \\
&\quad + \{1 - 2(+1)(x - 1)\}(+x)^2 \times 1 + (x - 1)(+x)^2 \times 1 \\
&= 0 + \{1 - 2(x - 1)\}x^2 + (x - 1)x^2 \\
&= (3 - 2x)x^2 + (x - 1)x^2 \times 1 \\
&= 3x^2 - 2x^3 + x^3 - x^2 \\
&= 2x^2 - x^3 \\
&= -x^3 + 2x^2
\end{aligned}$$

EXAMPLE 4.8.2. Using Hermite's polynomial, obtain the appropriate polynomial which meets the following specifications:

	x_k	y_k	y_k'
x_1	0	0	0
x_2	4	2	0

Here

$$H_1(x) = \frac{x - x_2}{x_1 - x_2} = \frac{x-4}{0-4} = -\frac{x}{4} + 1$$

$$\therefore H_1'(x) = -\frac{1}{4} \Rightarrow H_1'(x_1) = -\frac{1}{4}$$

$$H_2(x) = \frac{x - x_1}{x_2 - x_1} = \frac{x-0}{4-0} = \frac{x}{4}$$

$$\therefore H_2'(x) = \frac{1}{4} \Rightarrow H_2'(x_2) = \frac{1}{4}$$

Hence

$$\begin{aligned}\phi(x) &= \{1 - 2H_1'(x_1)(x - x_1)\}\{H_1(x)\}^2 y_1 + (x - x_1)\{H_1(x)\}^2 y_1' \\ &\quad + \{1 - 2H_2'(x)(x - x_2)\}\{H_2(x)\} y_2 + (x - x_2)\{H_2(x)\}^2 y_2' \\ &= \left\{1 - 2\left(-\frac{1}{4}\right)(x-0)\right\}\left(-\frac{x}{4}+1\right)^2 \times 0 + (x-0)\left\{-\frac{x}{4}+1\right\}^2 \times 0 \\ &\quad + \left\{1 - 2\left(\frac{1}{4}\right)(x-4)\right\}\left(\frac{x}{4}\right)^2 \times 2 + (x-4)\left(\frac{x}{4}\right)^2 \times 0 \\ &= \left\{1 - \frac{1}{2}(x-4)\right\}\frac{x^2}{16} \times 2 + \frac{x^2(x-4)}{16} \times 0 \\ &= \left(3 - \frac{1}{2}x\right)\frac{x^2}{8} = -\frac{x^3}{16} + \frac{3x^2}{8}\end{aligned}$$

EXERCISES

1. Prepare the divided difference table for the following data and hence obtain $f(50)$:

x	1	3	4	6	10
$f(x)$	0	18	58	190	920

2. Given

   $f(0) = 659,$

   $f(2) = 710,$

   $f(3) = 740,$

   $f(6) = 856$

   Applying Lagrange's formula, find the form of the function $f(x)$.

3. (a) By means of Lagrange's formula prove that, approximately

   $$u_0 = u_2 - 0.3(u_4 - u_{-4}) + 0.2(u_{-4} - u_{-6})$$

   [*Hints:* Shift the origin at 6; obtain u_0, u_2, u_8, u_{10}. Find u_6. Use Lagrange's formula. Finally, revert the origin at 0.]

   (b) By means of Lagrange's formula, find the value of u_1, given

   $u_0 = 0.4,$

   $u_2 = 0.128,$

   $u_3 = 0.224,$

   $u_4 = 0.376$

4. Given u_{-1}, u_0, u_1 and u_2. Using Lagrange's formula, show that

$$u_x = yu_0 + xu_1 + \frac{y(y^2-1)}{3!}\Delta^2 u_{-1} + \frac{x(x^2-1)}{3!}\Delta^2 u_0$$

where $x + y = 1$.

5. Using Lagrange's formula, prove that approximately

$$y_0 = \frac{1}{2}(y_1 + y_{-1}) + \frac{1}{16}\{(y_3 - y_1) - (y_{-1} - y_{-3})\}$$

6. Given $f(x) = 1/x$ for $x = a, b$ and c. Show that the second order divided difference $[a, b, c]$ is equal to $1/abc$.

7. Starting from the definition of divided difference, derive Newton's divided difference formula for unequal intervals. State the necessary assumptions. What modifications will you introduce in the formula if the argument be equispaced at an interval of h?

8. Find the fourth order divided difference with arguments 3, 5, 11, 13, 17 of the functions $f(x) = x^3 - 2x$. Show that $[3, 5, 11, 13, 17]$ is zero. Explain the reason.

9. (a) If $f(x) = 1/x$, show by constructing suitable difference table that

$$[a, b, c, d] = -\frac{1}{abcd}$$

(b) If $f(x) = 1/x^2$, then

$$[a, b, c, d] = -\frac{abc + abd + acd + bcd}{a^2b^2c^2d^2}$$

10. The values of u_x are given at $x = a, b$ and c. Show that under certain conditions (to be stated) the maximum or the minimum value is attained by

$$x = \frac{(b^2 - c^2)u_a + (c^2 - a^2)u_b + (a^2 - b^2)u_c}{2\{(b - c)u_a + (c - a)u_b + (a - b)u_c\}}$$

11. If $u_0, u_1, \ldots, u_6$ be the consecutive terms of a series, prove that

$$u_3 = 0.05(u_0 + u_6) - 0.3(u_1 + u_5) + 0.75(u_2 + u_4)$$

12. If $f(x)$ is given for

$x = n - 1, n - 2, \ldots, 1, 0$

show that $f(x)$ can be written as

$$\frac{x!}{(x-n)!(n-1)!}\left[\frac{f(x_{n-1})}{(x-n+1)} - \binom{n-1}{1}\frac{f(x_{n-2})}{(x-n+2)} + \binom{n-1}{2}\frac{f(x_{n-3})}{(x-x+3)} + \cdots + \binom{n-1}{n-1}\frac{f(x_0)}{x}\right]$$

13. Using definition of divided difference, prove that

(a) $\delta\{f(x) \pm \phi(x)\} = \delta\{f(x)\} \pm \delta\{\phi(x)\}$;

(b) $\delta\{c \cdot f(x)\} = c \cdot \delta\{f(x)\}$;

when c is a constant and δ stands for the operation of divided difference in this case.

14. Using the following table, find $f(x)$ as a polynomial in powers of $(x - 6)$:

x	−1	0	2	3	7	10
$f(x)$	−11	1	1	1	141	561

Also find $f'(6)$, $f''(6)$ and $f'''(6)$.

Chapter 5

Central Difference Interpolation Formulae

Summary

5.1 Introduction

In the earlier chapter, we have obtained Newton's General Divided difference and Lagrange's formula to interpolate $f(x)$ for any intermediate value of x in $x_0 < x < x_n$. Lagrange's formula involves a great deal of computational work, which is rather tedious and will not yield accurate results, if the values of the argument are wide apart. Newton's formulae are fundamental and are applicable to nearly all cases of interpolation. Where the values of the argument x proceed by unit interval, Newton-Gregory Formula can be applied to give satisfactory results. If, however, the value of $f(x)$ is required for some value of x between $x = 0$ and $x = 1$, we should obtain a better result, if our knowledge about the shape of curve $y = f(x)$ extends on both sides of the values of x between which we wish to interpolate $f(x)$. It may be useful, sometimes, to shift the origin to some convenient point so that the entries

$y_0, y_1, y_2, \dots$ become $y_{-3}, y_{-2}, y_{-1}, y_0, y_1, y_2, \dots$

By Newton-Gregory formulae, we may expand $y = f(x) \equiv y_x$ in terms of a given value of y_x on either side of the origin. These will have smaller coefficients and a more rapidly converging series. There are various Central Difference interpolation formulae.

Some of the important central difference interpolation formulae are:

- Newton-Gauss formulae (*Forward and Backward*),
- Stirling's formula,
- Bessel's formulae,
- Laplace-Everett formula.

5.2 Newton-Gauss forward formula (also called Gauss's first form G_1)

Let $y = f(x)$ be given for $(2n+1)$ equispaced values of argument

$x_0, x_0 \pm h, x_0 \pm 2h, \ldots, x_0 \pm nh$

The corresponding entries y_i's, $(i = 0, \pm1, \pm2, \ldots, \pm n)$ are also available. It is required to interpolate $f(x)$ for some value of x lying in $x_0 - h < x < x_0 + h$.

We shall start with Newton's General Divided Difference formula with $(2n+1)$ arguments $x_0, x_1, \ldots, x_{2n}$. [For our convenience, we shall omit operator δ before the box bracket denoting the divided differences.]

Let us set

$$\begin{aligned}
x_0 &= x_0 \text{ (origin)} \\
x_1 &= x_0 + h \\
x_2 &= x_0 - h \\
x_3 &= x_0 + 2h \\
&\ldots\ldots\ldots\ldots \\
&\ldots\ldots\ldots\ldots \\
x_{2n-1} &= x_0 + nh \\
x_{2n} &= x_0 - nh
\end{aligned}$$

The corresponding entries are

$y_0, y_1, y_{-1}, y_2, \ldots, y_n, y_{-n}$ respectively.

We shall show the construction of the difference table in this case with 9 entries as under:

Argument x	Entries y	Δy	$\Delta^2 y$	$\Delta^3 y$	$\Delta^4 y$
x_0-4h	y_{-4}				
		Δy_{-4}			
x_0-3h	y_{-3}		$\Delta^2 y_{-4}$		
		Δy_{-3}		$\Delta^3 y_{-4}$	
x_0-2h	y_{-2}		$\Delta^2 y_{-3}$		$\Delta^4 y_{-4}$
		Δy_{-2}		$\Delta^3 y_{-3}$	
x_0-h	y_{-1}		$\Delta^2 y_{-2}$		$\Delta^4 y_{-3}$
	Gauss's Backward (G_2)	Δy_{-1} (G_2)		$\Delta^3 y_{-2}$ (G_2)	
Origin ← x_0	y_0		$\Delta^2 y_{-1}$		$\Delta^4 y_{-2}$
	Gauss's Forward (G_1)	Δy_0 (G_1)		$\Delta^3 y_{-1}$ (G_1)	
x_0+h	y_1		$\Delta^2 y_0$		$\Delta^4 y_{-1}$
		Δy_1		$\Delta^3 y_0$	
x_0+2h	y_2		$\Delta^2 y_1$		$\Delta^4 y_0$
		Δy_2		$\Delta^3 y_1$	
x_0+3h	y_3		$\Delta^2 y_2$		
		Δy_3			
x_0+4h	y_4				

From Newton's Divided difference interpolation formula with $(2n + 1)$ equispaced arguments, we write

$$f(x) \simeq f(x_0) + (x - x_0)[x_0, x_0 + h] + (x - x_0)(x - x_0 - h)[x_0, x_0 + h, x_0 - h] + \cdots + (x - x_0)(x - x_0 - h)(x - x_0 + h)(x - x_0 + h) \cdots (x - x_0 - nh) [x_0, x_0 + h, \ldots, x_0 - nh]$$

Using the relation between the divided difference and the ordinary difference of the same order

$$[x_0, x_0 + h] = \frac{\Delta y_0}{h}$$ (Suffix of y will correspond to the lowest argument in the divided difference)

$$[x_0, x_0 + h, x_0 - h] = \frac{\Delta^2 y_{-1}}{2!h^2}$$

....

$$[x_0, x_0 + h, x_0 - h, x_0 + 2h, \ldots, x_0 - nh] = \frac{\Delta^{2n} y_{-n}}{(2n)!h^{2n}}$$

We write,

$$f(x) \simeq y_0 + (x - x_0)\frac{\Delta y_0}{h} + (x - x_0)(x - x_0 - h)\frac{\Delta^2 y_{-1}}{2!h^2} + (x - x_0)(x - x_0 - h)(x - x_0 + h)\frac{\Delta^3 y_{-1}}{3!h^3} + \cdots + (x - x_0)(x - x_0 - h) \cdots (x - x_0 - nh)\frac{\Delta^{2n} y_{-n}}{(2n)!h^{2n}}$$

writing $(x - x_0)/h = u$.

$$f(x) \simeq y_0 + u\Delta y_0 + \binom{u}{2}\Delta^2 y_{-1} + \binom{u+1}{3}\Delta^3 y_{-1} + \cdots + \binom{u+n-1}{2n}\Delta^{2n} y_{-n} \quad (5.2.1)$$

This is known as Gauss's Forward Formula (*Gauss's First Form* G_1).

Note: We have assumed that $f(x)$ can be approximated by a polynomial of degree $2n$.

5.3 Gauss's backward formula with $(2n + 1)$ equidistant arguments (also called Gauss's second form G_2)

We set the arguments as under:

$$
\begin{aligned}
x_0 &= x_0, \\
x_1 &= x_0 - h, \\
x_2 &= x_0 + h, \\
x_3 &= x_0 - 2h, \\
x_4 &= x_0 + 2h, \\
&\cdots\cdots\cdots \\
x_{2n} &= x_0 + nh
\end{aligned}
$$

Newton's Divided Difference Formula with the above arguments may be written as

$$
\begin{aligned}
f(x) \simeq y_0 &+ (x - x_0)[x_0, x_0 - h] + (x - x_0)(x - x_0 + h)[x_0, x_0 - h, x_0 + h] + \cdots \\
&+ (x - x_0)(x - x_0 + h) \cdots (x - x_0 - nh)[x_0, x_0 - h, \ldots, x_0 - nh, x_0 + nh]
\end{aligned}
$$

Expressing divided differences in terms of the ordinary differences of the corresponding orders, we write

$$
\begin{aligned}
f(x) \simeq y_0 &+ (x - x_0)\frac{\Delta y_{-1}}{h} + (x - x_0)(x - x_0 + h)\frac{\Delta^2 y_{-1}}{2!h^2} + \cdots \\
&+ (x - x_0)(x - x_0 + h) \cdots (x - x_0 + nh)\frac{\Delta^{2n} y_{-n}}{(2n)!h^{2n}}
\end{aligned}
$$

Writing $(x - x_0)/h = u$.

$$
f(x) \simeq y_0 + u\Delta y_{-1} + \binom{u+1}{2}\Delta^2 y_{-1} + \binom{u+1}{3}\Delta^3 y_{-2} + \cdots + \binom{u+n}{2n}\Delta^{2n} y_{-n} \qquad (5.3.1)
$$

This is *Gauss's Backward Formula.* This form (G_2) employs *odd* differences *below* the central line through y_0 and even differences on the central line. Gauss's Forward formula, however, employs odd differences above the central line through y_0 and even differences on the central line.

5.4 Stirling's central difference interpolation formula with $(2n + 1)$ arguments

This is obtained by taking arithmetic mean of Gauss's forward and Gauss's backward formulae in (5.3.1) and (5.3.2) as

$$f(x) \simeq y_0 + u\frac{\Delta y_0 + \Delta y_{-1}}{2} + \frac{\binom{u}{2} + \binom{u+1}{2}}{2}\Delta^2 y_{-1} + \binom{u+1}{3}\frac{\Delta^3 y_{-1} + \Delta^3 y_{-2}}{2}$$
$$+ \frac{\binom{u+1}{4} + \binom{u+2}{4}}{2}\Delta^4 y_{-2} + \cdots + \frac{\binom{u+n-1}{2n} + \binom{u+n}{2n}}{2}\Delta^{2n} y_{-n}$$
$$\Rightarrow f(x) \simeq y_0 + u\frac{\Delta y_0 + \Delta y_{-1}}{2} + \frac{u^2}{2!}\Delta^2 y_{-1} + \frac{u(u^2-1)}{3!}\cdot\frac{\Delta^3 y_{-1} + \Delta 3 y_{-2}}{2}$$
$$+ \frac{u^2(u^2-1)}{4!}\Delta^4 y_{-2} + \cdots + \frac{u^2(u^2-1^2)(u^2-2^2)\cdots(y^2 - \overline{n-1}^2)}{(2n)!}\Delta^{2n} y_{-n} \quad (5.4.1)$$

This is known as *Stirling's Interpolation formula.* It works on the means of the odd differences above and below the central line through y_0, and even differences on the central line.

5.5 Newton-Gauss's backward formula with pivotal entry y_1 (also called Gauss's third form G_3)

We write

$$\frac{x - x_1}{h} = u' \quad \text{and} \quad \frac{x - x_0}{h} = u \text{ (as before).}$$

$$\therefore x - x_1 = u'h = x - x_0 + x_0 - x_1 = uh - h = (uh - h) = (u-1)h$$

$$\therefore u' = u - 1$$

Thus, Gauss's Backward Formula with origin at x_1 will be obtained from (5.3.1) by advancing the subscripts of y's by 1 and replacing u by $u - 1$.

Thus, we obtain Gauss's backward formula with pivotal entry y_1 as

$$f(x) \simeq y_1 + (u-1)\Delta y_0 + \binom{u}{2}\Delta^2 y_0 + \binom{u}{3}\Delta^3 y_{-1} + \cdots + \binom{u+n-1}{2n}\Delta^{2n} y_{-n+1} \quad (5.5.1)$$

5.6 Bessel's interpolation formula with $(2n + 2)$ arguments

It works on Gauss's Forward formula (G_1) with $(2n + 2)$ arguments

$x_0 = x_0,$

$x_1 = x_0 = h,$

$x_2 = x_0 - h,$

. . . .

$x_{2n} = x_0 - nh,$

$x_{2n+1} = x_0 + \overline{n+1}\, h$

Formula (5.2.1) reduces to

$$f(x) \simeq y_0 + \binom{u}{1}\Delta y_0 + \binom{u}{2}\Delta^2 y_{-1} + \binom{u+1}{3}\Delta^3 y_{-1} + \binom{u+1}{4}\Delta^4 y_{-2} + \cdots$$
$$+ \binom{u+n-1}{2n}\Delta^{2n} y_{-n} + \binom{u+n}{2n+1}\Delta^{2n+1} y_{-n} \quad (5.6.1)$$

Gauss's Backward Formula with (2*n* + 2) arguments

$$x_0 = x_0,$$
$$x_1 = x_0 - h,$$
$$x_2 = x_0 + h,$$
$$x_3 = x_0 - 2h,$$
$$\dots\dots\dots\dots$$
$$x_{2n} = x_0 + nh,$$
$$x_{2n+1} = x_0 - \overline{n+1}h$$

and pivotal entry y_1

$$f(x) \simeq y_1 + (u-1)\Delta y_0 + \binom{u}{2}\Delta^2 y_0 + \binom{u}{3}\Delta^3 y_{-1} + \cdots + \binom{u+n-1}{2n}\Delta^{2n} y_{-n+1} + \binom{u+n-1}{2n+1}\Delta^{2n+1} y_{-n} \quad (5.6.2)$$

Bessel's Formula with (2*n* + 2) arguments will be obtained by taking arithmetic mean of (5.6.1) and (5.6.2) as

$$f(x) \simeq \frac{y_0 + y_1}{2} + \left(u - \frac{1}{2}\right)\Delta y_0 + \binom{u}{2}\frac{\Delta^2 y_0 + \Delta^2 y_{-1}}{2} + \frac{\left(u-\frac{1}{2}\right)(u-1)u}{3!}\Delta^3 y_{-1}$$
$$+ \binom{u+1}{4}\frac{\Delta^4 y_{-1} + \Delta^4 y_{-2}}{2} + \cdots + \binom{u+n-1}{2n}\cdot\frac{\Delta^{2n} y_{-n+1} + \Delta^{2n} y_{-n}}{2}$$
$$+ \frac{\binom{u+n}{2n+1} + \binom{u+n-1}{2n+1}}{2}\Delta^{2n+1} y_{-n}$$

Writing $u - 1/2 = v$,

$$f(x) \simeq \frac{y_0 + y_1}{2} + v\Delta y_0 + \frac{\left(v^2 - \frac{1}{4}\right)}{2!}\cdot\frac{\Delta^2 y_0 + \Delta^2 y_{-1}}{2} + \frac{v\left(v^2 - \frac{1}{4}\right)}{3!}\Delta^3 y_{-1}$$
$$+ \frac{\left(v^2 - \frac{1}{4}\right)\left(v^2 - \frac{9}{4}\right)}{4!}\cdot\frac{\Delta^4 y_{-1} + \Delta^4 y_{-2}}{2}$$
$$+ \frac{v\left(v^2 - \frac{1}{4}\right)\left(v^2 - \frac{9}{4}\right)}{5!}\cdot\Delta^5 y_{-2} + \cdots$$
$$+ \frac{v\left(v^2 - \frac{1}{4}\right)\left(v^2 - \frac{3^2}{4}\right)\cdots\left(v^2 - \frac{\overline{2n-1}^2}{4}\right)}{(2n+1)!}\Delta^{2n+1} y_{-n} \quad (5.6.3)$$

We shall show the employment of the finite differences in a difference table with eight entries for Bessel's formula.

Argument x	Entries y	Δy	$\Delta^2 y$	$\Delta^3 y$	$\Delta^4 y$	$\Delta^5 y$	$\Delta^6 y$
x_{-3}	y_{-3}						
		Δy_{-3}					
x_{-2}	y_{-2}		$\Delta^2 y_{-3}$				
		Δy_{-2}		$\Delta^3 y_{-3}$			
x_{-1}	y_{-1}		$\Delta^2 y_{-2}$		$\Delta^4 y_{-3}$		
		Δy_{-1}		$\Delta^3 y_{-2}$		$\Delta^4 y_{-3}$	
x_0	y_0		$\Delta^2 y_{-1}$		$\Delta^4 y_{-2}$		$\Delta^4 y_{-3}$
		Δy_0		$\Delta^3 y_{-1}$		$\Delta^4 y_{-2}$	
x_1	y_1		$\Delta^2 y_0$		$\Delta^4 y_{-1}$		$\Delta^4 y_{-2}$
		Δy_1		$\Delta^3 y_0$		$\Delta^4 y_{-1}$	
x_2	y_2		$\Delta^2 y_1$		$\Delta^4 y_0$		
		Δy_2		$\Delta^3 y_1$			
x_3	y_3		$\Delta^2 y_2$				
		Δy_3					
x_4	y_4						

If we put $u = 1/2$, or, $v = 0$ in (5.6.3), we obtain

$$f(x) \simeq \frac{y_0 - y_1}{2} - \frac{1}{8} \cdot \frac{\Delta^2 y_{-1} + \Delta^2 y_0}{2} + \frac{3}{128} \cdot \frac{\Delta^4 y_{-1} + \Delta^4 y_{-2}}{2} - \frac{5}{1024} \frac{\Delta^6 y_{-2} + \Delta^6 y_{-3}}{2} + (-1)^n \cdot \frac{\{1.3.5 \cdots (2n - 1)\}^2}{2^{2n} \cdot (2n)!} \cdot \frac{\Delta^{2n} y_{-n} + \Delta^{2n} y_{-n+1}}{2} \quad (5.6.4)$$

This is a special case of Bessel's formula which is used when we are required to compute $f(x)$ midway between the two given values. It uses only the even order differences.

REMARK 5.1. It may be asked which of these formulae, Stirling's or Bessel's would give more accurate result. Sometimes one give more accurate result and sometimes the other. For a given table of differences, the rapidity of convergence depends on the magnitude of u (or v in case of Bessel's formula).

If $-0.25 < u < +0.25$, the series in Stirling's formula will converge more rapidly. On the other hand, if $-0.25 < u - \frac{1}{2} = v < +0.25$, Bessel's formula will make the converge more rapidly.

We may however mention that Stirling's formula or Bessel's formula is applicable for both the cases, (a) odd number of arguments, (b) even number of arguments. In case of even number of arguments, Bessel's formula will exhaust all the possible finite differences in the table, the last odd order difference may be negligible in some cases. In that case, Stirling's formula will also yield fairly close result.

EXAMPLE 5.6.1. Compute the value

$$y = \frac{2}{\sqrt{\pi}} \int_0^x e^{-x^2} dx, \text{ when } x = 0.6538$$

x	y	Δy	$\Delta^2 y$	$\Delta^3 y$	$\Delta^4 y$	$\Delta^5 y$	$\Delta^6 y$
0.62	0.6194114						
		0.0076349					
0.63	0.6270463		−0.0000955				
		0.0075394		−0.0000004			
0.64	0.6345857		−0.0000959		0.0000001		
		0.0074435		−0.0000003		0.0000001	
0.65	0.6420292		−0.0000962		0.0000002		−0.0000004
		0.0073473		−0.0000001		−0.0000003	
0.66	0.6493765		−0.0000963		−0.0000001		
		0.0072510		−0.0000002			
0.67	0.6566275		−0.0000965				
		0.0071545					
0.68	0.6637820						

Solution: Here

$$u = \frac{0.6538 - 0.65}{0.01} = 0.38$$

(a) Using Gauss's Forward Formula (G_1):

$$y = y_0 + u\Delta y_0 + \frac{u(u-1)}{2!}\Delta^2 y_{-1} + \frac{u(u^2-1)}{3!}\Delta^3 y_{-1} + \cdots$$

$$= 0.6420292 + 0.38 \times 0.0073475 + \frac{0.38(-0.62)}{2} \times (-0.0000962) + \cdots$$

$$= 0.6420292 + 0.0027920 + 0.00001133 + \cdots$$

$$= 0.6448325$$

(b) Using Gauss's Backward Formula (G_2):

$$y = y_0 + u\Delta y_{-1} + \frac{u(u+1)}{2!}\Delta^2 y_{-1} + \frac{u(u^2-1)}{3!}\Delta^3 y_{-2} + \cdots$$

$$= 0.6420292 + 0.38 \times 0.0074435 + \frac{0.38 \times (1.38)}{2} \times (-0.0000962) + \cdots$$

$$= 0.6420292 + 0.00282853 - 0.0000252$$

$$= 0.6448325$$

(c) Using Stirling's Formula [A.M. of (a) and (b)].

$$y = \frac{0.64448325 + 0.6448325}{2} = 0.6448325$$

(d) Using Bessel's Formula:

$$y = \frac{y_0 + y_1}{2} + v\Delta y_0 + \frac{\left(v^2 - \frac{1}{4}\right)}{2!} \cdot \frac{\Delta^2 y_{-1} + \Delta^2 y_0}{2} + \frac{v\left(v^2 - \frac{1}{4}\right)}{3!}\Delta^3 y_{-1} + \ldots,$$

where

$$v = u - \frac{1}{2} = -0.12$$

$$= \frac{0.6420292 + 0.6493765}{2} + (-0.12) \times 0.0073473 + (-0.1178)(-0.00009625) + \cdots$$

$$= 0.64570285 - 0.0008817 + 0.00001134 = 0.6448325$$

All the central difference interpolation formulae give the same result up to seven decimal places.

However, we advocate the use of Bessel's formula ($v = -0.12$) since it was sufficient to use only the three terms of the series for greater speed of convergence.

5.7 Laplace-Everett formula

This central difference interpolation formula is helpful in interpolating between a central interval while subdividing an interval.

We may derive this formula from Gauss's forward formula (G_1) in (5.2.1) by replacing all odd order differences by the lower order even differences.

Noting that

$$\Delta f(x_0) \equiv \Delta y_0 = y_1 - y_0$$

$$\Delta^3 f(x_0 - h) \equiv \Delta^3 y_{-1} = \Delta^2 y_0 - \Delta^2 y_{-1}, \quad \text{etc.}$$

We obtain from Gauss's forward formula (5.2.1)

$$f(x) \simeq y_0 + \binom{u}{1}\Delta y_0 + \binom{u}{2}\Delta^2 y_{-1} + \binom{u+1}{3}\Delta^3 y_{-1} + \binom{u+1}{4}\Delta^4 y_{-2} + \cdots + \binom{u+n-1}{2n}\Delta^{2n} y_{-n}$$

$$= y_0 + u(y_1 - y_0) + \binom{u}{2}\Delta^2 y_{-1} + \binom{u+1}{3}(\Delta^2 y_0 - \Delta^2 y_{-1}) + \binom{u+1}{4}\Delta^4 y_{-2} + \binom{u+2}{5}(\Delta^4 y_{-1} - \Delta^4 y_{-2}) + \cdots + \binom{u+n-1}{2n}\Delta^{2n} y_{-n}$$

$$= uy_1 + \binom{u+1}{3}\Delta^2 y_0 + \binom{u+2}{5}\Delta^4 y_{-1} + \cdots + (1-u)y_0 + \left\{\binom{u}{2} - \binom{u+1}{3}\right\}\Delta^2 y_{-1} + \left\{\binom{u+1}{4} - \binom{u+2}{5}\right\}\Delta^4 y_{-2} + \cdots \tag{5.7.1}$$

Noting that

$$\binom{u}{2} - \binom{u+1}{3} = \frac{u(u-1)}{6}(2-u) = \binom{\xi+1}{3}, \quad \text{where } \xi = 1 - u$$

Also

$$\binom{u+1}{4} - \binom{u+2}{5} = \frac{(u+1)u(u-1)(u-2)(3-u)}{120}$$

$$= \frac{(2-\xi)(1-\xi)(-\xi)(2_\xi)(-\xi-1)}{120}, \quad \xi = 1-u$$

$$= \frac{(\xi-2)(\xi-1)(\xi)(\xi+1)(\xi+2)}{120} = \binom{\xi+2}{5}$$

Then (5.7.1) gives

$$f(x) \simeq y_{0+u} = uy_1 + \binom{u+1}{3}\Delta^2 y_0 + \binom{u+2}{5}\Delta^4 y_{-1} + \cdots$$

$$+ \xi y_0 + \binom{\xi+1}{3}\Delta^2 y_{-1} + \binom{\xi+2}{5}\Delta^4 y_{-2} + \cdots \qquad (5.7.2)$$

This is the most common form of *Everett's formula.* Everett's formula uses *only even differences* of the function and therefore for a rapidly converging series of differences, we will have only a few number of terms in both the 'u-series' and the 'ξ-series' of (5.7.2). Again this formula can be conveniently used in sub-tabulation, i.e. sub-division of intervals. Here the terms are such that they may be used twice; they occur both in an "u" expansion and in the reverse order in the next "ξ" expansion.

For example, Everett's formula gives

$$y_{0+u} = uy_1 + \binom{u+1}{3}\Delta^2 y_0 + \binom{u+2}{5}\Delta^4 y_{-1} + \cdots$$

$$+ \xi y_0 + \binom{\xi+1}{3}\Delta^2 y_{-1} + \binom{\xi+2}{5}\Delta^4 y_{-2} + \cdots \qquad (5.7.3)$$

Also

$$y_{1+\xi} = \xi y_2 + \binom{\xi+1}{3}\Delta^2 y_1 + \binom{\xi+2}{5}\Delta^4 y_0 + \cdots$$

$$+ uy_1 + \binom{u+1}{3}\Delta^2 y_0 + \binom{u+2}{5}\Delta^4 y_{-1} + \cdots \qquad (5.7.4)$$

and the second line in $y_{1+\xi}$ is the same as the first line in y_{0+u}.

Suppose we have data, where entries are given at quinquennial points 0, 5, 10, . . . , etc. We are required to interpolate y_1, y_2, y_3, y_4 in between y_0 and y_5 and also to interpolate y_6, y_7, y_8, y_9 in between y_5 and y_{10}.

We may write $u = 0.2, \quad 0.4, \quad 0.6, \quad 0.8$

and $\xi = 0.8, \quad 0.6, \quad 0.4, \quad 0.2.$

The first line of $y_{0.2}$ will be the same as the second line for $y_{1.8}$, but in the reverse order and so on. Thus, the computational labour in sub-tabulations is reduced to a great extent.

To illustrate the above point, we may consider the practical computations of the problem given below:

EXAMPLE 5.7.1. Apply Everett's Formula to subdivide the intervals and find $f(41)$, $f(42)$, $f(43)$, $f(44)$, $f(46)$, $f(47)$, $f(48)$, $f(49)$ from the following data:

x	30	35	40	45	50	55	60
$f(x)$	761	852	998	1220	1560	2100	2980

Hence state the advantages of this method in sub-tabulation.

Table 5.7.1: *Diagonal difference table*

x	y_x	Δy	$\Delta^2 y$	$\Delta^3 y$	$\Delta^4 y$
$30 = x_{-2}$	$761 (= y_{-2})$				
		$91 (= \Delta y_{-2})$			
$35 = x_{-1}$	$852 (= y_{-1})$		$55 (= \Delta^2 y_{-2})$		
		$146 (= \Delta y_{-1})$		$21 (= \Delta^3 y_{-2})$	
$40 = x_0$	$998 (= y_0)$		$76 (= \Delta^2 y_{-1})$		$21 (= \Delta^4 y_{-2})$
		$222 (= \Delta y_0)$		$42 (= \Delta^3 y_{-1})$	
$45 = x_1$	$1220 (= y_1)$		$118 (= \Delta^2 y_0)$		$40 (= \Delta^4 y_{-1})$
		$340 (= \Delta y_1)$		$82 (= \Delta^3 y_0)$	
$50 = x_2$	$1560 (= y_2)$		$200 (= \Delta^2 y_1)$		$58 (= \Delta^4 y_0)$
		$540 (= \Delta y_2)$		$140 (= \Delta^3 y_1)$	
$55 = x_3$	$2100 (= y_3)$		$340 (= \Delta^2 y_2)$		
		$880 (= \Delta y_3)$			
$60 = x_4$	$2980 (= y_4)$				

We take $x_0 = 40$ as origin and $5 (= h)$ as unit of measurement. Transformed values of argument are: −2, −1, 0, 1, 2, 3, 4.

Here $y_{-2}, y_{-1}, y_0, y_1, y_2, y_3, y_4$ are given:

We are required to interpolate

$y_{0+0.2}$, $y_{0+0.4}$, $y_{0+0.6}$, $y_{0+0.8}$

So "u" values in these sub-intervals will be $u = 0.2$, $u = 0.4$, $u = 0.6$, $u = 0.8$ and corresponding ξ values $\xi = 0.8$, $\xi = 0.6$, $\xi = 0.4$, $\xi = 0.2$.

Table 5.7.2: *Computations of 'u-series' and 'ξ-series' in Everett's formula*

x	u	ξ	$\binom{u+1}{3}$	$\binom{u+2}{5}$	$\binom{\xi+1}{3}$	$\binom{\xi+1}{5}$	$\binom{u+1}{3}\Delta^2 y_0$	$\binom{u+2}{5}\Delta^4 y_{-1}$	'u-series' = uy_1 + col(8) + col(9)
(1)	(2)	(3)	(4)	(5)	(6)	(7)	(8)	(9)	(10)
40	0	1.0							
41	0.2	0.8	−0.032	0.006336	−0.048	0.008064	−3.776	0.25344	240.47744
42	0.4	0.6	−0.056	0.01752	−0.064	0.011648	−6.608	0.43008	481.82208
43	0.6	0.4	−0.064	0.011648	−0.056	0.010752	−7.552	0.46592	724.91392
44	0.8	0.2	−0.048	0.0080064	−0.032	0.006336	−5.664	0.32256	970.6585
45			*	*	*	*			*
46	0.8	0.2	−0.048	0.008064	−0.032	0.006336			970.65856
47	0.6	0.4	−0.064	0.011648	−0.056	0.010752			724.91392
48	0.4	0.6	−0.056	0.010752	−0.064	0.011648			481.82208
49	0.2	0.8	−0.032	0.006336	−0.048	0.0080064			240.47744

x	$\binom{\xi+1}{3}\Delta^2 y_{-1}$	$\binom{\xi+2}{5}\Delta^4 y_{-2}$	ξ-series $= \xi y_0$ + col(11) + col(12)	Interpolated y col(10) + col(13)
	(11)	(12)	(13)	(14)
41	−3.648	0.169344	794.92134	1035
42	−4.864	0.244608	594.1806	1076
43	−4.256	0.225792	395.16979	1120
44	−2.432	0.133056	197.30105	1168
45	$\binom{\xi+1}{3}\Delta^2 y_1$	$\binom{\xi+2}{5}\Delta^4 y_0$	ξ-series $= \xi y_2$ + col(15) + col(16)	Interpolated y col(10) + col(17)
	(15)	(16)	(17)	
46	−6.400	0.367488	305.96748	1277
47	−11.200	0.623616	613.42361	1338
48	−12.800	0.675584	923.87558	1406
49	−9.600	0.467712	1238.8677	1479

*Computations shown in these columns are copied in the reverse order from the respective overhead columns.

Note 5.7.1. 'u'-series for $x = 46, 47, 48, 49$, need not be calculated afresh. It is simply copied in the reverse order from 'u-series' for $x = 41, 42, 43$ and 44. Labour of computation is thus reduced to a great extent.

5.8 Sheppard's central difference operators

We have already used the following types of operators:

(1) The Shifting operator E $\Rightarrow$ $Ef(x) = f(x+h)$

(2) The forward difference operator Δ $\Rightarrow$ $\Delta f(x) = f(x+h) - f(x)$

(3) The backward difference operator ∇ (or Δ_1) $\Rightarrow$ $\nabla f(x) = f(x) - f(x-h)$ (or $= \Delta_1 f(x)$).

We shall now introduce the following set of central difference operators:

(4) The central difference operator δ:

$$\delta f(x) = f\left(x + \frac{h}{2}\right) - f\left(x - \frac{h}{2}\right)$$

$$\Rightarrow \delta \equiv (E^{1/2} - E^{-1/2})$$

(5) The mean value operator μ (average operator):

$$\mu f(x) = \frac{1}{2}\left[f\left(x + \frac{h}{2}\right) + f\left(x - \frac{h}{2}\right)\right]$$

$$\Rightarrow \mu \equiv \frac{1}{2}\left[E^{1/2} + E^{1/2}\right]$$

(6) The central sum operator σ:

defined as $\sigma \equiv \delta^{-1}$

$$\Rightarrow \sigma f(x) = \sigma f(x-h) + f\left(x - \frac{h}{2}\right)$$

$$\Rightarrow \quad f(x) = \sigma f\left(x + \frac{h}{2}\right) - \sigma f\left(x - \frac{h}{2}\right)$$

We now prove below some important results based on the above operators δ, μ, σ.

(a) Prove

$\delta \equiv E^{1/2} - E^{-1/2} \equiv \Delta E^{-1/2} \equiv \nabla E^{-1/2}$

Proof. We have

$$\delta f(x) = f\left(x + \frac{h}{2}\right) - f\left(x - \frac{h}{2}\right) \quad \text{[by definition]}$$

$$= E^{1/2} f(x) - E^{-1/2} f(x)$$

$$\therefore \text{ symbolically } \delta \equiv E^{1/2} - E^{-1/2} \tag{5.8.1}$$

$$\equiv \frac{E-1}{E^{1/2}}$$

$$\equiv \Delta E^{-1/2} \tag{5.8.2}$$

$$\equiv \nabla E^{-1/2} \tag{5.8.3}$$

(b) Prove that

$\mu\delta = \frac{1}{2}(\Delta E^{-1} + \Delta)$

Proof. The average operator μ is defined as

$$\mu \equiv \frac{1}{2}(E^{1/2} + E^{-1/2})$$

$$\therefore \mu f(x) = \frac{1}{2}\left[f\left(x + \frac{h}{2}\right) + f\left(x - \frac{h}{2}\right)\right]$$

Also

$$\delta f(x) = f\left(x + \frac{h}{2}\right) - f\left(x - \frac{h}{2}\right) = (E^{1/2} - E^{-1/2}) f(x)$$

$$\therefore \delta \equiv E^{1/2} - E^{-1/2}$$

$$\therefore \mu\delta \equiv \frac{1}{2}\left(E^{1/2} + E^{-1/2}\right)\left(E^{1/2} - E^{-1/2}\right)$$

$$= \frac{1}{2}(E - E^{-1}) = \frac{1}{2}\left(E - \frac{1}{E}\right) = \frac{1}{2}E^{-1}(E+1)(E-1)$$

$$= \frac{1}{2}(1 + E^{-1})\Delta = \frac{1}{2}(\Delta E^{-1} + \Delta) \tag{5.8.4}$$

(c) Prove that

$$\sigma = \frac{E^{1/2}}{E-1}$$

Proof. Central sum operator σ is defined by

$$\sigma f(x) = \sigma f(x-h) + f\left(x - \frac{h}{2}\right)$$

$$\Rightarrow f(x) = \sigma f\left(x + \frac{h}{2}\right) - \sigma f\left(x - \frac{h}{2}\right)$$

Also

$$(\sigma - \sigma E^{-1}) f(x) = \sigma f(x) - \sigma f(x-h) = \sigma f(x-h) f\left(x - \frac{h}{2}\right) - \sigma f(x-h)$$

$$= f\left(x - \frac{h}{2}\right) = E^{-1/2} f(x)$$

$$\therefore \sigma - \sigma E^{-1} \equiv E^{-1/2}, \text{ or, } \sigma(1 - E^{-1}) = E^{-1/2}$$

$$\therefore \sigma = \frac{E^{-1/2}}{1 - \frac{1}{E}} = \frac{E^{1/2}}{E-1} \qquad (5.8.5)$$

(d) To show that

$$\mu^2 \equiv 1 + \frac{1}{4}\delta^2$$

Proof.

$$\mu^2 f(x) = \left\{\frac{1}{2}\left(E^{1/2} + E^{-1/2}\right)\right\}^2 f(x) = \frac{1}{4}\left\{\left(E^{1/2} - E^{-1/2}\right)^2 + 4\right\} f(x)$$

$$= \frac{1}{4}\left(\delta^2 + 4\right) f(x) \Rightarrow \mu^2 \equiv \frac{1}{4}\left(4 + \delta^2\right)$$

EXAMPLE 5.8.1. If D, E, δ and μ be finite difference operators with usual meanings and if $hD = u$, where h is the interval of differencing, prove the following relations between the operators: (a) $E = e^u$, (b) $\delta = 2\sinh u/2$, (c) $\mu = \cosh u/2$, (d) $(1+E)\delta = 2(E-1)\mu$.

Proof. (a) The proof is given by the definition $e^{hD} = (1 + \Delta) = E \Rightarrow e^u = E$.

(b) We have

$$\delta f(x) = f\left(x + \frac{h}{2}\right) - f\left(x - \frac{h}{2}\right) = \left(E^{1/2} - E^{-1/2}\right) f(x)$$

$$\therefore \delta \equiv E^{1/2} - E^{-1/2} = e^{u/2} - e^{-u/2} \quad [\because E \equiv e^u]$$

$$= 2\sinh\frac{u}{2}$$

(c) Agian,

$$\mu = \frac{1}{2}\left\{f\left(x + \frac{h}{2}\right) + f\left(x - \frac{h}{2}\right)\right\} = \frac{1}{2}\left[E^{1/2} + E^{-1/2}\right] f(x)$$

$$\therefore \mu \equiv \frac{1}{2}\left[e^{u/2} + e^{-u/2}\right] = \cosh\frac{u}{2}$$

(d) Also

$$(E+1)\delta f(x) = (E+1)\left[f\left(x+\frac{h}{2}\right) - f\left(x-\frac{h}{2}\right)\right]$$

$$= Ef\left(x+\frac{h}{2}\right) - Ef\left(x-\frac{h}{2}\right) + f\left(x+\frac{h}{2}\right) - f\left(x-\frac{1}{2}\right)$$

$$= f\left(x+\frac{3h}{2}\right) - f\left(x+\frac{h}{2}\right) + f\left(x+\frac{h}{2}\right) - f\left(x-\frac{h}{2}\right)$$

$$= f\left(x+\frac{3h}{2}\right) - f\left(x-\frac{h}{2}\right) \tag{5.8.6}$$

Again

$$2(E-1)\mu f(x) = 2(E-1)\frac{1}{2}\left[f\left(x+\frac{h}{2}\right) + f\left(x-\frac{h}{2}\right)\right]$$

$$= Ef\left(x+\frac{h}{2}\right) + Ef\left(x-\frac{h}{2}\right) - f\left(x+\frac{h}{2}\right) - f\left(x-\frac{h}{2}\right)$$

$$= f\left(x+\frac{3h}{2}\right) + f\left(x+\frac{h}{2}\right) - f\left(x+\frac{h}{2}\right) - f\left(x-\frac{h}{2}\right)$$

$$= f\left(x+\frac{3h}{2}\right) - f\left(x-\frac{h}{2}\right) \tag{5.8.7}$$

From (5.8.6) and (5.8.7), we may write symbolically $(E+1)\delta \equiv 2(E-1)\mu$.

5.9 Summary

Various relations between operators $\delta, \mu, \sigma, \nabla, \Delta, E$ and $D \equiv d/dx$.

(a) By definition $\delta \equiv (E^{1/2} - E^{-1/2}) \equiv E^{-1/2}\Delta \equiv \Delta E^{-1/2} \Rightarrow \delta^n \equiv \Delta^n \cdot E^{-n/2}$.

(b) By definition $\mu \equiv \frac{1}{2}(E^{1/2} + E^{-1/2})$.

(c) $\mu\delta \equiv \frac{1}{2}(\Delta E^{-1} + \Delta)$.

Table 5.9.1: *Difference table in terms of δ*

x	y					
⋮						
−3	y_{-3}					
		$\delta y_{-5/2}$				
−2	y_{-2}		$\delta^2 y_{-2}$			
		$\delta y_{-3/2}$		$\delta^3 y_{-3/2}$		
−1	y_{-1}		$\delta^2 y_{-1}$		$\delta^4 y_{-1}$	
		$\delta y_{-1/2}$		$\delta^3 y_{-1/2}$		$\delta^5 y_{-/2}$
0	y_0		$\delta^2 y_0$		$\delta^4 y_0$	
		$\delta y_{1/2}$		$\delta^3 y_{1/2}$		$\delta^5 y_{1/2}$
1	y_1		$\delta^2 y_1$		$\delta^4 y_1$	
		$\delta y_{3/2}$		$\delta^3 y_{3/2}$		
2	y_2		$\delta^2 y_2$			
		$\delta y_{5/2}$				
3	y_3					
⋮						

Gauss's Forward formula (in this notation) is:

$$y_x = y_0 + \binom{x}{1}\delta y_{1/2} + \binom{x}{2}\delta^2 y_0 + \binom{x+1}{3}\delta^3 y_{1/2} + \binom{x+1}{4} y_0 + \cdots$$

and Stirling's formula is

$$y_x = y_0 + x\mu\delta y_0 + \frac{1}{2}x^2\delta^2 y_0 + \binom{x+1}{3}\mu\delta^3 y_0 + \cdots$$

EXAMPLE 5.9.1. Given y_0, y_1, y_2, y_3, y_4 and y_5 (fifth differences constant), prove that

$$y_{2\frac{1}{2}} = \frac{c}{2} + \frac{25(c-b)+3(a-c)}{256}$$

where $a = y_0 + y_5$, $b = y_1 + y_4$, $c = cy_2 + y_3$.

Proof. For arguments 0, 1, 2, 3, 4 and 5 (origin O), we may take

$$v = u - \frac{1}{2} = x - \frac{1}{2}$$

In Bessel's formula and write

$$x = v + \frac{1}{2}$$

$$\therefore\ y_{v+\frac{1}{2}} = \frac{y_9 + y_1}{2} + v\Delta y_0 + \frac{\left(v^2 - \frac{1}{4}\right)}{2!}\frac{\Delta^2 y_0 + \Delta^2 y_{-1}}{2} + \frac{v\left(v^2 - \frac{1}{4}\right)}{3!}\Delta^3 y_{-1}$$
$$+ \frac{\left(v^2 - \frac{1}{4}\right)\left(v^2 - \frac{9}{4}\right)}{4!}\cdot\frac{\Delta^4 y_{-2} + \Delta^4 3y_{-1}}{2}$$
$$+ \frac{v\left(v^2 - \frac{1}{4}\right)\left(v^2 - \frac{9}{4}\right)}{5!}\Delta^5 y_{-2}$$

Putting $v = 0$.

$$y_{1/2} = \frac{y_0 + y_1}{2} - \frac{1}{6}(\Delta^2 y_0 + \Delta^2 y_{-1}) + \frac{3}{256}(\Delta^4 y_{-2} + \Delta^4 y_{-1})$$

Shifting the origin at 2.

$$y_{2\frac{1}{2}} = \frac{y_2 + y_3}{2} - \frac{1}{16}(\Delta^2 y_2 + \Delta^2 y_1) + \frac{3}{256}(\Delta^4 y_0 + \Delta^4 y_1)$$
$$= \frac{c}{2} - \frac{1}{16}\{(y_4 - 2y_3 + y_2) + (y_3 - 2y_2 + y_1)\}$$
$$+ \frac{3}{256}\{y_4 - 4y_3 + 6y_2 - 4y_1 + y_0 + y_5 - 4y_4 + 6y_3 - 4y_2 + y_1\}$$
$$= \frac{c}{2} - \frac{1}{16}\{(y_1 + y_4) - (y_2 + y_3)\}$$
$$+ \frac{3}{256}\{(y_0 + y_5) - 3(y_1 + y_4) + 2(y_2 + y_3)\}.$$
$$= \frac{c}{2} - \frac{1}{16}(b - c) + \frac{3}{156}\{a - 3b + 2c\}$$
$$= \frac{c}{2} + \frac{1}{256}(3a - 9b + 6c + 16c - 16b)$$
$$= \frac{c}{2} + \frac{25(c-b) + 3a - 3c}{256}$$
$$= \frac{c}{2} + \frac{25(c-b) + 3(a-c)}{256} \quad \text{(Proved)}$$

Example 5.9.2. Prove that if third differences are assumed to be constant

$$y_x = uy_0 + \frac{u(u^2-1)}{3!}\Delta^2 y_{-1} + \frac{\xi(\xi^2-1)}{3!}\Delta^2 y_0$$

where $\xi = 1 - u$.

Proof. We may apply Everett's formula and noting that terms containing Δ^4 and higher order differences are zero, we obtain the result. (here $\xi = \frac{x-x_0}{h}$).

Starting from Gauss's Forward formula, we write

$$y = y_0 + \xi\Delta y_0 + \binom{\xi}{2}\Delta^2 y_{-1} + \binom{\xi+1}{3}\Delta^3 y_{-1} \ (\because \text{ third differences are constant})$$

$$= y_0 + \xi(y_1 - y_0) + \binom{\xi}{2}\Delta^2 y_{-1} + \binom{\xi+1}{3}(\Delta^2 y_0 - \Delta^2 y_{-1})$$

$$= (1-\xi)y_0 + \xi y_1 + \binom{\xi+1}{3}\Delta^2 y_0 + \left\{\binom{\xi}{2} - \binom{\xi+1}{3}\right\}\Delta^2 y_{-1}$$

$$= \xi y_1 + \binom{\xi+1}{3}\Delta^2 y_0 + uy_0 + \binom{u+1}{3}\Delta^2 y_{-1} \quad (u = 1-\xi)$$

$$\left[\because \binom{\xi}{2} - \binom{\xi+1}{3} = \frac{\xi(\xi-1)}{2} - \frac{\xi(\xi^2-1)}{6} = \frac{\xi(\xi-1)}{6}\{3-(\xi+1)\} = \xi(\xi-1)(2-\xi)\right.$$

$$\left. = \frac{(1-u)(-u)(1+u)}{6} = \frac{u(u^2-1)}{6} = \binom{u+1}{3}\right]$$

EXERCISES

1. Starting from the definition of divided differences, obtain Newton's General divided difference formula and give its remainder term.
2. Starting from Newton's General divided difference formula with equispaced values of argument (x), obtain Gauss's forward and backward formula. Hence derive Stirling's interpolation formula.
3. Derive Bessel's Central Difference Interpolation formula. Discuss the situations where Bessel's formula is to be preferred to Stirling's and other central difference interpolation formula.
4. Derive Everett's Central difference interpolation formula and explain its usefulness in sub-tabulations.
5. Given $u_{-1} = 110$, $u_0 = 105$, $u_1 = 107$, $u_2 = 111$; find $u_{1/2}$ by
   (a) Gauss's Forward Formula
   (b) Gauss's Backward Formula
   (c) Stirling's Interpolation Formula
   (d) Bessel's Interpolation Formula
   (e) Laplace-Everett's Formula.
   Which one seems to give the best result? [*Ans:* 106.56; 105.15; 105.86; 105.13; 105.52]

6. Given:

$\sqrt{12400} = 111.355287$

$\sqrt{12450} = 111.579508$

$\sqrt{12500} = 111.803399$

$\sqrt{12550} = 112.026783$

$\sqrt{12600} = 112.249722$

$\sqrt{12650} = 112.472219$

Apply any suitable interpolation formula to evaluate $\sqrt{125.32}$. [*Ans:* 111.9464157]

7. If the third differences are constant, prove that

$$u_{x+\frac{1}{2}} = \frac{1}{2}(u_x + u_{x+1}) - \frac{1}{16}(\Delta^2 u_{x-1} + \Delta^2 u_x)$$

8. Apply Laplace-Everett's Formula to find the suitable polynomial which takes the values u_x as given below:

x	4	6	8	10	12
u_x	1	3	8	20	26

[*Ans:* $y_x = 4(x-8) + \frac{7}{48}(x-6)(x-8)(x-10) - \frac{17}{32 \times 120}(x-4)(x-6)(x-8)(x-10) - 10(x-10) + \frac{1}{8}(x-8)(x-10)(x-12)$]

9. Apply Everett's Formula to sub-divide the intervals and find $f(41), f(42), f(43), f(44), f(46), f(47), f(48), f(49)$ from the following data:

x	30	35	40	45	50	55	60
$f(x)$	761	852	998	1220	1560	2100	2980

Hence state the advantages of this method in sub-divisions.

[*Ans:* $f(41) = 1035, f(42) = 1076, f(43) = 1120, f(44) = 1168, f(46) = 1277, f(47) = 1338, f(48) = 1406, f(49) = 1479.$]

10. Use Stirling's formula to find u_{36}, given $u_{20} = 512$, $u_{30} = 439$, $u_{40} = 346$, $u_{50} = 243$. [*Ans:* 385.6.]

11. Interpolate by means of Gauss's Backward Formula, the production ('000 tons) in 1936, given the following table:

Year	1900	1910	1920	1930	1940	1950
Productions ('000 tones)	120	150	200	270	390	520

[*Ans:* 332.2.]

12. Apply Bessel's formula to obtain u_{25}, given $u_{20} = 2864$, $u_{24} = 3186$, $u_{28} = 3587$, $y_{32} = 3970$. [*Ans:* 3282.6.]

[*Hints:* Take $x = 24$ as origin.]

13. (a) Define the operators E, Δ, μ, δ, D and ∇ used in calculus of finite differences and prove that

i. $\delta \equiv \Delta E^{-1/2} \equiv \nabla E^{1/2}$

ii. $\mu^2 \equiv 1 + \frac{1}{4}\delta^2$

iii. $\delta^n y_x = \Delta^n E^{-3/2} y_x = \nabla^n E^{3/2} y_x$ [use result in (i)]

iv. $E^{1/2} f(x) = \mu f(x) + \frac{1}{2}\delta^2 f(x)$

(b) Show that $\delta\sigma \neq \sigma\delta$

14. (a) Symbols having their usual meanings, show that

i. $\delta^2 y_0 = y_1 - 2y_0 + y_{-1}$

ii. $\delta^2 y_{1/2} = y_2 - 3y_1 + 3y_0 - y_{-1}$

iii. If, further, $h.D = u$, then $\delta = 2\sinh\frac{u}{2}$

(b) i. $\left(\frac{u}{\delta}\right)^2 = 1 - \frac{\delta^2}{12} + \frac{\delta^4}{90} - \frac{\delta^6}{560} + 0(\delta)^8.$

[*Hints:* First show that $\frac{u}{\delta} = \frac{2}{\delta}\sinh^{-1}\frac{\delta}{2}$, then expand RHS in power series of δ^2.]

ii. $\left(\frac{u}{\delta}\right) = 1 - \frac{\delta^2}{24} + \frac{3}{640}\delta^4 + \frac{5}{7168}\delta^6 + 0(\delta)^8.$

(c) $\frac{u}{\delta\mu} = 1 - \frac{\delta^2}{6} + \frac{\delta^4}{30} - \frac{\delta^6}{140} + 0(\delta)^8.$

(d) Show that in general

$$\delta^n f(x) = \sum_{k=0}^{n}(-1)^k \frac{n!}{k!(n-k)!} f\left(x + \frac{n}{2} - k\right)$$

Hence find $\delta^2 f(x)$ and $\delta^3 f(x)$.

[*Hints:* $\delta^n = (E^{1/2} - E^{-1/2})^n = E^{-n/2}(E-1)^n$.

$$\therefore \delta^n f(x) = E^{-n/2}(E-1)^n f(x) = (E-1)^n f\left(x - \frac{n}{2}\right)$$

$$= \sum_{k=0}^{n}(-1)^k \cdot \binom{n}{k} \cdot E^{n-k} f\left(x - \frac{n}{2}\right)$$

$$= \sum_{k=0}^{n}(-1)^k \binom{n}{k} f\left(x + \frac{n}{2} - k\right)]$$

15. (a) Show that

$$\Delta\,[f(x-1)\Delta g(x-1)] = \nabla\,[f(x)\Delta g(x)]$$
$$= \Delta\,[f(x-1)\nabla g(x-1)]$$
$$= \Delta\,[f(x) - \nabla g(x)]$$

(b) $\Delta\nabla f(x) = \nabla\Delta f(x) = \delta^2 f(x)$.

(c) If $\nabla f(x) = f(x) - f(x-1)$, show that

$$\nabla^n f(x) = f(x) - \binom{n}{1} f(x-1) + \binom{n}{2} f(x-2) + \cdots + (-1)^n f(x-n)$$

16. Find $\Delta^3(1-x)(1-2x)(1-3x)$, when the interval of differencing is unity.

17. $\Delta \log f(x) = \log\left\{1 + \frac{\Delta f(x)}{f(x)}\right\}$.

Chapter 6

The Accuracy of Interpolation Formulae

Summary

6.1 Introduction

In the problems of interpolation, we approximated the unknown function, $f(x)$ by a polynomial $\phi(x) \cdot f$ suitable degree in x, such that

$f(x_i) = \phi(x_i), \quad i = 0, 1, 2, \ldots, n$

The values of arguments $x_0, x_1, x_2, \ldots, x_n$ are given and the corresponding entries $f(x_0), f(x_1), f(x_2), \ldots, f(x_n)$ are also given. At all other points within the range (x_0, x_n), $f(x) \neq \phi(x)$.

We write

$f(x) = \phi(x) + R(x)$

where $R(x)$ in the error term or remainder term. In the following sections we shall try to derive theoretical expressions of the error terms of the different interpolation formulae.

6.2 Remainder term in Newton's forward formula and in Lagrange's interpolation formula

We shall assume that the unknown function $f(x)$ is continuous and possesses continuous derivatives of all orders within the interval $[x_0, x_n]$. We may define an arbitrary function

$F(z)$ given by

$$F(z) = f(z) - \phi(z) - \{f(x) - \phi(x)\} \cdot \frac{(z - x_0)(z - x_1)\cdots(z - x_n)}{(x - x_0)(x - x_1)\cdots(x - x_n)}$$

where $f(x)$ denotes the given function and $\phi(x)$ denotes the interpolating polynomial.

So $F(z)$ is also continuous, possesses continuous derivatives of all orders in $[x_0, x_n]$ and vanishes for the $(n + 2)$ values, $z = x, x_0, x_1, x_2, \ldots, x_n$.

Thus, $F(z)$ satisfies all the conditions of Rolle's theorem in each of the $(n+1)$ sub-intervals given by the points $x_1, x_0, x_1, \ldots, x_n$. Hence $F'(z)$ vanishes at least once in each of these $(n + 1)$ sub-intervals. Therefore, within the entire interval (x_0, x_n), $F'(x)$ vanishes at least $(n + 1)$ times; $F''(z)$ vanishes at least n times; $F'''(z)$ vanishes at least $(n - 1)$ times, etc. Hence $F^{(n+1)}(z)$ vanishes at least once in (x_0, x_n), say, at ξ; $(x_0 < \xi < x_n)$.

We may note that the expression $(z - x_0), (z - x_1), \ldots, (z - x_n)$ is a polynomial of degree $(n + 1)$ in z with coefficient of $z^{n+1} = 1$. Therefore, $(n + 1)^{th}$ derivative of this polynomial will be $(n + 1)!$.

$$\therefore F^{(n+1)}(\xi) = f^{(n+1)}(\xi) - 0 - \{f(x) - \phi(x)\} \frac{(n + 1)!}{(x - x_0), (x - x_1), \ldots, (x - x_n)}$$

$$= 0$$

$$\therefore f(x) - \phi(x) = \frac{f^{(n+1)}(\xi)}{(n + 1)!}(x - x_0), (x - x_1), \ldots, (x - x_n) \tag{6.2.1}$$

$$\therefore \text{Error} = R_n(x) = \frac{f^{(n+1)}(\xi)}{(n + 1)!}(x - x_0), (x - x_1), \ldots, (x - x_n)$$

where ξ is some value of n between x_0 and x_n.

This remainder term in (6.2.1) is true for Lagrange's interpolation formula.

For equispaced intervals of arguments $x_0, x_1, \ldots, x_n$, we may note that

$$x - x_0 = hu,$$
$$x - x_1 = h(u - 1),$$
$$x - x_2 = h(u - 2),$$
$$\ldots\ldots\ldots\ldots\ldots$$
$$x - x_n = h(u - n)$$

Then

$$R(x) = \frac{h^{(n+1)} f^{(n+1)} \xi}{(n + 1)!} u(u - 1), (u - 2), \ldots, (u - n)$$

The above is the analytical form of the error in Newton's Forward Formula. Using the relation between $D \equiv d/dx$ and Δ and noting that

$$f^{(n+1)}(\xi) \equiv D^{n+1}(\xi) \simeq \frac{\Delta^{n+1} f(x_0)}{h^{n+1}} \text{ (practically)}$$

we may write

$$R(x) = \frac{\Delta^{n+1} y_0}{(n + 1)!} u(u - 1), (u - 2), \ldots, (u - n) \tag{6.2.2}$$

The smaller the interval of h is taken, the more nearly does (6.2.2) give the actual error.

6.3 Remainder term in Newton's backward formula

Here we take the auxiliary function $f(z)$ in the form:

$$F(z) = f(z) - \phi(z) - \{f(x) - \phi(x)\} \cdot \frac{(z - x_n), (z - x_{n-1}), \ldots, (z - x_0)}{(x - x_n), (x - x_{n-1}), \ldots, (x - x_0)}$$

Proceeding as in (6.2.2), we have

$$F^{n+1}(\xi) = f^{(n+1)}(\xi) - 0 - R_n(x) \cdot \frac{(n+1)!}{(x - x_n), (x - x_{n-1}), \ldots, (x - x_0)}$$

$$\therefore R_n(x) = \frac{f^{(n+1)}(\xi)}{(n+1)!}(x - x_n), (x - x_{n-1}), \ldots, (x - x_0)$$

where $x_0 < \xi < x_n$.

Noting that

$$\begin{aligned} x - x_n &= hu, \\ x - x_{n-1} &= h(u+1), \\ x - x_{n-2} &= h(u+2), \\ &\ldots\ldots\ldots\ldots\ldots \\ x - x_0 &= h(u+n), \end{aligned}$$

we write

$$R_n(x) \equiv f(x) - \phi(x) = \frac{h^{n+1} f^{n+1}(\xi)}{(n+1)!} u(u+1), (u+2), \ldots, (u+n) \tag{6.3.1}$$

When the analytical form of the function $f(x)$ is unknown, we may replace $f^{n+1}(\xi)$ by

$h^{n+1}\Delta_{n+1}f(x_n)$, [or $h^{n+1}\nabla^{n+1}f(x_n)$]

and write

$$R_n(x) = \frac{\Delta_{n+1} y_n}{(n+1)!} u(u+1), (u+2), \ldots, (u+n) \text{ (notation of horizontal difference table)}$$

$$= \frac{\nabla^{n+1} y_n}{(n+1)!} u(u+1), (u+2), \ldots, (u+n) \text{ (notation of backward differences).}$$

6.4 Remainder term in Stirling's formula

Here we assume that $f(x)$ is continuous and possesses continuous derivatives of all orders in $[x_{-n}, x_n]$.

The auxiliary function $F(z)$ is given by

$$F(z) = f(z) - \phi(z) - \{f(x) - \phi(x)\} \cdot \frac{(z - x_0)(z - x_1)(z - x_{-1}) \cdots (z - x_n)(z - x_{-n})}{(x - x_0)(x - x_1)(x - x_{-1}) \cdots (x - x_n)(x - x_{-n})}$$

Here also $F(z)$ is continuous and possesses continuous derivatives of all orders in (x_{-n}, x_n) and vanishes for the $(2n+2)$ values of $z = x$, $z = x_0$, $z = x_1$, $z = x_{-1}$, $\ldots$, $z = x_n$, $z = x_{-n}$.

Noting that $\phi(x)$ is a polynomial of degree $2n$, we write

$$F^{(2n+1)}(\xi) = f^{(2n+1)}(\xi) - 0$$
$$- R(x) \cdot \frac{(2n+1)!}{(x-x_0)(x-x_1)(x-x_{-1})\cdots(x-x_n)(x-x_{-n})} = 0$$

$$\Rightarrow R(x) = \frac{f^{(2n+1)}(\xi)}{(2n+1)!}(x-x_0)(x-x_1)\cdots(x-x_n)(x-x_{-n})$$

Writing $x - x_0 = hu$ and $x_1 = x_0 + h$, $x_{-1} = x_0 - h$, etc.

$$R(x) = \frac{f^{(2n+1)}(\xi)}{(2n+1)!} h^{2n+1} u(u^2 - 1^2)\cdots(u^2 - n^2) \tag{6.4.1}$$

where $x_{-n} < \xi < x_n$.

If the analytical form of the function $f(x)$ is unknown, we replace $f^{(2n+1)}(\xi)$ by m_{2n+1}, where

$$m_{2n+1} = \frac{\Delta^{2n+1} y_{-n-1} + \Delta^{2n+1} y_{-n}}{2}$$

and the error term in Stirling's formula becomes

$$R(x) = \frac{\Delta^{2n+1} y_{-n-1} + \Delta^{2n+1} y_{-n}}{2} \cdot \frac{1}{(2n+1)!} u(u^2 - 1)\cdots(u^2 - n^2) \tag{6.4.2}$$

In formulae (6.4.1) and (6.4.2), n is the number of intervals on each side of the origin (x_0).

6.5 Remainder term in Bessel's formula

Here the arbitrary function $F(z)$ may be written as

$$F(z) = f(z) - \phi(z) - \{f(x) - \phi(x)\}$$
$$\times \frac{(z-x_0)(z-x_1)(z-x_{-1})\cdots(z-x_n)(z-x_{-n})(z-x_{n+1})}{(x-x_0)(x-x_1)(x-x_{-1})\cdots(x-x_n)(x-x_{-n})(x-x_{n+1})}$$

$F(z)$ is continuous and possesses continuous derivatives of all orders in the interval $[x_{-n}, x_{n+1}]$. $F(z)$ also vanishes at the $(2n + 3)$ points $z = x, x_0, x_1, x_{-1}, \cdots x_{+n}, x_{-n}, x_{n+1}$. Since $\phi(z)$ is a polynomial of degree $(2n + 1)$, its $(2n + 2)$th derivative is zero.

$$\therefore F^{(2n+2)}(\xi) = f^{(2n+2)}(\xi) - 0 - \{f(x) - \phi(x)\}$$
$$\times \frac{(2n+2)!}{(x-x_0)(x-x_1)(x-x_{-1})\cdots(x-x_n)(x-x_{-n})(x-x_{n+1})} = 0$$

$$\Rightarrow R_n(x) = f(x) - \phi(x)$$
$$= \frac{f^{(2n+2)}(\xi)}{(2n+2)!}(x-x_0)(x-x_1)(x-x_{-1})\cdots(x-x_n)(x-x_{-n})(x-x_{n+1})$$

Putting

$(x - x_0) = hu$, $(x - x_1) = h(u - 1)$, $(x - x_{-1}) = h(u + 1)$

$(x - x_2) = h(u - 2)$, $(x - x_{-2}) = h(u + 2)$, etc.

we may write

$$R_n(x) = \frac{f^{(2n+2)}(\xi) \cdot h^{2n+2}}{(2n+2)!} \cdot u(u^2 - 1)(u^2 - 2^2)\cdots(u^2 - n^2)(u - n - 1) \tag{6.5.1}$$

This is the error term, when the analytical form of the unknown function $f(x)$ is known If, however, $f(x)$ is not known we may write

$$R_n(x) = \frac{m_{2n+2}}{(2n+2)!} u(u^2-1)(u^2-2^2)\cdots(u^2-n^2)(u-n-1)$$

where

$$m_{2n+2} = \frac{\Delta^{2n+2} y_{-n-1} + \Delta^{2n+2} y_{-n}}{2} \tag{6.5.2}$$

On putting

$$u = \left(v + \frac{1}{2}\right)$$

in (6.5.1) and (6.5.2), we obtain

$$R_n(x) = \frac{h^{2n+2} f^{(2n+2)}(\xi)}{(2n+2)!} \left(v^2 - \frac{1}{4}\right)\left(v^2 - \frac{9}{4}\right)\cdots\left(v^2 - \frac{\overline{2n+1^2}}{4}\right) \tag{6.5.3}$$

$$R_n(x) = \frac{m_{2n+2}}{(2n+2)!} \left(v^2 - \frac{1}{4}\right)\left(v^2 - \frac{9}{4}\right)\cdots\left(v^2 - \frac{\overline{2n+1^2}}{4}\right) \tag{6.5.4}$$

where

$$m_{2n+2} = \frac{\Delta^{2n+2} y_{-n-1} + \Delta^{2n+2} y_{-n}}{2}$$

Putting $v = 1/2$ in (6.5.3) and (6.5.4), we get the remainder terms in the Bessel's formula for interpolating two halves, namely,

$$R_n(x) = \frac{h^{2n+2} f^{(2n+2)}(\xi)}{(2n+2)!} (-1)^{n+1} \frac{\{1.3.5\cdots(2n+1)\}^2}{2^{2n+2}} \tag{6.5.5}$$

$$R_n(x) = \frac{m_{2n+2}}{(2n+2)!} (-1)^{n+1} \frac{\{1.3.5\cdots(2n+1)\}^2}{2^{2n+2}} \tag{6.5.6}$$

6.6 Remainder term in Newton's divided difference formula

Newton's general divided difference formula gives

$$\begin{aligned} f(x) = f(x_0) &+ (x-x_0)\delta[x_0,x_1] + (x-x_0)(x-x_1)\delta[x_0,x_1,x_2] + \cdots \\ &+ (x-x_0)(x-x_1)\cdots(x-x_{n-1})\delta[x_0,x_1,x_2,\ldots,x_n] \\ &+ (x-x_0)(x-x_1)\cdots(x-x_n)\delta[x,x_0,x_1,x_2,\ldots,x_n] \end{aligned}$$

The interpolating polynomial of nth degree is given by

$$\begin{aligned} \phi(x) &= f(x_0) + (x-x_0)\delta[x_0,x_1] + \cdots \\ &\quad + (x-x_0)(x-x_1)\cdots(x-x_{n-1})\delta[x_0,x_1,\ldots,x_n] \\ \therefore R_n(x) &= f(x) - \phi(x) = (x-x_0)(x-x_1)\cdots(x-x_n)\delta[x,x_0,x_1,\ldots,x_n] \end{aligned}$$

If the arguments be equispaced at an interval of h

$$\begin{aligned} R_n(x) &= u(u-1)\cdots(u-n)h^{n+1} \frac{\Delta^{n+1} y_0}{(n+1)!\,h^{n+1}} \\ &= u(u-1)(u-2)\cdots(u-n)\frac{\Delta^{n+1} y_0}{(n+1)!} \end{aligned} \tag{6.6.1}$$

6.7 The accuracy of linear interpolation from tables

We shall now try to obtain a simple expression for the maximum error inherent in linear interpolation from tables.

We write the auxiliary function $F(z)$ as

$$F(z) = f(z) - \phi(z) - \{f(x) - \phi(x)\} \cdot \frac{(z - x_0)(z - x_1)}{(x - x_0)(x - x_1)}$$

We assume $f(x)$ to be continuous and continuously differentiable in $[x_0, x_1]$ and $\phi(x)$ is the interpolating polynomial of degree 1 in x, hence $F(z)$ is also continuous and continuously differentiable in $[x_0, x_1]$ and $F(z)$ vanishes at three points $z = x, x_0, x_1$.

$\therefore$ by Rolle's theorem $F'(z)$ vanishes at least two times in $[x_0, x_1]$ and $F''(z)$ vanishes at least once in $[x_0, x_1]$.

$$\therefore F''(\xi) = f''(\xi) - 0 - \{f(x) - \phi(x)\} \cdot \frac{2!}{(x - x_0)(x - x_1)} = 0$$

$$\therefore f(x) - \phi(x) = R_1 = \frac{f''(\xi)}{2!}(x - x_0)(x - x_1)$$

where $x_0 < \xi < x_1$.

Let M be the maximum absolute value of the second derivative of $f(x)$ in $[x_0, x_1]$ and we write $(x - x_0)/h = u$, where $h = x_1 - x_0$. Then

$$R_1 = \frac{1}{2}h^2 Mu(u - 1) = \frac{1}{2}h^2 M(u^2 - u)$$

$$\frac{dR_1}{du} = \frac{1}{2}h^2 M(2u - 1)$$

To find the maximum value of R_1, we take

$$\frac{dR_1}{du} = 0 \Rightarrow u = \frac{1}{2}$$

$$\therefore |R_1(\max)| = \frac{1}{2}h^2 M\left|\frac{1}{4} - \frac{1}{2}\right| = \frac{1}{8}h^2 M$$

Thus,

$$R_1(\max) \leq \frac{1}{8}h^2 M \tag{6.7.1}$$

Note: The student should always bear in mind that linear interpolation is permissible only when first differences are constant or nearly so. We should therefore always compute a few first differences and verify if they are nearly constant before we use linear interpolation formula.

EXAMPLE 6.7.1. In Barlow's Tables, the function $1/N$ is tabulated at unit intervals from 1 to 12,500. Find the possible error in the linear interpolation of this function when $N = 574$.

Solution:

$$f(N) = \frac{1}{N}, \quad f'(N) = -\frac{1}{N^2}. \quad \therefore f''(N) = \frac{2}{N^3}$$

Taking $h = 1$, $N = 574$ and substituting in (6.7.1)

$$R \leq \frac{1}{4(574)^3} = \frac{1}{756476896}, \quad \text{or,} \quad R_{(\max)} < 0.00000000132$$

6.8 Recapitulation of formula for the remainder

We may collect all the remainder terms in different interpolation formula for easy reference.

6.8.1 Newton's forward formula

- $R_n = \dfrac{h^{n+1}f^{(n+1)}(\xi)}{(n+1)!}u(u-1)(u-2)\cdots(u-n)$ (analytical form).
- $R_n\dfrac{\Delta^{n+1}y_0}{(n+1)!}u(u-1)(u-2)\cdots(u-n)$.

6.8.2 Newton's backward formula

- $R_n = \dfrac{h^{n+1}f^{(n+1)}(\xi)}{(n+1)!}u(u+1)(u+2)\cdots(u+n)$ (analytical form).
- $R_n\dfrac{\Delta^{n+1}y_0}{(n+1)!}u(u+1)(u+2)\cdots(u+n)$ (notation of horizontal difference table).

  or, $\dfrac{\nabla^{n+1}y_n}{(n+1)!}u(u+1)(u+2)\cdots(u+n)$ (notation of Backward difference).

6.8.3 Stirling's formula

- $R_n = \dfrac{h^{2n+1}f^{(2n+1)}(\xi)}{(2n+1)!}u(u^2-1^2)(u^2-2^2)\cdots(u^2-n^2)$ (analytical form).
- $R_n\dfrac{m_{2n+1}}{(2n+1)!}u(u^2-1^2)(u^2-2^2)\cdots(u^2-n^2)$,

  where $m_{2n+1} = \dfrac{\nabla^{2n+1}y_{-n-1} + \nabla^{2n+1}y_{-n}}{2}$.

6.8.4 Bessel's formula

- $R_n = \dfrac{h^{2n+2}f^{2n+2}(\xi)}{(2n+2)!}\left(v^2-\dfrac{1}{4}\right)\left(v^2+\dfrac{9}{4}\right)\cdots\left(v^2-\dfrac{\overline{2n+1}^2}{4}\right)$ (analytical form).
- $R+n = \dfrac{m_{2n+2}}{(2n+2)!}\left(v^2-\dfrac{1}{4}\right)\left(v^2+\dfrac{9}{4}\right)\cdots\left(v^2-\dfrac{\overline{2n+1}^2}{4}\right)$,

  where $m_{2n+2} = \dfrac{\Delta^{2n+2}y_{-n-1} + \Delta^{2n+2}y_{-n}}{2}$.

6.8.5 Lagrange's formula

- $R_n = \dfrac{f^{(n+1)}(\xi)}{(n+1)!}(x-x_0)(x-x_1)\cdots(x-x_n)$ (analytical form).
- $R_n = [xx_0x_1\cdots x_n](x-x_0)(x-x_1)\cdots(x-x_n)$.

Chapter 7

Inverse Interpolation

Summary

.1 Introduction

general interpolation problems discussed in the earlier chapters (Chapters 4 and 5), we ere given values of y corresponding to various values of the argument x (independent ariable), and we were required to interpolate a value of the entry y (dependent variable) orresponding to a value of x, intermediate between two given values. However, on the asis of the given (x_i, y_i) $(i = 1, 2, \ldots, n)$, if it is required to interpolate the value of the rgument corresponding to an intermediate value of the *entry*, the process is just reverse nd it is called inverse interpolation.

When the form of the function $y = f(x)$ is known and the inverse function $x = f^{-1}(y)$ xists, the problem is easy to solve. But when the form of the function is not known and the ata give only a set of (x, y) values, the problem is slightly complicated. By replacing $f(x)$ y a suitable polynomial $\phi(x)$, we may express y in terms of the interpolating polynomial

$$y = y_0 + u\Delta y_0 + \binom{u}{2}\Delta^2 y_0 + \cdots + \binom{u}{n}\Delta^n y_0$$

where

$$u = \frac{x - x_0}{h}, \quad (h = \text{unit of differencing}).$$

Here for a given y, we shall have an equation in x (of degree n), which is a difficult problem to solve. If, however, we assume that the second order differences are constant, we shall have a quadratic equation in x which is not a laborious problem. The assumption that the second order differences are constant, however, will not always be admissible. The solution also gives more than one value of x corresponding to a given y.

We now propose a number of procedures to solve the problem of inverse interpolation.

- Use of Lagrange's interpolation formula
- Method of successive approximations or iteration
- Reversion of series.

7.2 Use of Lagrange's formula for inverse interpolation

In Lagrange's interpolation formula

$$y = \frac{(x - x_1)(x - x_2)\cdots(x - x_n)}{(x_0 - x_1)(x_0 - x_2)\cdots(x_0 - x_n)} y_0 + \cdots + \frac{(x - x_0)(x - x_1)\cdots(x - x_{n-1})}{(x_n - x_0)(x_n - x_1)\cdots(x_n - x_{n-1})} y_n$$

where y is expressed as a function of x.

If we interchange x and y we may write

$$\begin{aligned} x = {} & \frac{(y - y_1)(y - y_2)\cdots(y - y_n)}{(y_0 - y_1)(y_0 - y_2)\cdots(y_0 - y_n)} x_0 + \frac{(y - y_0)(y - y_2)\cdots(y - y_n)}{(y_1 - y_0)(y_1 - y_2)\cdots(y_1 - y_n)} x_1 + \cdots \\ & + \frac{(y - y_0)(y - y_1)\cdots(y - y_{r-1})(y - y_{r+1})\cdots(y - y_n)}{(y_{r-1} - x_0)(y_r - y_1)\cdots(y_r - y_{r-1})(y_r - y_{r+1})\cdots(y_r - y_n)} x_r + \cdots \\ & + \frac{(y - y_0)(y - y_1)\cdots(y - y_{n-1})}{(y_n - y_0)(y_n - y_1)\cdots(y_n - y_{n-1})} x_n \end{aligned} \tag{7.2.1}$$

It may be said that in expression (7.2.1), x is expressed as a function of y. We may use the above formula for inverse interpolation.

7.3 Method of successive approximation

We shall consider the case, where the values of x are given for equispaced interval of h as $x_0, x_1, x_2, \ldots, x_n$; $(x_i = x_0 + ih)$ $(i = 0, 1, 2, \ldots, n)$. The corresponding values of y are y_0, $y_1, y_2, \ldots, y_n$. We are required to interpolate x, for some y :

- near the beginning of the tabulated values of y.
- near the end of the tabulated values of y.
- somewhere near the middle of the tabulated values of y.

Our starting formula will be respectively: Newton's forward formula, Newton's backward formula and any of the suitable central difference interpolation formula.

We first consider Newton's forward formula given by

$$\begin{aligned} y = {} & y_0 + u\Delta y_0 + \frac{u(u-1)}{2!}\Delta^2 y_0 + \frac{u(u-1)(u-2)}{3!}\Delta^3 y_0 + \cdots \\ & + \frac{u(u-1)(u-2)\cdots(u-n+1)}{n!}\Delta^n y_0 \end{aligned}$$

where $u = (x - x_0)/h$.

Transposing and dividing by Δy_0,

$$\begin{aligned} \frac{y - y_0}{\Delta y_0} = {} & u + \frac{u(u-1)}{2!} \cdot \frac{\Delta^2 y_0}{\Delta y_0} + \frac{u(u-1)(u-2)}{3!} \cdot \frac{\Delta^3 y_0}{\Delta y_0} + \cdots \\ & + \frac{u(u-1)\cdots(u-n+1)}{n!} \cdot \frac{\Delta^n y_0}{\Delta y_0} \end{aligned} \tag{7.3.1}$$

To get the first approximation of u, we neglect all differences higher than the first, and write

$$u^{(1)} = \frac{y - y_0}{\Delta y_0}$$

For subsequent approximations of u, we write the expression (7.3.1) as

$$u = \frac{y - y_0}{\Delta y_0} - \frac{u(u-1)}{2!} \cdot \frac{\Delta^2 y_0}{\Delta y_0} - \frac{u(u-1)(u-2)}{3!} \cdot \frac{\Delta^3 y_0}{\Delta y_0} - \cdots - \frac{u(u-1)\cdots(u-n+1)}{n!} \cdot \frac{\Delta^n y_0}{\Delta y_0}$$

Hence

$$u^{(2)} = \frac{y - y_0}{\Delta y_0} - \frac{u^{(1)}(u^{(1)}-1)}{2!} \cdot \frac{\Delta^2 y_0}{\Delta y_0} - \frac{u^{(1)}(u^{(1)}-1)(u^{(2)}-2)}{3!} \cdot \frac{\Delta^3 y_0}{\Delta y_0} \cdots - \frac{u^{(1)}(u^{(1)}-1)\cdots(u^{(1)}-n+1)}{n!} \cdot \frac{\Delta^n y_0}{y_0} \cdots$$

$$u^{(n)} = \frac{y - y_0}{\Delta y_0} - \frac{u^{(n-1)} - (u^{(n-1)}-1)}{2!} \cdot \frac{\Delta^2 y_0}{\Delta y_0} \cdots - \frac{u^{(n-1)}(u^{(n-1)}-1)\cdots(u^{(n-1)}-n+1)}{n!} \cdot \frac{\Delta^n y_0}{\Delta y_0}$$

We continue the approximations, till the two successive approximations of u agree up to the required number of decimal places.

Suppose for the given y, the interpolated value of u is $u^{(n)}$. Then $u^{(n)} = (x - x_0)/h$ gives $x = x_0 + hu^{(n)}$.

For the other two cases (2) and (3), we use the same technique with Newton's backward formula and central difference interpolation formula respectively.

We shall illustrate this method with the following example where we use Gauss's forward interpolation formula for our iteration process.

EXAMPLE 7.3.1. The following table gives the values of the integral

$$\frac{2}{\sqrt{\pi}} \int_0^x e^{-x^2} dx$$

corresponding to certain values of x. Interpolate the value of x for which the above integral $= 1/2$.

Solution:

$y = \frac{2}{\sqrt{\pi}} \int_0^x e^{-x^2} dx$	x
0.4846555	0.46
0.4937452	0.47
0.5027498	0.48
0.5116683	0.49

x	y	Δy	$\Delta^2 y$	$\Delta^3 y$
0.46	0.4846555			
		$0.0090897 = (\Delta y_{-1})$		
0.47	0.4937452		$-0.0000851 = (\Delta^2 y_{-1})$	
		$0.0090046 = (\Delta y_0)$		$0.0000345 = (\Delta^3 y_{-1})$
0.48	0.5027498		$-0.0001196 = (\Delta^2 y_0)$	
		$0.0088850 = (\Delta y_1)$		
0.49	0.5116683			

If $x_0 = 0.47$, $x_{-1} = 0.46$, $x_1 = 0.48$ and $x_2 = 0.40$ and $h = 0.01$, we have $y_0 = 0.4937452$, $y_{-1} = 0.4846555$ $y_1 = 0.5027498$ and $y_2 = 0.5116683$.

Gauss's forward formula gives

$$y = y_0 + u\Delta y_0 + \frac{u(u-1)}{2!}\Delta^2 y_{-1} + \frac{u(u^2-1)}{3!}\Delta^3 y_{-1}$$

$$\therefore \frac{y - y_0}{\Delta y_0} = u + \frac{u(u-1)}{2!}\cdot\frac{\Delta^2 y_{-1}}{\Delta y_0} + \frac{u(u^2-1)}{3!}\cdot\frac{\Delta^3 y_{-1}}{\Delta y_0}$$

$$u^{(1)} = \frac{y - y_0}{\Delta y_0} = \frac{.5 - 0.4937452}{0.0090046} = \frac{0.0062548}{0.0090046} = 0.694623$$

Again

$$u = \frac{y - y_0}{\Delta y_0} - \frac{u(u-1)}{2!}\cdot\frac{\Delta^2 y_{-1}}{\Delta y_0} - \frac{u(u^2-1)}{3!}\cdot\frac{\Delta^3 y_{-1}}{\Delta y_0}$$

$$u^{(2)} = 0.694623 - \frac{0.694623 \times (-0.305377)}{2} \times \frac{-0.0000851}{0.0090046}$$

$$- \frac{0.694623(0.694623^2 - 1)}{6} \times \frac{-0.0000345}{0.0090046}$$

$$= 0.694623 - 0.0010024 - 0.0002295$$

$$= 0.693391$$

$$u^{(3)} = 0.694623 - \frac{0.693391(0.693391 - 1)}{2} \times \frac{-0.0000851}{0.0090046}$$

$$- \frac{0.693391(0.693391^2 - 1)}{6} \times \frac{-0.0000345}{0.0090046}$$

$$= 0.694623 - 0.0010046 - 0.0002299$$

$$= 0.693389$$

Up to 5 decimal places $u = 0.69339$.

$\therefore x = x_0 + 0.01 \times 0.69339 = 0.4769$

7.4 Method of reversion of series

The apparent method of inverse interpolation is the reversion of series. We have seen that all the interpolation formulae so far discussed are in the form of a power series and any convergent power series can be reverted.

Any interpolation formula can be expressed as a power series in k and so we write

$$y = a_0 + a_1 u + a_2 u^2 + a_3 u^3 + \cdots \tag{7.4.1}$$

where $u = (x - x_0)/h$ and the constants $a_0, a_1, a_2, \cdots$ can be expressed in terms of the differences.

Let $\omega = (y - a_0)/a_1$ and we assume that the given power series is convergent, so that u can be written as a power series in ω.

$$u = c_1\omega + c_2\omega^2 + c_2\omega^3 + c_4\omega^3 + \cdots \tag{7.4.2}$$

$\therefore$ from (7.4.1) and (7.4.2), we write

$$\begin{aligned} y = a_0 &+ a_1(c_1\omega + c_2\omega^2 + c_3\omega^3 + c_4\omega^4 + \cdots) \\ &+ a_2(c_1\omega + c_2\omega^2 + c_3\omega^3 + c_4\omega^4 + \cdots)^2 \\ &+ a_3(c_1\omega + c_2\omega^2 + c_3\omega^3 + c_4\omega^4 + \cdots)^3 \\ &+ a_4(c_1\omega + c_2\omega^2 + c_3\omega^3 + c_4\omega^4 + \cdots)^4 + \cdots \end{aligned}$$

$$\Rightarrow \frac{y - a_0}{a_1} = (c_1\omega + c_2\omega^2 + c_3\omega^3 + \cdots) + \frac{a_2}{a_1}(c_1\omega + c_2\omega^2 + \cdots)^2 + \cdots$$

$$\begin{aligned} \Rightarrow \omega = (c_1\omega &+ c_2\omega^2 + c_3\omega^3 + c_4\omega^4 + \cdots) \\ &+ \frac{a_2}{a_1}(c_1\omega + c_2\omega^2 + c_3\omega^3 + c_4\omega^4 + \cdots)^2 \\ &+ \frac{a_3}{a_1}(c_1\omega + c_2\omega^2 + c_3\omega^3 + c_4\omega^4 + \cdots)^3 \\ &+ \frac{a_4}{a_1}(c_1\omega + c_2\omega^2 + c_3\omega^3 + c_4\omega^4 + \cdots)^4 \\ &+ \frac{a_5}{a_1}(c_1\omega + c_2\omega^2 + c_3\omega^3 + c_4\omega^4 + \cdots)^5 + \cdots \end{aligned} \tag{7.4.3}$$

Equating coefficients of like powers of ω from both sides:

Equating coefficient of ω : $c_1 = 1$.

Equating coefficient of ω^2 : $c_2 + \frac{a_2}{a_1}c_1^2 = 0 \Rightarrow c_2 = -\frac{a_2}{a_1}$.

Equating coefficient of ω^3 : $c_3 + \frac{a_2}{a_1} \cdot 2c_1c_2 + \frac{a_3}{a_1} \cdot c_1^3 = 0 \Rightarrow c_3 = -\frac{a_3}{a_1} + 2\left(\frac{a_2}{a_1}\right)^2$.

Equating coefficient of ω^4 : $c_4 + 2\frac{a_2}{a_1} \cdot c_1c_3 + \frac{a_2}{a_1}c_2^2 + 3\frac{a_3}{a_1} \cdot c_1^2c_2 + \frac{a_4}{a_1} \cdot c_1^4 = 0$

$$\Rightarrow c_4 = -\frac{a_4}{a_1} + 5\frac{a_2a_3}{a_1^2} - 5\left(\frac{a_2}{a_1}\right)^3.$$

Equating coefficient of

$$\omega^5 : c_5 + 2\frac{a_2}{a_1}c_1c_4 + 2\frac{a_2}{a_1}c_2c_3 + 3\frac{a_3}{a_1}c_1^2c_3 + 3\frac{a_3}{a_1}c_1c_2^2 + 4\frac{a_4}{a_1}c_1^3c_2 + \frac{a_5}{a_1}c_1^5 = 0$$

$$\Rightarrow c_5 = -\frac{a_5}{a_1} + 6\frac{a_2a_4}{a_1^2} + 3\left(\frac{a_3}{a_1}\right)^2 - 21\frac{a_2^2a_3}{a_1^3} + 14\left(\frac{a_2}{a_1}\right)^4$$

and so on.

When reverting a series with numerical coefficients, we may follow the following steps:

Step 1: The appropriate interpolation formula should be written as a power series in u: $\left(u = \frac{x - x_0}{h}\right)$

The constants $a_0, a_1, a_2, \ldots$, etc. are written in terms of differences of different orders.

Step 2: Evaluate $c_1, c_2, c_3, c_4, c_5, \ldots$, in terms of the values of $a_0, a_1, a_2, \cdots$.

Step 3: Evaluate

$$u = c_1\omega + c_2\omega^2 + c_3\omega^3 + c_4\omega^4 + c_5\omega^5 + \cdots$$

$$= c_1\left(\frac{y - a_0}{a_1}\right) + c_2\left(\frac{y - a_0}{a_1}\right)^2 + \cdots$$

Step 4: From u, we obtain $x = x_0 + hu$.

We shall now write Newton's forward, Gauss's forward and backward, Stirling's and Bessel's formulae in the form of power series and then write down the values of a_0, a_1, a_3, a_4, etc. in each case. We shall stop our computations with the fourth differences, since higher order differences are usually small. If required, one may proceed to compute higher differences.

7.4.1 Newton's forward formula

$$y = y_0 + u\Delta y_0 + \frac{u^2 - u}{2}\Delta^2 y_0 + \frac{u^3 - 3u^2 + 2u}{6}\Delta^3 y_0 + \frac{u^4 - 6u^3 + 11u^2 - 6u}{24}\Delta^4 y_0 + \cdots$$

$$= y_0 + \left(\Delta y_0 - \frac{\Delta^2 y_0}{2} + \frac{1}{3}\Delta^3 y_0 - \frac{1}{4}\Delta^4 y_0\right) u + \left(\frac{1}{2}\Delta^2 y_0 - \frac{1}{2}\Delta^3 y_0 + \frac{11}{24}\Delta^4 y_0\right) u^2$$
$$+ \left(\frac{1}{6}\Delta^3 y_0 - \frac{1}{4}\Delta^4 y_0\right) u^3 + \left(\frac{1}{24}\Delta^4 y_0\right) u^4 + \cdots$$

Hence

$$a_0 = y_0$$

$$a_1 = \Delta y_0 - \frac{1}{2}\Delta^2 y_0 + \frac{1}{3}\Delta^3 y_0 - \frac{1}{4}\Delta^4 y_0$$

$$a_2 = \frac{1}{2}\Delta^2 y_0 - \frac{1}{2}\Delta^3 y_0 + \frac{11}{24}\Delta^4 y_0$$

$$a_3 = \frac{1}{6}\Delta^3 y_0 - \frac{1}{4}\Delta^4 y_0$$

$$a_4 = \frac{1}{24}\Delta^4 y_0$$

7.4.2 Gauss's forward formula

$$y = y_0 + u\Delta y_0 + \frac{u^2 - u}{2}\Delta^2 y_{-1} + \frac{u^3 - u}{6}\Delta^3 y_{-1} + \frac{u^4 - 2u^3 - u^2 + 2u}{24}\Delta^4 y_{-2}$$

$$= y_0 + \left(\Delta y_0 - \frac{1}{2}\Delta^2 y_{-1} - \frac{1}{6}\Delta^3 y_{-1} + \frac{1}{12}\Delta^4 y_{-2}\right) u + \left(\frac{1}{2}\Delta^2 y_{-1} - \frac{1}{24}\Delta^4 y_{-2}\right) u^2$$
$$+ \left(\frac{1}{6}\Delta^3 y_{-1} - \frac{1}{12}\Delta^4 y_{-2}\right) u^3 + \left(\frac{1}{24}\Delta^4 y_{-1}\right) u^4 + \cdots$$

Hence

$$
\begin{aligned}
a_0 &= y_0 \\
a_1 &= \Delta y_0 - \frac{1}{2}\Delta^2 y_{-1} - \frac{1}{6}\Delta^3 y_{-1} + \frac{1}{12}\Delta^4 y_{-2} \\
a_2 &= \frac{1}{2}\Delta^2 y_{-1} - \frac{1}{24}\Delta^4 y_{-2} \\
a_3 &= \frac{1}{6}\Delta^3 y_{-1} - \frac{1}{12}\Delta^4 y_{-2} \\
a_4 &= \frac{1}{24}\Delta^4 y_{-2}
\end{aligned}
$$

7.4.3 Gauss's backward formula

$$
\begin{aligned}
y &= y_0 + u\Delta y_{-1} + \frac{u^2+u}{2}\Delta^2 y_{-1} + \frac{u^3-u}{6}\Delta^3 y_{-2} + \frac{u^4+2u^3-u^2-2u}{24}\Delta^4 y_{-2} + \cdots \\
&= y_0 + \left(\Delta y_{-1} + \frac{1}{2}\Delta^2_{-1} - \frac{1}{6}\Delta^3 y_{-2} - \frac{1}{12}\Delta^4 y_{-2}\right)u + \left(\frac{1}{2}\Delta^2 y_{-1} - \frac{1}{24}\Delta^4 y_{-2}\right)u^2 \\
&\quad + \left(\frac{1}{6}\Delta^3 y_{-2} + \frac{1}{12}\Delta^4 y_{-2}\right)u^3 + \left(\frac{1}{24}\Delta^4 y_{-2}\right)u^4 + \cdots
\end{aligned}
$$

Hence

$$
\begin{aligned}
a_0 &= y_0 \\
a_1 &= \Delta y_{-1} + \frac{1}{2}\Delta^2 y_{-1} - \frac{1}{6}\Delta^3 y_{-2} - \frac{1}{12}\Delta^4 y_{-2} \\
a_2 &= \frac{1}{2}\Delta^2 y_{-1} - \frac{1}{24}\Delta^4 y_{-2} \\
a_3 &= \frac{1}{6}\Delta^3 y_{-2} + \frac{1}{12}\Delta^4 y_{-2} \\
a_4 &= \frac{1}{24}\Delta^4 y_{-2}
\end{aligned}
$$

7.4.4 Stirling's formula

$$
\begin{aligned}
y &= y_0 + u\frac{\Delta y_{-1} + \Delta y_0}{2} + \frac{u^2}{2}\Delta^2 y_{-1} + \frac{u^3-u}{6}\frac{\Delta^3 y_{-2} + \Delta^3 y_{-1}}{2} + \frac{u^4-u^2}{24}\Delta^4 y_{-2} + \cdots \\
&= y_0 + \left\{\frac{\Delta y_{-1} + \Delta y_0}{2} - \frac{1}{12}(\Delta^3 y_{-2} + \Delta^3 y_{-1})\right\}u + \left\{\frac{1}{2}\Delta^2 y_{-1} - \frac{1}{24}\Delta^4 y_{-2}\right\}u^2 \\
&\quad + \left\{\frac{1}{12}(\Delta^3 y_{-2} + \Delta^3 y_{-1})\right\}u^3 + \left(\frac{1}{24}\Delta^4 y_{-2}\right)u^4 + \cdots
\end{aligned}
$$

Hence

$$a_0 = y_0$$
$$a_1 = \frac{1}{2}(\Delta y_{-1} + \Delta y_0) - \frac{1}{12}(\Delta^3 y_{-2} + \Delta^3 y_{-1})$$
$$a_2 = \frac{1}{2}\Delta^2 y_{-1} - \frac{1}{24}\Delta^4 y_{-2}$$
$$a_3 = \frac{1}{12}(\Delta^3 y_{-2} + \Delta^3 y_{-1})$$
$$a_4 = \frac{1}{24}\Delta^4 y_{-2}$$

7.4.5 Bessel's formula

$$\begin{aligned} y &= \frac{y_0 + y_1}{2} + v\Delta y_0 + \frac{\left(v^2 - \frac{1}{4}\right)}{2} \cdot \frac{\Delta^2 y_{-1} + \Delta^2 y_0}{2} + \frac{v\left(v^2 - \frac{1}{4}\right)}{6} \cdot \Delta^3 y_{-1} \\ &\quad + \frac{\left(v^2 - \frac{1}{4}\right)\left(v^2 - \frac{9}{4}\right)}{24} \cdot \frac{\Delta^4 y_{-2} + \Delta^4 y_{-1}}{2} \\ &= \frac{y_0 + y_1}{2} - \frac{1}{16}(\Delta^2 y_{-1} + \Delta^2 y_0) + \frac{3}{256}(\Delta^4 y_{-2} + \Delta^4 y_{-1}) + \left(\Delta y_0 - \frac{1}{24}\Delta^3 y_{-1}\right) v \\ &\quad + \left\{\frac{\Delta^2 y_{-1} + \Delta^2 y_0}{4} - \frac{5(\Delta^4 y_{-2} + \Delta^4 y_{-1})}{96}\right\} v^2 + \left(\frac{1}{6}\Delta^3 y_{-1}\right) v^3 \\ &\quad + \left\{\frac{1}{48}(\Delta^4 y_2 + \Delta^4 y_{-1})\right\} v^4 \ldots \end{aligned}$$

where

$$v = \frac{x - x_0}{h} - \frac{1}{2}$$

Hence

$$a_0 = \frac{y_0 + y_1}{2} - \frac{1}{16}(\Delta^2 y_{-1} + \Delta^2 y_0) + \frac{3}{256}(\Delta^4 y_{-2} + \Delta^4 y_{-1})$$
$$a_1 = \Delta y_0 - \frac{1}{24}\Delta^3 y_{-1}$$
$$a_2 = \frac{\Delta^2 y_{-1} + \Delta^2 y_0}{4} - \frac{5(\Delta^4 y_{-2} + \Delta^4 y_{-1})}{96}$$
$$a_3 = \frac{1}{6}\Delta^3 y_{-1}$$
$$a_4 = \frac{1}{48}(\Delta^4 y_{-2} + \Delta^4 y_{-1})$$

We shall now work out the Example using Bessel's formula.

Here

$$a_0 = \frac{0.4937452 + 0.5027498}{2} - \frac{1}{16}(-0.0000851 - 0.0001196)$$
$$= 0.4982475 - 0.0000128 = 0.4982603$$
$$a_1 = 0.0090046 + \frac{0.0000345}{24} = 0.0090060$$
$$a_2 = \frac{-0.0000851 - 0.0001196}{4} = -0.0000512$$
$$a_3 = \frac{-0.0000345}{6} = -0.0000058$$

$a_4 \simeq 0$ (practically negligible).

Using $y = 0.5$, we have

$$\omega = \frac{y - a_0}{a_1} = \frac{0.5 - 0.4982603}{0.0090060} = 0.1931712$$
$$c_1 = 1$$
$$c_2 = -\frac{a_2}{a_1} = -\frac{-0.0000512}{0.0090060} = 0.0056851$$
$$c_3 = -\frac{a_3}{a_1} + 2\left(\frac{a_2}{a_1}\right)^2 = -\frac{-0.0000058}{0.0090060} + 2\left(-\frac{0.0000512}{0.0090060}\right)^2$$
$$= 0.0006440 + 2(0.0056851)^2 = 0.0006440 + 0.0000646$$
$$= 0.0007086$$
$$c_4 = -\frac{a_4}{a_1} + 5\frac{a_2 a_3}{a_1^2} - 5\left(\frac{a_2}{a_1}\right)^3$$
$$= 5\frac{(-0.0000512) \times (-0.0000058)}{(0.0090060)^2} - 5\left(\frac{-0.0000512}{0.0090060}\right)^3$$
$$= \frac{0.0000000014848}{0.000081108} + 5(0.0056851)^3$$
$$= 0.00001831 + 0.00000092 = 0.00001923$$
$$v = c_1\omega_0 + c_2\omega^2 + c_3\omega^3 + c_4\omega^4 + \cdots$$
$$= (0.1931712) + (0.0056851)(0.1931712)^2 + (0.0007086)(0.1931712)^3$$
$$+ (0.00001923)(0.1931712)^4$$
$$= 0.1931712 + 0.0002121 + 0.00000511 + 0.0000000052$$
$$= 0.1933884$$
$$\therefore u = v + \frac{1}{2} = 0.6933884$$
$$\therefore x = x_0 + 0.01 \times 0.6933884 = 0.4769339$$

7.5 Use of inverse interpolation method to find the root of an algebraic equation

Let the equation be given by $f(x) = 0$. Given that the root of the given equation lies between x_0 and x_1. If $x_1 - x_0$ be equal to h, we may take some equidistant values of x on both sides of x_0 and evaluate corresponding values of $y = f(x)$. We may prepare difference tables and any suitable central difference interpolation formula. The method boils down to finding x, when y is given to be equal to zero, which is in fact a problem of inverse interpolation. A simple example will make the position clear.

EXAMPLE 7.5.1. Find the real root of the equation $x^3 + x - 4 = 0$.

Solution: It is seen that the root lies between 1.3 and 1.4, since

$$f(1.4) = +0.144$$

and $f(1.3) = -0.503$

We take
$$x_0 = 1.3,\quad x_1 = 1.4,\quad x_{-1} = 1.2,\quad x_2 = 1.5,\quad x_{-2} = 1.1,\quad x_3 = 1.6,\quad x_{-3} = 1.0$$

x	$y = x^3 + x - 4$	Δy	$\Delta^2 y$	$\Delta^3 y$
1.0	−2			
		0.431		
1.1	−1.569		0.066	
		0.497		0.006
1.2	−1.072		0.072	
		0.569		0.006
1.3	−0.503		0.078	
		0.647		0.006
1.4	0.144		0.084	
		0.731		0.006
1.5	0.875		0.090	
		0.821		
1.6	1.696			

Taking origin at 1.3, interval of differencing 0.1, and using Bessel's formula

$$y = \frac{-0.503 + 0.144}{2} + v(0.647) + \frac{v^2 - 0.25}{2} \times \frac{0.078 + 0.084}{2} + \frac{v(v^2 - 0.25)}{6} \times 0.006$$

$$= \left\{\frac{-0.359}{2} - 0.125 \times \frac{0.162}{2}\right\} + (0.647 - 0.00025)v + \frac{0.162}{4}v^2 + 0.001v^3$$

$$= \{-0.1795 - 0.01025\} + 0.64675v + 0.0405v^2 + 0.001v^3$$

$$= -0.18975 + 0.64675v + 0.0405v^2 + .001v^3$$

$$\omega = \frac{y + 0.18975}{0.64675} = 0.293391(\text{ for } y = 0)$$

$$c_2 = -\frac{0.0405}{0.64675} = -0.062621$$

$$c_3 = -\frac{a_3}{a_1} + 2\left(\frac{a_2}{a_1}\right)^2 = -\frac{0.001}{0.64675} + 2\left(\frac{0.0405}{0.64675}\right)^2$$

$$= -0.0015462 + 0.00784273 = 0.00629653$$

$$c_4 = 5\frac{a_2 a_3}{a_1^2} - 5\left(\frac{a_2}{a_1}\right)^2$$

$$= 5(0.0009687 - 0.00392136) = -0.0147633$$

$$v = 1 \times (0.293391) + (-0.062621)(0.293391)^2 + 0.00629653(0.293391)^3$$

$$- 0.0147633(0.293391)^4$$

$$= 0.293391 - 0.00539031 + 0.0001590 - 0.0001094 = 0.2880503$$

$$u = v + \tfrac{1}{2} = 0.7880503$$

$$\therefore x = x_0 + hu = 1.3 + 0.1 \times 0.7880503 = 1.3788$$

EXERCISES

1. The equation $x^3 - 6x - 11 = 0$ has a root between 3 and 4. Obtain the root by inverse interpolation correct to three places of decimals. [*Ans:* 3.091.]
2. Find to 3 places of decimals the smallest positive root of the equation $x^3 - 7x + 5$ by making a difference table when $x = 0, 1/2 + 1$. [*Ans:* 0.783.]
3. Given the following values:

x	0	1	2
y	1	3	26

Obtain x, when $y = 18$. [*Ans:* 1.741.]

4. Write a note on inverse interpolation and show how it can be used to find a real root of an algebraic equation.
5. Apply Lagrange's formula to find a root of $f(x) = 0$, when $f(26) = -34$, $f(30) = -12$, $f(34) = 4$, $f(38) = 14$. [*Ans:* 32.663.]
6. Find the root of the equation $10x^3 - 15x + 3 = 0$, which lies between 1 and 2. [*Ans:* 1.109.]
7. Find the real root of $2x - \log_{10} x = 7$ by the method of successive approximations (*iteration*). [*Ans:* 3.7893.]
8. Apply the method of iteration of find the real root of the equation $xe^x - 2 = 0$, which lies between 0.8 and 0.9. (Give the root correct to 4 decimal places). [*Ans:* 0.8526.]
9. If Cosh $x = 1.286$, find x, using inverse interpolation, using the following data:

x	Cosh x
0.735	1.2824937
0.736	1.2832974
0.737	1.2841023
0.738	1.2849085
0.739	1.2857159
0.740	1.2865247
0.741	1.2873348
0.742	1.2881461

[*Ans:* 0.7393514.]

10. Explain how Newton-Gregory formula can be used to carry out inverse interpolation of second order. Show that a good approximation to the interpolating factor x will be given by

$$x = \frac{f_x - f_0}{\Delta f_0}\left[1 - \left\{\frac{f_x - f_0}{2(\Delta f_0)^2}\right\}\Delta^2 f_0\right]$$

Chapter 8

Interpolation with Two Independent Variables

Summary

8.1 Introduction

In chapters 3, 4 and 5 we have considered the problems of interpolation for a function of one independent variable (argument). Sometimes we may encounter problems where the function depends on two independent variables (arguments) and we may be required to interpolate function for both the arguments. For example, the hour angle Z of the Sun depends on the altitude (x) and declinations (y) at a place in a certain latitude. We may express this relation by

$$Z = f(x, y) \tag{8.1.1}$$

The form of this functional relationship may not always be given. Instead, the values of Z may be given for

$$x = x_0, x_1, x_2, \ldots, x_m$$

and $y = y_0, y_1, y_2, \ldots, y_n$

The corresponding Z values may be given in the following tabular form:

x/y	x_0	x_1	x_2	$\cdots$	$\cdots$	$\cdots$	$\cdots$	x_m
y_0	Z_{00}	Z_{10}	Z_{20}	$\cdots$	$\cdots$	$\cdots$	$\cdots$	Z_{m0}
y_1	Z_{01}	Z_{11}	Z_{21}	$\cdots$	$\cdots$	$\cdots$	$\cdots$	Z_{m1}
y_2	Z_{02}	Z_{12}	Z_{22}	$\cdots$	$\cdots$	$\cdots$	$\cdots$	Z_{m2}
$\vdots$	$\vdots$	$\vdots$	$\vdots$	$\vdots$	$\vdots$	$\vdots$	$\vdots$	$\vdots$
y_n	Z_{0n}	Z_{1n}	Z_{2n}	$\cdots$	$\cdots$	$\cdots$	$\cdots$	Z_{mn}

We are required to interpolate Z for some given x and y in

$x_0 < x < x_m$

and $y_0 < y < y_n$

8.2 Repeated use of interpolation formula for one argument

We may solve such problems by first interpolating with respect to one argument, say x, and then interpolation with respect to the other. There may be three different situations:

- both the arguments may be equispaced;
- one of the arguments may be equispaced;
- both are not equispaced.

We first take up case (i). Let the argument x be given at an equispaced interval of h,

i.e. $x_i = x_0 + ih, \quad (i = 0, 1, 2, \ldots, m)$

and the argument y be given at an interval of k,

i.e. $y_i = y_0 + jk, \quad (j = 0, 1, 2, \ldots, n)$

We first interpolate with respect to the argument x, say, keeping y fixed at different points

$y_0, y_1, y_2, \ldots, y_n$

we have to use $(n + 1)$ times the appropriate interpolation formula for one argument. The appropriate interpolation formula may be Newton's Forward, Newton's Backward or any of the Central Difference Interpolation Formula according as the given value of x lies near the beginning, near the end, or near the middle of the tabulated values of x. Having obtained the interpolated values

$Z_{x,0}; Z_{x,1}; Z_{x,2}, \ldots, Z_{x,n}$

corresponding to the values $y_0, y_1, y_2, \ldots, y_n$ (for fixed x) of the argument y, we again repeat the process, using an appropriate interpolation formula for interpolating $Z_{x,y}$ for $y_0 < y < y_n$.

EXAMPLE 8.2.1. The following table gives the values of the function $F(0.05, \nu_1, \nu_2)$ for different values of ν_1 and ν_2

ν_1 \ ν_2	10	20	30	40
5	4.74	4.56	4.50	4.46
10	2.98	2.77	2.70	2.66
15	2.54	2.33	2.25	2.20
20	2.35	2.12	2.04	1.99

Find the value $F(0.05, 38, 12)$ taking $1/\nu_1$ and $1/\nu_2$ as the arguments.

Solution: Since the arguments in this case are $1/\nu_1$ and $1/\nu_2$ and they are not equispaced, we use Newton's General Divided Difference formula to interpolate $F(0.05, 38, \nu_2)$ for $\nu_2 = 5, 10, 15$ and 20.

We first obtain $F(0.05, 38, 5)$. For this we prepare the following divided difference table:

$\frac{1}{\nu_1}$	$F(0.05, \nu_1, 5)$	1st order divided difference	2nd order divided difference	3rd order divided difference
0.1000	4.74			
		$\frac{-0.18}{-0.05} = 3.6$		
0.0500	4.56		0	
		$\frac{-0.06}{-1/16} = 3.6$		$\frac{-48}{-3/40} = 640$
0.0333	4.50		$\frac{1.2}{-1/40} = -48$	
		$\frac{-0.04}{-1/120} = 4.8$		
0.0250	4.46			

$$F(0.05, 38, 5) = 4.74 + (x - 0.1) \times 3.6 + (x - 0.1)(x - 0.05)\left(x - \frac{1}{30}\right) \times 640$$

(where $x = 1/38$)

$$= 4.74 - 0.2653 + (-0.07368)(-0.023684)(-0.0070175) \times 640$$

$$= 4.74 - 0.2653 - 0.0078373 = 4.47$$

We next find $F(0.05, 38, 10)$

$\frac{1}{\nu_1}$	$F(0.05, \nu_1, 10)$	1st order divided difference	2nd order divided difference	3rd order divided difference
0.1000	2.98			
		$\frac{-0.21}{-0.05} = 4.2$		
0.0500	2.77		0	
		$\frac{-0.07}{-1/60} = 4.2$		320
0.0333	2.70		−24	
		$\frac{-0.04}{-1/120} = 4.8$		
0.0250	2.66			

$$F(0.05, 38, 10) = 2.98 + (x - 0.1) \times 4.2 + (x - 0.1)(x - 0.05) \times 0$$

$$+ (x - 0.01)(x - 0.05)\left(x - \frac{1}{30}\right) \times 320 \text{ (where } x = 1/38)$$

$$= 2.98 - 0.30947 + (-0.073684)(-0.023684)(-0.0070175) \times 320$$

$$= 2.98 - 0.30947 - 0.00392 = 2.67$$

To find $F(0.05, 38, 15)$, we prepare the following difference table:

$\frac{1}{v_1}$	$F(0.05, v_1, 15)$	1st order divided difference	2nd order divided difference	3rd order divided difference
0.1000	2.54			
		$\frac{-0.21}{-0.05} = 4.2$		
0.0500	2.33		$\frac{0.6}{-1/15} = -9$	
		$\frac{-0.08}{-1/60} = 4.8$		$\frac{-39}{-3/40} = 520$
0.0333	2.25		$\frac{1.2}{-1/20} = -48$	
		$\frac{-0.05}{-1/120} = 6.0$		
0.025	2.20			

$$
\begin{aligned}
F(0.05, 38, 15) &= 2.54 + (x - 0.1) \times 4.2 + (x - 0.1)(x - 0.05)(-9) \\
&\quad + (x - 0.1)(x - 0.05)(x - 0.03333) \times (520) \text{ (where } x = 1/38) \\
&= 2.54 - 0.30947 - 0.0157053 - 0.0063682354 \\
&= 2.208 \simeq 2.21
\end{aligned}
$$

To find $F(0.05, 38, 20)$, we prepare the following table of divided differences:

$\frac{1}{v_1}$	$F(0.05, v_1, 20)$	1st order divided difference	2nd order divided difference	3rd order divided difference
0.1000	2.35			
		$\frac{-0.23}{-0.05} = 4.6$		
0.0500	2.12		$\frac{0.2}{-1/15} = -3$	
		$\frac{-0.08}{-1/60} = 4.8$		$\frac{-45}{-3/40} = 600$
0.03333	2.04		$\frac{1.2}{-1/40} = -48$	
		$\frac{-0.05}{-1/120} = 6.0$		
0.0250	1.99			

$$
\begin{aligned}
F(0.05, 38, 20) &= 2.35 + (x - 0.1) \times 4.6 + (x - 0.1)(x - 0.05)(-3) \\
&\quad + (x - 0.1)(x - 0.05)(x - 0.03333)(600) \text{ (where } x = 1/38) \\
&= 2.35 - 0.33895 + (-0.073684)(-0.023684)(-3) \\
&\quad + (0.073684)(-0.023684)(-0.0070175) \times (600) \\
&= 2.35 - 0.33895 - 0.0052354 - 0.007348 \\
&= 1.9985 \simeq 2.00
\end{aligned}
$$

Our next problem is to interpolate $F(0.05, 38, 12)$ from the series of entries obtained by earlier interpolations. We again use Newton's General Divided Difference formula:

$\frac{1}{\nu_2}$	$F(0.5, \nu_2, 38)$	1st order divided difference	2nd order divided difference	3rd order divided difference
1/5	4.47			
		$\frac{-1.80}{-1/10} = 18$		
1/10	2.67		$\frac{-4.2}{-2/15} = 31.5$	
		$\frac{-0.46}{-1/30} = 13.8$		$\frac{-19.5}{-0.15} = 185$
1/15	2.21		$\frac{-0.6}{-1/20} = 12$	
		$\frac{-0.22}{-1/60} = 13.2$		
1/20	2.00			

$$
\begin{aligned}
F(0.05, 38, 12) &= 4.47 + (x - 0.2) \times 18 + (x - 0.2)(x - 0.1)(31.5) \\
&\quad + (x - 0.2)(x - 0.1)\left(x - \frac{1}{15}\right)(185) \text{ (where } x = 1/38) \\
&= 4.47 - 2.1 + (-0.11667)(-0.016667) \times (31.5) \\
&\quad + (-0.11667)(-0.016667)(0.0166666) \times 185 \\
&= 4.47 - 2.1 + 0.062517 + 0.005994 \\
&= 2.4372 \simeq 2.44
\end{aligned}
$$

NOTE 8.2.1. We have intentionally taken an example which comes under case (iii) of section 8.2, where both the arguments $1/\nu_1$ and $1/\nu_2$ are not equispaced, so the use of Divided Difference Interpolation is made. If, however, the arguments be equispaced, we may use Newton's Backward Formula for the first step and then any central difference interpolation in the second step.

8.3 Use of bivariate interpolation formula: Construction of two-way difference table

Let the function be denoted by $Z = f(x, y)$, where x and y are two arguments. We shall consider the case of equidistant arguments with intervals h and k respectively.

We now define two-way differences

$$
\begin{aligned}
\Delta^{1,0} f(x, y) &= f(x + h, y) - f(x, y) = E_x f(x, y) - f(x, y) \\
&\equiv (E_x - 1) f(x, y)
\end{aligned}
$$

and
$$
\begin{aligned}
\Delta^{0,1} f(x, y) &= f(x, y + k) - f(x, y) = E_y f(x, y) - f(x, y) \\
&\equiv (E_y - 1) f(x, y)
\end{aligned}
$$

Similarly, we define

$$E_x^u E_y^v f(x, y) = f(x + uh, y + vk)$$

Thus, we have the following equivalence relation

$$\Delta^{m,n} \equiv \Delta_x^m \Delta_y^n$$

where $\Delta_x \equiv E_x - 1; \Delta_y \equiv E_y - 1$.

We now write the general bivariate interpolation formula for the case of equidistant values of the two arguments:

Let

$$u = \frac{x - x_0}{h}$$
$$v = \frac{y - y_0}{k}$$

$$\begin{aligned}
f(x, y) &= f(x_0 + hu, y_0 + kv) \\
&\equiv E_x^u E_y^v f(x_0, y_0) = (1 + \Delta x)^u (1 + \Delta y)^v f(x_0, y_0) \\
&= \left[1 + u\Delta x + \binom{u}{2}\Delta^2 x + \binom{y}{3}\Delta_x^3 + \cdots\right] \\
&\quad \times \left[1 + v\Delta y + \binom{v}{2}\Delta^2 y + \binom{v}{3}\Delta^3 y + \cdots\right] f(x_0, y_0) \\
&= \left[1 + u\Delta x + \binom{u}{2}\Delta^2 x + v\Delta y + \binom{v}{2}\Delta^2 y + uv(\Delta x)(\Delta y)\right] f(x_0, y_0) \\
&= \left[1 + u\Delta x + v\Delta y + \binom{u}{2}\Delta^2 x + \binom{v}{2}\Delta^2 y + uv\Delta x\Delta y + \cdots\right] f(x_0, y_0) \\
&= f(x_0, y_0) + u\Delta^{1,0} f(x_0, y_0) + v\Delta^{0,1} f(x_0, y_0) \\
&\quad + \binom{u}{2}\Delta^{2,0} f(x_0, y_0) + uv\Delta^{1,1} f(x_0, y_0) + \cdots
\end{aligned} \tag{8.3.1}$$

If u and v are small and higher order differences above $\Delta^{2,0}$ and $\Delta^{0,2}$ are negligible, the computational labour will not be prohibition.

EXAMPLE 8.3.1. Given the following table of $Z_{x,y} = f(x, y)$:

x \ y	40	45	50
40	$(Z_{0,0})$	12.880 $(Z_{1,0})$	11.898 $(Z_{2,0})$
45	13.133 $(Z_{0,1})$	12.432 (Z_{11})	(Z_{21})
50	12.450 $(Z_{0,2})$	$(Z_{1,2})$	$(Z_{2,2})$

Find $Z_{42,43} = f(42, 43)$.

Solution: Taking the origin at 40, for both x and y, and unit 5 in each case, we are required to find

$$
\begin{aligned}
Z_{0.4,0.6} &= E_x^{2/5} E_y^{3/5} Z_{0,0} \\
&= \left[1 + 0.4\Delta x - 0.12\Delta^2 x + \cdots\right]\left[1 + 0.6\Delta y - 0.12\Delta^2 y + \cdots\right] Z_{0,0} \\
&= \left[1 + 0.4\Delta x + 0.6\Delta y + 0.24\Delta x\Delta y - 0.12\Delta^2 x - 0.12\Delta^2 y + \cdots\right] Z_{0,0} \\
&= Z_{0,0} + 0.4(Z_{1,0} - Z_{0,0}) + 0.6(Z_{0,1} - Z_{0,0}) \\
&\quad + 0.24(Z_{1,1} - Z_{0,1} - Z_{1,0} + Z_{0,0}) \\
&\quad - 0.12(Z_{2,0} - 2Z_{1,0} + Z_{0,0}) - 0.12(Z_{0,2} - 2Z_{0,1} + Z_{0,0}) \\
&= Z_{1,0}(0.4 + 0.24 - 0.24) + Z_{0,1}(0.6 + 0.24 - 0.24) \\
&\quad - Z_{2,0}(0.12) - Z_{0,2}(0.12) + Z_{1,1}(0.24) \\
&= 0.4Z_{1,0} + 0.6Z_{0,1} - 0.12Z_{2,0} - 0.12Z_{0,2} + 0.24Z_{1,1} \\
&= 13.094
\end{aligned}
$$

EXERCISES

1. Given $Z_{0,0} = f(0, 0)$,
   $Z_{0,1} = f(0, 1)$,
   $Z_{1,0} = f(1, 0)$,
   $Z_{1,1} = f(1, 1)$.

   Obtain the function $Z_{x,y}$ where $0 < x < 1, 0 < y < 1$.

   [*Hints:* $Z_{x,y} = E_x^x E_y^y Z_{0,0}$, (noting that origin is (0, 0) and unit for both x and y is 1)
   $= (1 + \Delta x)^x (1 + \Delta y)^y Z_{0,0}$
   $= [(1 + x\Delta x)(1 + y\Delta y)]\, Z_{0,0}$
   $= (1 + x\Delta x + y\Delta y + xy\Delta x\Delta y)Z_{0,0}$ (assuming $\Delta^2 x$, $\Delta^2 y$, etc. to be zero)
   $= Z_{0,0} + x(Z_{1,0} - Z_{0,0} + y(Z_{0,1} - Z_{0,0}) + xy(Z_{1,1} - Z_{1,0} - Z_{0,1} + Z_{0,0})$, etc.]

2. Show that if $Z_{1,0} = Z_{0,1}$ and $Z_{2,0} = Z_{0,2}$, then

$$Z_{x,y} = Z_{0,0} + (x + y)\left[\Delta x + \frac{x + y - 1}{2}\Delta^2 x\right] Z_{0,0} + xy[Z_{1,1} - Z_{2,0}]$$

3. Given $Z_{20,15} = 7.004$,
   $Z_{20,20} = 5.304$,
   $Z_{20,25} = 4.325$,
   $Z_{25,15} = 7.029$,
   $Z_{25,20} = 5.346$,
   $Z_{30,15} = 7.075$;

   find $f_{23,17}$ as accurately as you can. [*Ans:* 6.389]

4. The following table gives the hour angle (t) of the sun corresponding to certain altitudes (a) and declination (b) at a certain place in a certain latitude. Find the hour angle corresponding to
   (a) $a = 16°$, $d = 12°$
   (b) $a = 12°$, $d = 16°$.

	$a = 10°$	$a = 14°$	$a = 18°$	$a = 22°$
$d = 20°$	$6^h11^m20^s$	$5^h50^m17^s$	5^h29m27^s	$5^h8^m48^s$
15°	$5^h55^m41^s$	$5^h35^m5^s$	$5^h14^m39^s$	$4h54^m17^s$
10°	$5^h40^m16^s$	$5^h19^m56^s$	$4^h59^m3^s$	$4^h39^m17^s$
5°	$5^h24^m50^s$	$5^h4^m30^s$	$4^h44^m4^s$	$4^h23^m29^s$
0°	$5^h9^m5^s$	$4^h48^m29^s$	$4^h27^m39^s$	$4^h6^m28^s$

[*Ans:* (a) $5^h15^m50^s$, (b) $5^h48^m25^s$.]

Chapter 9

Numerical Differentiation

Summary

9.1 Introduction

A function $f(x)$, which is generally unknown, is given for $x = x_0, x_1, x_2, \ldots, x_n$. The corresponding entries are $f(x_0), f(x_1), \ldots, f(x_n)$. On the basis of these data, we are required to find the derivative or successive derivatives of $f(x)$. The technique or procedure which helps us to accomplish this job, is known as Numerical Differentiation.

First we approximate the unknown function $f(x)$ by an interpolation formula and then we differentiate the interpolating polynomial as many times as desired. The interpolating polynomial to be selected for this purpose, depends on the following situations:

1. The value of the function is given for equispaced values of the argument, and the derivative of the function is to be found for some given x lying near the beginning of the tabulated values. We may recommend Newton's Forward Interpolation Formula in such cases.
2. The value of the function $f(x)$ is given for equispaced values of the argument, and the derivative is to be found at some point lying near the end of the tabulated values. We may use Newton's Backward Formula in such cases.
3. The value of the function in given for equispaced values of the argument and the derivative is to be found at some point lying near the middle of the tabulated values. We may use any suitable Central Difference Interpolation Formula.
4. The function $f(x)$ is given for the values of the argument which are not equispaced. In that case, we may use Newton's Divided Difference Formula to approximate the unknown function $f(x)$. We may illustrate these by solving a few examples:

Example 9.1.1. Find the first two derivatives of $f(x)$ at $x = 1.0$ from the given data:

x	y	Δy	$\Delta^2 y$	$\Delta^3 y$	$\Delta^4 y$
1.0	−14.000				
		3.968			
1.2	−10.032		0.768		
		4.736		0.048	
1.4	−5.296		0.816		0
		5.552		0.048	
1.6	+0.256		0.864		0
		6.416		0.048	
1.8	6.672		0.912		
		7.328			
2.0	14.000				

Solution: Here the argument is given for equispaced interval of length $h = 0.2$ and we are required to find the derivative of $f(x)$ at $x = 1.0$, which is at the start of the tabulated values. Therefore, we approximate $f(x)$ by Newton's Forward Formula [nth degree polynomial of u, where $u = (x - x_0)/h = 0$.]

Noting that the third difference is constant and

$$\frac{d}{dx}f(x) = \frac{d}{du}\cdot f(x)\cdot\frac{du}{dx} = \frac{1}{h}\cdot\frac{d}{du}f(x) = (0.2)^{-1}\frac{d}{du}f(x)$$

we write

$$f(x) \simeq f(1.0) + u\Delta f(1.0) + \frac{u(u-1)}{2!}\Delta^2 f(1.0) + \frac{u(u-1)(u-2)}{3!}\Delta^3 f(1.0)$$

$$f'(x) = (0.2)^{-1}\left[\Delta f(1.0) + \frac{2u-1}{2}\Delta^2 f(1.0) + \frac{3u^2-6u+2}{6}\Delta^3 f(1.0)\right]$$

$$\therefore f'(x)\Big|_{x=1.0} = (0.2)^{-1}\left[\Delta f(1.0) - \frac{1}{2}\Delta^2 f(1.0) + \frac{1}{3}\Delta^3 f(1.0)\right], \quad \text{[where } u = 0]$$

$$= (0.2)^{-1}\left[3.968 - \frac{1}{2}\cdot 0.768 + \frac{1}{3}\cdot 0.048\right]$$

$$= (0.2)^{-1}[3.968 - 0.384 + 0.016] = 18$$

Again

$$\frac{d^2}{dx^2}f(x) = \frac{d}{dx}\left[h^{-1}\frac{d}{du}f(x)\right] = \frac{d}{du}\cdot\frac{du}{dx}\left[h^{-1}\frac{d}{du}f(x)\right] = \frac{1}{h^2}\cdot\frac{d^2}{du^2}f(x)$$

$$\therefore f''(x)\Big|_{x=1.0} = \frac{1}{h^2}\left[\Delta^2 f(1.0) + (u-1)\Delta^3 f(1.0)\right]_{u=0} = \frac{1}{0.04}[0.768 - 0.048]$$

$$= \frac{1}{0.04}\times 0.720 = 18$$

Example 9.1.2. Find the first derivative of the function $y = f(x)$ at $x = 6$ and $x = 3$ given in the following table:

x	$y = f(x)$	Δy	$\Delta^2 y$	$\Delta^3 y$	$\Delta^4 y$	Δ^5
1	2.7173					
		0.6027				
2	3.3200		0.1315			
		0.7342		0.0321		
3	4.0542		0.1636		0.0031	
		0.8978		0.0352		0.0068
4	4.9520		0.1988		0.0099	
		1.0966		0.0451		
5	6.0486		0.2439			
		1.3405				
6	7.3891					

Solution: Here the arguments are equispaced at an interval of $h = 1$. To find the derivative at $x = 6$, we may first approximate $f(x)$ by a fifth degree polynomial in u, where

$$u = \frac{x - x_n}{h} = \frac{6-6}{h} = 0$$

Also

$$\frac{du}{dx} = \frac{1}{h} = 1 \quad (\text{for } h = 1)$$

$$\therefore f'(x) \simeq \frac{d}{du}\left[y_n + u\Delta_1 y_n + \frac{u(u+1)}{2}\Delta^2 y_n + \frac{u(u+1)(u+2)}{6}\Delta_3 y_n + \frac{u(u+1)(u+2)(u+3)}{24}\Delta_4 y_n + \frac{u(u+1)(u+2)(u+3)(u+4)}{120}\Delta_5 y_n\right]$$

$$\simeq \left[\Delta_1 y_n + \frac{2u+1}{2}\Delta_2 y_n + \frac{3u^2+6u+2}{6}\Delta_3 y_n + \frac{4u^3+18u^2+22u+6}{24}\Delta_4 y_n + \frac{5u^4+40u^3+105u^2+76u+24}{120}\Delta_5 y_n\right]$$

$$\therefore f'(x)\Big|_{x=6} = \left[\Delta_1 y_n + \frac{1}{2}\Delta_2 y_n + \frac{1}{3}\Delta_3 y_n + \frac{1}{4}\Delta_4 y_n + \frac{1}{5}\Delta_5 y_n\right] \quad (\text{for } u = 0)$$

$$= \left[1.3405 + \frac{1}{2}\times 0.2439 + \frac{1}{3}\times 0.0451 + \frac{1}{4}\times 0.0099 + \frac{1}{5}\times 0.0068\right]$$

$$= [1.3405 + 0.12195 + 0.015033 + 0.002475 + 0.00136] = 1.4813$$

To find $f'(x)$ at $x = 3$, we may take the origin x_0 at 3, and unit of x equal to 1 as before, and approximate $f(x)$ by the interpolating polynomial given by Bessel's Formula.

Here

$$v = u - \frac{1}{2} = \frac{x - x_0}{h} - \frac{1}{2}. \quad \therefore \frac{dv}{dx} = \frac{1}{h} = 1 \quad (\text{in this case})$$

when $x = 3$, $v = -1/2$.

$$\frac{d}{du}f(x) = \frac{d}{dv}f(x)\cdot\frac{dv}{dx} \simeq \frac{d}{dv}\cdot\phi(x)$$

$$= \frac{d}{dv}\left[\frac{y_0+y_1}{2} + v\Delta y_0 + \frac{\left(v^2-\frac{1}{4}\right)}{2}\cdot\frac{\Delta^2 y_0+\Delta^2 y_{-1}}{2} + \frac{v\left(v^2-\frac{1}{4}\right)}{6}\cdot\Delta^3 y_{-1}\right.$$

$$\left. + \frac{\left(v^2-\frac{1}{4}\right)\left(v^2-\frac{9}{4}\right)}{24}\left(\frac{\Delta^4 y_{-1}+\Delta^4 y_{-2}}{2}\right) + \frac{v\left(v^2-\frac{1}{4}\right)\left(v^2-\frac{9}{4}\right)}{120}\Delta^5 y_{-2}\right]$$

$$= \left[\Delta y_0 + v\cdot\frac{\Delta^2 y_0+\Delta^2 y_{-1}}{2} + \frac{\left(3v^2-\frac{1}{4}\right)}{6}\Delta^3 y_{-1}\right.$$

$$\left. + \frac{4v^3-5v}{24}\cdot\frac{\Delta^4 y_{-1}+\Delta^4 y_{-2}}{2} + \frac{\left(5v^4-\frac{15}{2}v^2\right)}{120}\Delta^5 y_{-2}\right]$$

$$\therefore f'(x)\Big|_{x=3} = \left[\Delta y_0 - \frac{1}{4}\left(\Delta^2 y_0+\Delta^2 y_{-1}\right) + \frac{1}{12}\Delta^3 y_{-1} + \frac{1}{24}\left(\Delta^4 y_{-1}+\Delta^4 y_{-2}\right)\right.$$

$$\left. - \frac{5}{384}\Delta^5 y_{-2}\right] \quad \text{(for } v = -1/2\text{)}$$

$$= \left[0.8978 - \frac{1}{4}\times(0.136+0.198) + \frac{1}{12}\times 0.0352 + \frac{1}{24}(0.0031+0.0091)\right.$$

$$\left. - \frac{5}{384}\times 0.0068\right]$$

$$= [0.8978 - 0.0906 + 0.002933 + 0.000508 - 0.0000885] = 0.8106$$

EXAMPLE 9.1.3. To find $f'(10)$ from the following data:

x	$y = f(x)$	1st order divided difference	2nd order divided difference	3rd order divided difference	4th order divided difference
3	−11				
		18			
5	25		15.9792		
		145.8333		0.9993	
11	900		39.9620		−.02992
		1025		0.0120	
27	17300		40.3333		
		2033.3333			
36	35600				

Solution: Here,

$$f(x) \simeq f(3) + (x-3)\times 18 + (x-3)(x-5)\times 15.9792 + (x-3)(x-5)(x-11)$$
$$\times 0.9993 + (x-3)(x-5)(x-11)(x-27)\times(-0.02992)$$

Differentiating w.r.t. x,

$$f'(x) = 18 + (2x-8)\times 15.9792 + (3x^2 - 38x + 103)\times 0.9993$$
$$+\Big\{(x-5)(x-11)(x-27) + (x-3)(x-11)(x-27)$$
$$+ (x-3)(x-5)(x-27) + (x-3)(x-5)(x-11)\Big\}\times(-0.02492)$$

$$\therefore f'(10) = 18 + 191.7504 + 22.9839 + (85 + 119 - 595 - 35)\times(-0.02992)$$
$$= 18 + 191.7504 + 22.9839 + 12.74592 = 245.4802$$

[*Note:* Since the arguments are not equispaced, we have approximated $f(x)$ by Newton's Divided Difference Formula.]

9.2 To find the maximum or the minimum value of the function $f(x)$

The *unknown* function $f(x)$ may be given for some equispaced (or not) values of the arguments $x_0, x_1, x_2, \ldots, x_n$. The corresponding entries $y_0, y_1, y_2, \ldots, y_n$, $[y_i = f(x_i)]$ are also available. On the basis of the given data, we are required to find the maximum or the minimum value of $f(x)$.

In these type of problems, we first approximate the unknown function $f(x)$ by the appropriate interpolating polynomial. Next we find the maximum or the minimum value of the interpolating polynomial. This may be taken as the approximate maximum or minimum value of the unknown function.

EXAMPLE 9.2.1. Find the *minimum* value of $y = f(x)$ from the following data:

x :	0.2	0.3	0.4	0.5	0.6	0.7
$f(x)$:	0.9180	0.8972	0.8870	0.8860	0.8930	0.9085

Solution: We see that $f(x)$ reaches its minimum value in the neighbourhood of $x = 0.5$. Hence we may approximate $f(x)$ by any Central Difference Interpolation Formula, as the arguments are also equispaced.

x	$f(x)$	$\Delta f(x)$	$\Delta^2 f(x)$	$\Delta^3 f(x)$	$\Delta^4 f(x)$
$0.2(x_{-3})$	0.9180				
		−0.0208			
$0.3(x_{-2})$	0.8972		0.0106		
		$-0.0102(\Delta y_{-2})$		0.0014	
$0.4(x_{-1})$	0.8870		$0.0092(\Delta^2 y_{-2})$		$-0.0026(\Delta^4 y_{-3})$
		$-0.0010(\Delta y_{-1})$		$-0.0012(\Delta^3 y_{-2})$	
$0.5(x_0)$	0.8860		$0.0080(\Delta^2 y_{-1})$		$0.0017(\Delta^4 y_{-2})$
		$0.0070(\Delta y_0)$		$0.0005(\Delta^3 y_{-1})$	
$0.6(x_1)$	0.8930		$0.0085(\Delta^2 y_0)$		
		$0.0155(\Delta y_1)$			
$0.7(x_2)$	0.9085				

We may use Stirling's Formula to obtain the interpolating polynomial.

Let $u = (x - 0.5)/0.1$. We note that

$$\frac{du}{dx} = \frac{1}{0.1} = 10$$

$$f(x) \simeq \phi(x) = f(0.5) + u\frac{\Delta y_0 + \Delta y_{-1}}{2} + \frac{u^2}{2}\Delta^2 y_{-1} + \frac{u(u^2-1)}{6}\frac{\Delta^3 y_{-1} + \Delta^3 y_{-2}}{2}$$

$$+ \frac{u^2(u^2-1)}{24}\Delta^4 y_{-2}$$

$$\therefore f'(x) = \frac{d}{du}\cdot\frac{du}{dx}\phi(x)$$

$$= 10\left[\frac{\Delta y_0 + \Delta y_{-1}}{2} + u\Delta^2 y_{-1} + \frac{3u^2-1}{6}\left(\frac{\Delta^3 y_{-1} + \Delta^3 y_{-2}}{2}\right) + \frac{4u^3 - 2u}{24}\Delta^4 y_{-2}\right]$$

$$= 10\left[0.0030 + u \times 0.0080 + \frac{u^2}{2} \times (-0.00035) + \frac{u^3}{6} \times 0.0017\right.$$

$$\left. - u \times 0.000142 - 0.00006\right]$$

$$= 10[0.00294 + 0.007858u - 0.00017u^2 + 0.0003u^3]$$

$\therefore$ maximum or minimum,

$$f'(x) = 0 \Rightarrow 0.007858u = -0.0003u^3 + 0.000175u^2 - 0.00294$$

$$\Rightarrow u = -0.0382u^3 + 0.02227u^2 - 0.37414$$

$$u^{(1)} = -\frac{0.00294}{0.007858} = -0.3741$$

$$u^{(2)} = -0.0382u^{(1)^3} + 0.02227u^{(1)^2} - 0.37414 = -0.3690$$

$$u^{(3)} = -0.0382u^{(2)^3} + 0.02227u^{(2)^2} - 37414 = -0.3692$$

$$\therefore f(x)\Big|_{\min} = \left[0.8860 - 0.3692 \times 0.0030 + \frac{(-0.3692)^2}{2} \times 0.0080\right.$$

$$\left. + \frac{(-0.3692)\{(-0.3692)^2 - 1\}}{6}(-0.00035)\right]$$

$$= [0.8860 - 0.001108 + 0.000545 - 0.000019] = 0.8854$$

[*Note:* It may be shown that $f''(x)$ for $u = -0.3692$ will be positive.]

EXAMPLE 9.2.2. Deduce from Bessel's formula the following approximations:

(a) $\frac{d}{dx}f(x) = \Delta f\left(x - \frac{1}{2}\right) - \frac{1}{24}\Delta^3 f\left(x - \frac{3}{2}\right)$;

(b) $\frac{d^2}{dx^2}f(x) = \frac{1}{2}\left[\Delta^2 f\left(x - \frac{3}{2}\right) + \Delta^2 f\left(x - \frac{1}{2}\right)\right]$;

(c) $\frac{d^3}{dx^3}f(x) = \Delta^3 f\left(x - \frac{3}{2}\right)$.

Solution: (a) Noting that the origin is at 0 and unit of x is 1, we write

$$v = x - \frac{1}{2} \Rightarrow x = v + \frac{1}{2} \quad \text{(note that } dx = dv\text{)}$$

$$\therefore f\left(v + \frac{1}{2}\right) = \frac{f(0) + f(1)}{2} + v\Delta f(0) + \frac{\left(v^2 - \frac{1}{4}\right)}{2}\left[\frac{\Delta^2 f(-1) + \Delta^2 f(0)}{2}\right]$$

$$+ \frac{v\left(v^2 - \frac{1}{4}\right)}{6}\Delta^3 f(-1)$$

Differentiating w.r.t. v.

$$f'\left(v+\frac{1}{2}\right) = \Delta f(0) + v\frac{\Delta^2 f(-1) + \Delta^2(0)}{2} + \frac{\left(3v^2 - \frac{1}{4}\right)}{6}\Delta^3 f(-1) \tag{9.2.1}$$

Putting $v = 0$,

$$f'\left(\frac{1}{2}\right) = \Delta f(0) - \frac{1}{24}\Delta^3 f(-1)$$

Shifting the origin to $\left(x - \frac{1}{2}\right)$,

$$f'(x) = \Delta f\left(x - \frac{1}{2}\right) - \frac{1}{24}\Delta^3 f\left(x - \frac{3}{2}\right) \quad \text{(Proved)}$$

(b) Differentiating again (9.2.1) w.r.t. v,

$$f''\left(v+\frac{1}{2}\right) = \frac{\Delta^2 f(-1) + \Delta^2 f(0)}{2} + v\Delta^3 f(-1) \tag{9.2.2}$$

Putting $v = 0$,

$$f''\left(\frac{1}{2}\right) = \frac{\Delta^2 f(-1) + \Delta^2 f(0)}{2}$$

Shifting the origin to $\left(x - \frac{1}{2}\right)$,

$$f''(x) = \frac{\Delta^2 f\left(x - \frac{3}{2}\right) + \Delta^2 f\left(x - \frac{1}{2}\right)}{2} \quad \text{(Proved)}$$

(c) Differentiating again (9.2.2) w.r.t. v,

$$f'''\left(v+\frac{1}{2}\right) = \Delta^3 f(-1)$$

Putting $v = 0$,

$$f'''\left(\frac{1}{2}\right) = \Delta^3 f(-1)$$

Shifting the origin at $\left(x - \frac{1}{2}\right)$,

$$f'''(x) = \Delta^3 f\left(x - \frac{3}{2}\right) \quad \text{(Proved)} \tag{9.2.3}$$

Example 9.2.3. Taking unit of differencing as 1, show that the third derivative of $f(x)$ is approximately.

(a) $\Delta^3 f(0) + \left(x - \frac{3}{2}\right)\Delta^4 f(0)$;

(b) $\Delta^3 f(-1) + \left(x - \frac{1}{2}\right) \cdot \frac{\Delta^4 f(-2) + \Delta^4 f(-1)}{2}$.

Solution: (a) We take the origin at 0 and unit of differencing as 1. From Newton's forward formula,

$$f(x) = f(0) + x\Delta f(0) + \frac{x^2 - x}{2}\Delta^2 f(0) + \frac{x^3 - 3x^2 + 2x}{6}\Delta^3 f(0) + \frac{x^4 - 6x^3 + 11x^2 - 6x}{24}\Delta^4 f(0)$$

Differentiating w.r.t. x, successively

$$f'(x) = \Delta f(0) + \frac{2x-1}{2}\Delta^2 f(0) + \frac{3x^2-6x+2}{6}\Delta^3 f(0) + \frac{4x^3-18x^2+22x-6}{24}\Delta^4 f(0)$$

$$f''(x) = \Delta^2 f(0) + (x-1)\Delta^3 f(0) + \frac{12x^2-36x+22}{24}\Delta^4 f(0)$$

$$f'''(x) = \Delta^3 f(0) + \frac{24x-36}{24}\Delta^4 f(0) = \Delta^3 f(0) + \left(x-\frac{3}{2}\right)\Delta^4 f(0) \quad \text{(proved)}$$

(b) With the same origin and unit as in (a), we obtain from Bessel's formula

$$f(x) \simeq \frac{f(0)+f(1)}{2} + v\Delta f(x) + \frac{\left(v^2-\frac{1}{4}\right)}{2}\cdot\frac{\Delta^2 f(-1)+\Delta^2 f(0)}{2} + \frac{v\left(v^2-\frac{1}{4}\right)}{6}\cdot\Delta^3 f(-1) + \frac{\left(v^2-\frac{1}{4}\right)\left(v^2-\frac{1}{4}\right)}{24}\cdot\frac{\Delta^4 f(-2)+\Delta^4 f(-1)}{2}$$

(where $v = x - 1/2$)

$$\therefore f'(x) = \Delta f(x) \simeq \Delta f(0) + v\cdot\frac{\Delta^2 f(-1)+\Delta^2 f(0)}{2} + \frac{\left(3v^2-\frac{1}{4}\right)}{6}\Delta^3 f(-1) + \frac{\left(4v^3-\frac{10}{2}v\right)}{24}\cdot\frac{\Delta^4 f(-2)+\Delta^4 f(-1)}{2}$$

$$f''(x) = \frac{\Delta^2 f(-1)+\Delta^2 f(0)}{2} + v\Delta^3 f(-1) + \frac{12v^2-5}{24}\cdot\frac{\Delta^4 f(-2)+\Delta^4 f(-1)}{2}$$

$$f'''(x) = \Delta^3 f(-1) + v\cdot\frac{\Delta^4 f(-1)+\Delta^4 f(-1)}{2} \quad (\because v = x - 1/2)$$

$$f'''(x) = \Delta^3 f(-1) + \left(x-\frac{1}{2}\right)\cdot\frac{\Delta^4 f(-2)+\Delta^4 f(-1)}{2} \quad \text{(Proved)}$$

EXERCISES

1. Find dy/dx and d^2y/dx^2 at $x = 1$ from the following table:

x :	1.0	1.2	1.4	1.6	1.8	2.0
$y = f(x)$:	−14.000	−10.300	−5.320	0.260	6.670	13.880

[*Ans:* 17.19; 0.73.]

2. Find the first, second and the third derivatives of the function tabulated below, at the point $x = 3.1$.

x :	3.0	3.2	3.4	3.6	3.8	4.0
$y = f(x)$:	13.375	17.000	23.625	34.0000	48.875	69.125

[*Ans:* 18.01; 84.844; −106.25.]

3. Find dy/dx at $x = 1$ from the following table:

x :	0.8	0.9	1.0	1.1	1.2
$y = f(x)$:	0.7174	0.7835	0.8415	0.8913	0.9376

[*Ans:* −0.0655.]

4. Find the first derivative of $f(x)$ at $x = 4$.

x :	1	2	4	8	10
$y = f(x)$:	0	1	8	27	34

[*Ans:* 4.6059.]

[*Hints:* Use Newton's Divided Difference formula.]

5. Given $y_6 = 1.567$; $y_7 = 1.742$; $y_9 = 1.938$; $y_{12} = 2.458$; find dy/dx at $x = 8$.

[*Ans:* 0.091.]

6. Find dy/dx and d^2y/dx^2 at $x = 1$ from the following table by constructing a central difference table:

x :	1	2	3	4	5	6
$y = f(x)$:	198650	296423	389672	489561	567542	648237

[*Ans:* 134260; −137826.]

7. Show that as far as the third differences

$$\frac{dy_x}{dx} = \frac{2}{3}(y_{x+1} - y_{x-1}) - \frac{1}{12}(y_{x+2} - y_{x-2})$$

[*Hints:* Use Stirling's formula.]

8. Prove that

$$\frac{d}{dx}y_x = \frac{1}{h}(y_{x+h} - y_{x-h}) - \frac{1}{2h}(y_{x-2h} - y_{x-2h}) + \frac{1}{3h}(y_{x+3h} - y_{x-3h}) + \cdots$$

9. Deduce from Bessel's formula:

(a) $$\frac{d}{dx} = \Delta y_{x-(1/2)} - \frac{1}{24}\Delta^3 y_{x-(3/2)}$$

(b) $$\frac{d^2y_x}{dx^2} = \frac{1}{2}(\Delta^2 y_{x-3/2} + \Delta^2 y_{x-1/2})$$

(c) $$\frac{d^3y_x}{dx^3} = \Delta^3 y_{x-(3/2)}.$$

10. Find the first and the second derivatives of $\log_e x$ at $x = 340$

x :	300	310	320	330	340
$\log_e x$:	5.703782	5.736572	5.768321	5.799093	5.828946

[*Ans:* 0.029408; −0.0008665.]

11. The following table gives the values of an empirical function $f(x)$ for certain values of x. Find (a) $f'(93)$, (b) the value of x for which $f(x)$ is maximum, (c) the maximum value of $f(x)$ in the range of x.

x :	60	75	90	105	120
$y = f(x)$:	38.2	48.2	53.2	50.8	47.8

[*Ans:* $f(x)$ is minimum for $x = 91.062$ and the minimum value of $f(x) = 53.2458$.]

12. For what value of x is the following tabulated function a minimum? What is that minimum value?

x :	0.2	0.3	0.4	0.5	0.6	60.7
$y = f(x)$:	0.9182	0.8975	0.8873	0.8862	0.8935	0.9086

[*Ans:* y minimum when $x = 0.4623$.]

Chapter 10

Numerical Integration

Summary

10.1 Introduction

In evaluating a definite integral

$$\int_a^b f(x)dx$$

two problems may be encountered. The first is the form of the integrand $f(x)$ which must be known. The second problem is that the function $f(x)$ must be integrable in $[a, b]$.

In the theory of life contingencies, however, the above two conditions are rarely satisfied. The exact mathematical form of the function $f(x)$ may not be known. We may be given some observations on $f(x)$ for a number of given values of x in $[a, b]$. On the given values of $f(x)$ for $x = x_0(= a), x_1, x_2, \cdots x_n(= b)$, we may find the area of the curve $y = f(x)$ between the ordinates at $x = x_0$ and $x = x_n$. The technique by which this task is accomplished, is known as the *Numerical Integration*. When this device is applied to the integration of a function of a single variable, the process is known as *Numerical Quadrature.*

In this special device of integration, the integrand is first approximated by a polynomial with the help of an appropriate interpolation formula and then it is integrated between the given limits.

Suppose we are required to evaluate the definite integral

$$I = \int_a^b f(x)dx$$

The first step is to approximate $f(x)$ by a polynomial $\phi(x)$ of suitable degree. We then integrate $\phi(x)$ within the limits (a, b), i.e.

$$\int_a^b f(x)dx \simeq \int_a^b \phi(x)dx$$

The difference

$$\left[\int_a^b f(x)dx - \int_a^b \phi(x)dx\right]$$

is called the *Error of approximation.*

10.2 A general quadrature formula

Suppose an unknown function $f(x)$ is given for certain equidistant values of the argument x, say, $x_0, x_0 + h, x_0 + 2h, \cdots$. Let the integral to be evaluated be

$$I = \int_a^b f(x)dx$$

Let the range of integration (a, b) be divided into n equal parts, each of length h by $(n+1)$ equispaced values of x, say, $a = x_0, x_0 + h = x_1, x_0 + 2h = x_2, \ldots, x_0 + nh = x_n = b$, so that $b - a = nh$, and the value of the integrand at the given points be $y_0, y_1, \ldots, y_n$.

$$\therefore I = \int_a^b f(x)dx = \int_{x_0}^{x_0+nh} f(x)dx$$

Let us write $u = (x - x_0)/h$; $du = (1/h)dx$.

$$\therefore I = h\int_0^n f(x_0 + hu)du$$

Now $f(x_0 + hu)$ may be approximated by Newton's forward formula as

$$f(x_0 + hu) = E^u f(x_0) = (1 + \Delta)^u f(x_0)$$

$$= \left[1 + u\Delta + \binom{u}{2}\Delta^2 + \binom{u}{3}\Delta^3 + \cdots\right] f(x_0)$$

$$= f(x_0) + u\Delta f(x_0) + \binom{u}{2}\Delta^2 f(x_0) + \binom{u}{3}\Delta^3 f(x_0) + \cdots$$

$$\therefore I = h\int_0^n \left\{ f(x_0) + u\Delta f(x_0) + \frac{u^2 - u}{2}\Delta^2 f(x_0) + \frac{u^3 - 3u^2 + 2u}{6}\Delta^3 f(x_0) + \cdots \right\} du$$

$$= h\left[uy_0 + \frac{u^2}{2}\Delta y_0 + \frac{\frac{u^3}{3} - \frac{u^2}{2}}{2}\Delta^2 y_0 + \frac{\frac{u^4}{4} - u^3 + u^2}{6}\Delta^3 y_0 + \cdots \right]_0^n$$

$$= h\left[ny_0 + \frac{n^2}{2}\Delta y_0 + \frac{\frac{n^3}{3} - \frac{n^2}{2}}{2}\Delta^2 y_0 + \frac{\frac{n^4}{4} - n^3 + n^2}{6}\Delta^3 y_0 + \cdots \right] \quad (10.2.1)$$

The expression in (10.2.1) is called *General Gauss Legendre Quadrature Formula* for equidistant ordinates.

We shall now develop a number of quadrature formulae based on different assumptions on the degree of the interpolating polynomial.

10.3 Trapezoidal rule

We divide the range of integration (a, b) into n equal parts of width h, by ordinates at $a = x_0, a + h = x_1, a + 2h = x_2, \ldots, a + nh = x_n \equiv b$.

We assume that in each subinterval of width h, the function $f(x)$ may be approximated by a polynomial of degree 1, in x. Hence in the interpolation formula, we shall retain terms up to the first order of difference. Taking the interval (x_0, x_1), we have

$$I_1 = \int_{x_0}^{x_0+h} f(x)dx$$

Writing $(x - x_0)/h = u$; $dx = h\,du$,

$$I_1 = h\int_0^1 f(x_0 + hu)du = h\int_0^1 [f(x_0) + u\Delta f(x_0)]du = h\left[uf(x_0) + \frac{u^2}{2}\Delta f(x_0)\right]_0^1$$

$$= h\left[y_0 + \frac{1}{2}\Delta y_0\right] = \frac{h}{2}[2y_0 + (y_1 - y_0)] = \frac{h}{2}[y_0 + y_1]$$

Similarly,

$$I_2 = \int_{x_1}^{x_2} f(x)dx = \int_{x_0+h}^{x_0+2h} f(x)dx = h\int_1^2 f(x_0+hu)du = \frac{h}{2}[y_1 + y_2]$$

(by advancing the subscripts of y by 1)

....

$$I_n = \int_{x_{n-1}}^{x_n=b} f(x)dx = h\int_{n-1}^{n} f(x_0+hu)du = \frac{h}{2}[y_{n-1} + y_n]$$

Adding $I_1, I_2, \ldots, I_n$, we have

$$I = I_1 + I_2 + \cdots + I_n = \int_a^b f(x)dx = h\int_0^n f(x_0+hu)du$$

$$= \frac{h}{2}[(y_0 + y_n) + 2(y_1 + y_2 + \cdots + y_{n-1})] \qquad (10.3.1)$$

This rule is known as the *Trapezoidal Rule.*

[*Note:* We have in fact taken $n = 1$ in (10.3.1) and neglected terms containing Δ^2 and higher order of differences.]

10.4 Simpson's one-third rule

Here we assume that within any two consecutive subintervals of width h each, the interpolating polynomial $\phi(x)$ is of degree 2. Hence we may put $n = 2$ in (10.3.1) and neglect all differences above the second order. We write

$$I_1 = \int_{x_0}^{x_2} f(x)dx = h\int_0^2 \left\{y_0 + u\Delta y_0 + \frac{u^2-u}{2}\Delta^2 y_0\right\} du$$

$$= h\left[2y_0 + \frac{2^2}{2}\Delta y_0 + \frac{\frac{2^3}{3} - \frac{2^2}{2}}{2}\Delta^2 y_0\right]$$

$$= h\left[2y_0 + 2(y_1 - y_0) + \frac{1}{3}(y_2 - 2y_1 + y_0)\right]$$

$$\therefore I_1 = \frac{h}{3}[6y_0 + 6(y_1 - y_0) + (y_2 - 2y_1 + y_0)] = \frac{h}{3}[y_0 + 4y_1 + y_2] \qquad (10.4.1)$$

Similarly,

$$I_2 = \int_{x_2}^{x_4} f(x)dx = h\int_2^4 \left\{y_0 + u\Delta y_0 + \frac{u^2-u}{2}\Delta^2 y_0\right\} du = \frac{h}{3}[y_2 + 4y_3 + y_4]$$

(by advancing the subscripts by two units)

....

$$I_{n/2} = \int_{x_{n-2}}^{x_n} f(x)dx = \frac{h}{3}[y_{n-2} + 4y_{n-1} + y_n]$$

Adding $I_1, I_2, \ldots, I_{n/2}$, we have

$$I = I_1 + I_2 + \cdots + I_{n/2} = \int_{x_0}^{x_0+nh} f(x)dx$$

$$= \frac{h}{3}[(y_0 + y_n) + 4(y_1 + y_3 + y_5 + \cdots + y_{n-1})$$

$$+ 2(y_2 + y_4 + y_6 + \cdots + y_{n-2})] \qquad (10.4.2)$$

The above quadrature formula is known as *Simpson's One-Third Rule.*

10.5 Simpson's three-eighths rule

Here we assume that within any three consecutive sub-intervals of width h each, the interpolating polynomial approximating $f(x)$ is of degree 3. Hence in General Gauss Legendre Quadrature formula (10.2.1), we may write $n = 3$ and neglect all differences above Δ^3. Thus,

$$I_1 = \int_{x_0}^{x_0+3h} f(x)dx = h\int_0^3 \left[y_0 + u\Delta y_0 + \frac{u^2-u}{2}\Delta^2 y_0 + \frac{u^3-3u^2+2u}{6}\Delta^3 y_0\right] du$$

$$= h\left[3y_0 + \frac{3^2}{2}\Delta y_0 + \frac{\frac{3^3}{3}-\frac{3^2}{2}}{2}\Delta^2 y_0 + \frac{\frac{3^4}{4}-3^3+3^2}{6}\Delta^3 y_0\right]$$

$$= h\left[3y_0 + \frac{9}{2}(y_1 - y_0) + \frac{9}{4}(y_2 - 2y_1 + y_0) + \frac{\frac{81}{4}-18}{6}(y_3 - 3y_2 + 3y_1 - y_0)\right]$$

$$= h\left[3y_0 + \frac{9}{2}(y_1 - y_0) + \frac{9}{4}(y_2 - 2y_1 + y_0) + \frac{3}{8}(y_3 - 3y_2 + 3y_1 - y_0)\right]$$

$$= \frac{3h}{8}[8y_0 + 12(y_1 - y_0) + 6(y_2 - 2y_1 + y_0) + (y_3 - 3y_2 + 3y_1 - y_0)$$

$$= \frac{3h}{8}[y_0 + 3y_1 + 3y_2 + y_3] \qquad (10.5.1)$$

Similarly,

$$I_2 = \int_{x_3}^{x_6} f(x)dx = \frac{3h}{8}[y_3 + 3y_4 + 3y_5 + y_6]$$

(by advancing the subscripts of y by 3 units).

$$\therefore I_{n/3} = \int_{x_{n-3}}^{x_n} f(x)dx = \frac{3h}{8}[y_{n-3} + 3y_{n-2} + 3y_{n-1} + y_n]$$

Adding $I_1, I_2, \ldots, I_{n/3}$, we have

$$I = \frac{3h}{8}[(y_0 + y_n) + 3(y_1 + y_2 + y_4 + y_5 + \cdots + y_{n-2} + y_{n-1})$$

$$+ 2(y_3 + y_6 + \cdots + y_{n-3})] \qquad (10.5.2)$$

This formula (10.5.2) is known as *Simpson's Three-Eighths Rule.*

[*Note:* 1. Simpson's one-third rule can be applied when the range (a, b) is divided into even number of sub-intervals. 2. Simpson's three-eighths rule can be applied when the range (a, b) is divided into a number of sub-intervals, which must be a multiple of 3. 3. Trapezoidal rule, however, can be applied for any number of sub-intervals, odd or even.]

10.6 Weddle's rule

Here we assume that within any six consecutive sub-intervals of width h, each, the interpolating polynomial approximating $f(x)$ will be of degree 6 and we may neglect all differences above Δ^6. Thus, putting $n = 6$ in Gauss Legendre Quadrature Formula (10.2.1) and neglecting all differences above Δ^6, we may write

$$I_1 = \int_{x_0}^{x_0+6h} f(x)dx = h\int_0^6 \left[y_0 + u\Delta y_0 + \frac{u^2-u}{2}\Delta^2 y_0 + \frac{u^3-3u^2+2u}{6}\Delta^3 y_0 + \frac{u^4-6u^3+11u^2-6u}{24}\Delta^4 y_0 + \frac{u^5-10u^4+35u^3-50u^2+24u}{120}\Delta^5 y_0 + \frac{u^6-15u^5+85u^4-22u^3+27u^2-120u}{720}\Delta^6 y_0\right]du$$

$$= h\left[uy_0 + \frac{u^2}{2}\Delta y_0 + \frac{\frac{u^3}{3}-\frac{u^2}{2}}{2}\Delta^2 y_0 + \frac{\frac{u^4}{4}-u^3+u^2}{6}\Delta^3 y_0 + \frac{\frac{u^5}{5}-\frac{3}{2}u^4+\frac{11}{3}u^3-3u^2}{24}\Delta^4 y_0 + \frac{\frac{u^6}{6}-2u^5+\frac{35}{4}u^4-\frac{50}{3}u^3+12u^2}{120}\Delta^5 y_0 + \frac{\frac{u^7}{7}-\frac{5}{2}u^6+17u^5-\frac{225}{4}u^4+\frac{274}{3}u^3-60u^2}{720}\Delta^6 y_0\right]_{u=0}^{u=6}$$

$$= h\left[6y_0 + 18\Delta y_0 + 27\Delta^2 y_0 + 24\Delta^3 y_0 + \frac{123}{10}\Delta^4 y_0 + \frac{33}{10}\Delta^5 y_0 + \frac{41}{140}\Delta^6 y_0\right]$$

If we replace the last term in the expression above by $3/10\Delta^6 y_0$, the error committed will be $h/140\Delta^6 y_0$, which is negligible, if the value of h be such that the sixth order differences are small. Under these assumptions, we may replace the last term by $3/10\Delta^6 y_0$ and write

$$I_1 = h\left[6y_0 + 18\Delta y_0 + 27\Delta^2 y_0 + 24\Delta^3 y_0 + \frac{123}{10}\Delta^4 y_0 + \frac{33}{10}\Delta^5 y_0 + \frac{3}{10}\Delta^6 y_0\right]$$

$$= \frac{3h}{10}\left[20y_0 + 60\Delta y_0 + 90\Delta^2 y_0 + 80\Delta^3 y_0 + 41\Delta^4 y_0 + 11\Delta^5 y_0 + \Delta^6 y_0\right]$$

$$\therefore I_1 = \int_{x_0}^{x_0+6h} f(x)dx = \frac{3h}{10}[20y_0 + 60(y_1 - y_0) + 90(y_2 - 2y_1 + y_0) + 80(y_3 - 3y_2 + 3y_1 - y_0) + 41(y_4 - 4y_3 + 6y_2 - 4y_1 + y_0) + 11(y_5 - 5y_4 + 10y_3 - 10y_2 + 5y_1 - y_0) + (y_6 - 6y_5 + 15y_4 - 20y_3 + 15y_2 - 6y_1 + y_0)]$$

$$= \frac{3h}{10}[y_0 + 5y_1 + y_2 + 6y_3 + y_4 + 5y_5 + y_6] \qquad (10.6.1)$$

Similarly,

$$I_2 = \int_{x_0+6h}^{x_0+12h} f(x)dx = h\int_{6}^{12} f(x_0 + hu)du$$

$$= \frac{3h}{10}[y_6 + 5y_7 + y_8 + 6y_9 + y_{10} + 5y_{11} + y_{12}]$$

....

$$I_{n/6} = \int_{x_0+(n-6)h}^{x_0+nh} f(x)dx = \frac{3h}{10}[y_{n-6} + 5y_{n-5} + y_{n-4} + 6y_{n-3} + y_{n-2} + 5y_{n-1} + y_n]$$

(Number of sub-intervals is a multiple of 6)

Adding $I_1, I_2, \ldots, I_{n/6}$, we get

$$I = \int_{x_0}^{x_0+nh} f(x)dx$$

$$= \frac{3h}{10}[y_0 + 5y_1 + y_2 + 6y_3 + y_4 + 5y_5 + 2y_6 + 5y_7 + y_8 + 6y_9$$
$$+ y_{10} + 5y_{11} + 2y_{12} + \cdots$$
$$+ y_{n-6} + 5y_{n-5} + y_{n-4} + 6y_{n-3} + y_{n-2} + 5y_{n-1} + y_n$$

$$= \frac{3h}{10}[(y_0 + y_n) + (y_2 + y_4 + y_8 + y_{10} + y_{14} + y_{16} + \cdots + y_{n-4} + y_{n-2})$$
$$+ 5(y_1 + y_5 + y_7 + y_{11} + \cdots + y_{n-5} + y_{n-1})$$
$$+ 6(y_3 + y_9 + y_{15} + \cdots + y_{n-3}) + 2(y_6 + y_{12} + \cdots + y_{n-6})] \quad (10.6.2)$$

This is known as *Weddle's rule*, and is generally, more accurate than Simpson's rules. But it requires at least seven consecutive equispaced ordinates within the interval (a, b). It may be noted that if the integrand $f(x)$ is actually a polynomial of degree 5 or lower, Weddle's rule gives an exact result. However, judging the relative accuracy of the above quadrature formulae, we shall show in the next section that Simpson's one-third rule is better.

10.7 Relative accuracy of the three quadrature formulae: Trapezoidal rule, Simpson's rules and Weddle's rule

To compare the relative accuracy of the different quadrature formulae discussed so far, we consider six sub-intervals of length h, obtained by seven ordinates, namely, $x_0 - 3h$, $x_0 - 2h$, $x_0 - h$, x_0, $x_0 + h$, $x_0 + 2h$, $x_0 + 3h$.

We assume that integrand $f(x)$ is finite and continuous in the interval $(x_0 - 3h, x_0 + 3h)$ and has continuous derivatives of all orders up to, and including the sixth in that interval. Furthermore, we suppose that $F(x)$ is the integral of $f(x)$, so that $F'(x) = f(x)$ and

$$I = \int_{x_0-3h}^{x_0+3h} f(x)dx = \int_{x_0-3h}^{x_0+3h} F'(x)dx = F(x_0 + 3h) - F(x_0 - 3h) \quad (10.7.1)$$

10.7.1 The inherent error in trapezoidal rule

Let the value of the integral I obtained by the Trapezoidal Rule be denoted by I_T. Now,

$$I_T = \frac{h}{2}[f(x_0 - 3h) + f(x_0 + 3h) + 2\{f(x_0 + 2h) + f(x_0 - 2h) + f(x_0 + h) + f(x_0 - h) + f(x_0)\}] \tag{10.7.2}$$

Actual value of the integral I is given by

$$I = F(x_0 + 3h) - F(x_0 - 3h)$$

Expanding by Taylor's series,

$$I = 2\left\{(3h)F'(x_0) + \frac{(3h)^3}{3!}F'''(x_0) + \frac{(3h)^5}{5!}F^{v}(x_0) + \frac{(3h)^7}{7!}F^{vii}(x_0) + \cdots\right\}$$

Noting that $F'(x_0) = f(x_0)$ and $F''(x_0) = f'(x_0)$, etc. we have

$$I = 2\left[(3h)f(x_0) + \frac{27h^3}{6}f''(x_0) + \frac{243h^5}{120}f^{iv}(x_0) + \frac{729h^7}{1680}f^{vi}(x_0) + \cdots\right]$$

$$= 6hf(x_0) + 9h^3f''(x_0) + \frac{81}{20}h^5f^{iv}(x_0) + \frac{243}{280}h^7f^{vi}(x_0) + \cdots \tag{10.7.3}$$

From (10.7.2),

$$I_T = \frac{h}{2}\left[2\left\{f(x_0) + \frac{(3h)^2}{2!}f''(x_0) + \frac{(3h)^4}{4!}f^{iv}(x_0) + \frac{(3h)^6}{6!}f^{vi}(x_0) + \cdots\right\}\right.$$
$$+ 4\left\{f(x_0) + \frac{(2h)^2}{2!}f''(x_0) + \frac{(2h)^4}{4!}f^{iv}(x_0) + \frac{(2h)^6}{6!}f^{vi}(x_0) + \cdots\right\}$$
$$\left.+ 4\left\{f(x_0) + \frac{h^2}{2!}f''(x_0) + \frac{h^4}{4!}f^{iv}(x_0) + \frac{h^6}{6!}f^{vi}(x_0) + \cdots\right\} + 2f(x_0)\right]$$
$$= \frac{h}{2}\left[12f(x_0) + 19h^2f''(x_0) + \frac{115}{12}h^4f^{iv}(x_0) + \frac{859}{360}h^6f^{vi}(x_0) + \cdots\right]$$
$$= 6hf(x_0) + \frac{19}{2}h^3f''(x_0) + \frac{115}{24}h^5f^{iv}(x_0) + \frac{859}{720}h^7f^{vi}(x_0) + \cdots \tag{10.7.4}$$

$\therefore$ error in Trapezoidal Rule in six consecutive intervals of width h, each, is

$$E_T = I - I_T = -\frac{1}{2}h^3f''(x_0) - \frac{89}{120}h^5f^{iv}(x_0) - \frac{1639}{5040}h^7f^{vi}(x_0) - \cdots$$

Since in Trapezoidal rule, the second and the third terms in E_T are negligible in comparison to the first term, we may write, $E_T = -\frac{1}{2}h^3f''(\xi)$, for six sub-intervals.

Hence for n sub-intervals of length h,

$$E_T = -\frac{n}{12}h^3f''(\xi)$$

where $x_0 - 3h < \xi < x_0 + 3h$.

$$\therefore E_T \leq -\frac{nh^3}{12}f''(x_m) \tag{10.7.5}$$

where $f''(x_m)$ is the greatest value of $f''(x)$ in $x_0 - 3h < x < x_0 + 3h$.

The expression in (10.7.5) is the analytical form of the error term in the Trapezoidal rule.

10.7.2 The inherent error in Simpson's one-third rule

Here $I_{1/3}$ is given by

$$I_{1/3} = \frac{h}{3}[f(x_0 - 3h) + 4f(x_0 - 2h) + 2f(x_0 - h) + 4f(x_0) + 2f(x_0 + h) + 4f(x_0 + 2h) + f(x_0 + 3h)]$$

$$= \frac{h}{3}[\{f(x_0 + 3h) + f(x_0 - 3h)\} + 4\{f(x_0 + 2h) + f(x_0 - 2h)\} + 2\{f(x_0 + h) + f(x_0 - h)\} + 4f(x_0)]$$

$$= \frac{h}{3}\Bigg[2\left\{f(x_0) + \frac{(3h)^2}{2!}f''(x_0) + \frac{(3h)^4}{4!}f^{iv}(x_0) + \frac{(3h)^6}{6!}f^{vi}(x_0) + \cdots\right\} + 8\left\{f(x_0) + \frac{(2h)^2}{2!}f''(x_0) + \frac{(2h)^4}{4!}f^{iv}(x_0) + \frac{(2h)^6}{6!}f^{vi}(x_0) + \cdots\right\} + 4\left\{f(x_0) + \frac{h^2}{2!}f''(x_0) + \frac{h^4}{4!}f^{iv}(x_0) + \frac{h^6}{6!}f^{vi}(x_0) + \cdots\right\} + 4f(x_0)\Bigg]$$

$$= \frac{h}{3}\left[18f(x_0) + 27h^2f''(x_0) + \frac{49}{4}h^4f^{iv}(x_0) + \frac{329}{40}h^6f^{vi}(x_0) + \cdots\right]$$

$$= 6hf(x_0) + 9h^3f''(x_0) + \frac{49}{12}h^5f^{iv}(x_0) + \frac{329}{120}h^7f^{vi}(\xi)$$

$$\therefore E_{1/3} = I - I_{1/3} = -\frac{1}{30}h^5f^{iv}(x_0) - \frac{3}{35}h^7f^{vi}(\xi), \quad (x_0 - 3h < \xi < x_0 + 3h)$$

In Simpson's one-third rule $f^{vi}(\xi)$ is taken to be negligibly small and the inherent error in six sub-intervals is

$$E_{1/3} \simeq -\frac{1}{30}h^5f^{iv}(x_0)$$

Thus, in n sub-intervals of length h each, the total error in one-third formula is

$$E_{1/3} \simeq -\frac{n}{180}h^5f^{iv}(x_0)$$

If $f^{iv}(x_m)$ denotes the greatest value of the fourth derivative of $f(x)$ in the whole range of x, the maximum error in Simpson's one-third formula is

$$E_{1/3} \leq -\frac{nh^5}{180}f^{iv}(x_m) \tag{10.7.6}$$

10.7.3 The inherent error in Simpson's three-eighths rule

Here

$$I_{3/8} = \frac{3h}{8}\Big[\{f(x_0 + 3h) + f(x_0 - 3h)\} + 3\{f(x + 2h) + f(x - 2h)\} + 3\{f(x + h) + f(x - h)\} + 2f(x_0)\Big]$$

$$= \frac{3h}{8}\left[2\left\{f(x_0) + \frac{(3h)^2}{2!}f''(x_0) + \frac{(3h)^4}{4!}f^{iv}(x_0) + \frac{(3h)^6}{6!}f^{vi}(x_0) + \cdots\right\}\right.$$

$$+ 6\left\{f(x_0) + \frac{(2h)^2}{2!}f''(x_0) + \frac{(2h)^4}{4!}f^{iv}(x_0) + \frac{(2h)^6}{6!}f^{vi}(x_0) + \cdots\right\}$$

$$\left. + 6\left\{f(x_0) + \frac{h^2}{2!}f''(x_0) + \frac{h^4}{4!}f^{iv}(x_0) + \frac{h^6}{6!}f^{vi}(x_0) + \cdots\right\} + 2f(x_0)\right]$$

$$= \frac{3h}{8}\left[16f(x_0) + 24h^2f''(x_0) + 11h^4f^{iv}(x_0) + \frac{77}{30}h^6f^{vi}(x_0) + \cdots\right]$$

$$= 6hf(x_0) + 9h^3f''(x_0) + \frac{33}{8}h^5f^{iv}(x_0) + \frac{77}{80}h^7f^{vi}(x_0)$$

$$\therefore E_{3/8} = I - I_{3/8} = -\frac{3}{40}h^5f^{iv}(x_0) - \frac{53}{560}h^7f^{vi}(x_0) + \cdots$$

In Simpson's rule, we may take the sixth derivative of $f(x)$ to be negligibly small and write

$$E_{3/8} = -\frac{3}{40}h^5f^{iv}(x_0)$$

for six sub-intervals of width h each.

$\therefore$ total error in all the n sub-intervals of width h each is

$$E_{3/8} = -\frac{nh^5}{80}f^{iv}(x_0)$$

If $f^{iv}(x_m)$ is the maximum value of the fourth derivative in the entire range of x, we may write

$$E_{3/8} \le -\frac{nh^5}{80}f^{iv}(x_m) \tag{10.7.7}$$

COROLLARY 10.7.1. Show that errors in Simpson's one-third rule and three-eighths rule are approximately in the ratio 4 : 9.

Proof. From (10.7.6) and (10.7.7), we find that

$$E_{1/3} : E_{3/8} = -\frac{nh^5}{180} : -\frac{nh^5}{80} = \frac{1}{180} : \frac{1}{80} = \frac{1}{9} : \frac{1}{4} = 4 : 9$$

We may therefore conclude that Simpson's one-third rule gives more accurate result than his three-eighths rule.

10.7.4 The inherent error in Weddle's rule

Here,

$$I_w = \frac{3h}{10}[f(x_0 - 3h) + 5f(x_0 - 2h) + f(x_0 - h) + 6f(x_0)$$
$$+ f(x_0 + h) + 5f(x_0 + 2h) + f(x_0 + 3h)]$$

$$= \frac{3h}{10}[\{f(x_0 + 3h) + f(x_0 - 3h)\} + 5\{f(x_0 + 2h) + f(x_0 - 2h)\}$$
$$+ \{f(x_0 + h) + f(x_0 - h)\} + 6f(x_0)]$$

$$= \frac{3h}{10}\left[2\left\{f(x_0) + \frac{(3h)^2}{2!}f''(x_0) + \frac{(3h)^4}{4!}f^{iv}(x_0) + \frac{(3h)^6}{6!}f^{vi}(x_0) + \cdots\right\}\right.$$

$$+ 10\left\{f(x_0) + \frac{(2h)^2}{2!}f''(x_0) + \frac{(2h)^4}{4!}f^{iv}(x_0) + \frac{(2h)^6}{6!}f^{vi}(x_0) + \cdots\right\}$$

$$\left. + 2\left\{f(x_0) + \frac{h^2}{2!}f''(x_0) + \frac{h^4}{4!}f^{iv}(x_0) + \frac{h^6}{6!}f^{vi}(x_0) + \cdots\right\} + 6f(x_0)\right]$$

$$= \frac{3h}{10}\left[20f(x_0) + 30h^2 f''(x_0) + \frac{27}{2}h^4 f^{iv}(x_0) + \frac{35}{12}h^6 f^{vi}(x_0) + \cdots\right]$$

$$= 6hf(x_0) + 9h^3 f''(x_0) + \frac{81}{20}h^5 f^{iv}(x_0) + \frac{7}{8}h^7 f^{vi}(x_0) + \cdots \qquad (10.7.8)$$

$$\therefore E_w = I - I_w = -\frac{1}{140}h^7 f^{vi}(x_0)$$

for six sub-intervals, each of length h.

$\therefore$ total error in n such sub-intervals

$$E_w = -\frac{nh^7}{840}f^{vi}(x_0)$$

If $f^{vi}(x_m)$ be the greatest value of the sixth derivative of $f(x)$ in the entire range of x, then

$$E_w \leq -\frac{nh^7}{840}f^{vi}(x_m)$$

(Here E_w is the total error in the whole range by Weddle's rule).

[*Note* 1. In terms of difference, E_w can be written as $E_w = -(1/140)h^7\Delta^6 y_0$. (This is the total error in range $6h$ by Weddle's rule). 2. If $f(x)$ is a polynomial of degree 5 or lower, Weddle's Rule gives an exact result.]

Example 10.7.1. On the use of Trapezoidal, Simpson's and Weddle's Rule: Evaluate

$$\int_0^1 \frac{1}{1+x^2}dx$$

by taking 7 ordinates at $x = 0, 1/6, 2/6, 3/6, 4/6, 5/6, 6/6$. Here $h = 1/6$.

Solution:

x	$f(x) = 1/(1+x^2)$
0	1.0000000
1/6	0.9729730
2/6	0.9000000
3/6	0.8000000
4/6	0.6923077
5/6	0.5901639
1	0.5000000

$$I_T = \frac{1}{12}\left[f(0) + f(1) + 2\left\{f\left(\frac{1}{6}\right) + \cdots + f\left(\frac{5}{6}\right)\right\}\right] = \frac{1}{12}[9.4108892] = 0.7842408$$

$$I_{1/3} = \frac{1}{18}[1.5000000 + 9.4525476 + 3.1846154] = 0.7853979$$

$$I_{3/8} = \frac{3}{6 \times 8}[1 + 0.5000000 + 3\{0.9729730 + 0.9000000 + 0.6923077 + 0.5901639\}$$
$$+ 2 \times 0.8000000]$$
$$= \frac{1}{16}[1.50000000 + 9.4663338 + 1.6000000] = 0.7853959$$

$$I_w = \frac{3}{6 \times 10}[1 + 5 \times 0.9729730 + 0.9 + 6 \times 0.8 + 0.6923077 + 5 \times 0.5901639 + 0.5]$$
$$= \frac{1}{20}[1.5 + 7.8156845 + 0.9 + 4.8 + 0.6923077] = 0.7853996$$

Actual value

$$I = \int_0^1 \frac{1}{1 + x^2}dx = \left[\tan^{-1} x\right]_0^1 = \frac{\pi}{4} = 0.7853982$$

True value = 0.7853982.

Formula	Value	Error
Trapezoidal rule	0.7842408	0.0011574
Simpson's one-third	0.7853979	0.0000003
Simpson's three-eighths	0.7853959	0.0000023
Weddle's rule	0.7853996	−0.0000014

It is seen from the above that Simpson's one-third rule gives result closest to the actual value.

10.8 Hardy's formula

Certain approximate results of definite integral, due to *G.F. Hardy*, were used quite extensively at some time in actuarial work.

We assume $f(x)$ to be a polynomial of degree 5 given by

$$y_x \equiv f(x) = a_0 + a_1 x + a_2 x^2 + a_3 x^3 + a_4 x^4 + a_5 x^5$$

so that

$$\int_{-3}^{+3} f(x)dx = \left[a_0 x + a_1\frac{x^2}{2} + a_2\frac{x^3}{3} + a_3\frac{x^4}{4} + a_4\frac{x^5}{5} + a_5\frac{x^6}{6}\right]_{-3}^{+3}$$

$$= 2\left(3a_0 + 9a_2 + \frac{243}{5}a_4\right) = 6a_0 + 18a_2 + \frac{486}{5}a_4 \qquad (10.8.1)$$

$$f(0) = a_0, f(-2) + f(2) = 2a_0 + 8a_2 + 32a_4$$
$$f(-3) + f(3) = 2a_0 + 18a_2 + 162a_4$$

We now try to solve for a_0, a_2 and a_4, and find

$$a_0 = f(0) = y_0$$

$$a_2 = \frac{9}{40}(y_{-2} + y_2) - \frac{13}{36}y_0 - \frac{2}{45}(y_{-3} + y_3)$$

$$a_4 = -\frac{1}{40}(y_{-2} + y_2) + \frac{1}{36}y_0 + \frac{1}{90}(y_{-3} + y_3)$$

Substituting in (10.8.1),

$$\int_{-3}^{+3} y_x\,dx = 6y_0 + \frac{81}{20}(y_{-2} + y_2) - \frac{13}{2}y_0 - \frac{4}{5}(y_{-3} + y_3)$$

$$- \frac{486}{200}(y_{-2} + y_2) + \frac{486}{180}y_0 + \frac{486}{450}(y_{-3} + y_3)$$

$$= \frac{11}{5}y_0 + \frac{81}{50}(y_{-2} + y_2) + \frac{14}{50}(y_{-3} + y_3)$$

$$= 2.2y_0 + 1.62(y_{-2} + y_2) + 0.28(y_{-3} + y_3) \tag{10.8.2}$$

$$\Rightarrow \int_0^6 y_x\,dx = 0.28(y_0 + y_6) + 1.62(y_1 + y_5) + 2.2y_3 \tag{10.8.3}$$

This is known as *Hardy's formula* (37).

If the interval of differencing be h, this becomes

$$\int_0^{6h} y_x\,dx = h\{0.28(y_0 + y_{6h}) + 1.62(y_h + y_{5h}) + 2.2y_{3h}\}$$

Similarly,

$$\int_{6h}^{12h} y_x\,dx = h\{0.28(y_{6h} + y_{12h}) + 1.62(y_{7h} + y_{11h}) + 2.2y_{9h}\}, \text{ and so on.}$$

Since

$$\int_0^{\infty} y_x\,dx = \int_0^{6h} y_x\,dx + \int_{6h}^{12h} y_x\,dx + \int_{12h}^{18h} y_x\,dx + \cdots$$

$$\therefore \int_0^{\infty} y_x\,dx = h\{0.28(y_0 + 2y_{6h} + 2y_{12h} + \cdots) + 1.62(y_n + y_{5h} + y_{7h} + \cdots)$$

$$+ 2.2(y_{3h} + y_{9h} + \cdots)\} \tag{10.8.4}$$

This is called *Hardy's formula* (38).

Hardy's formulae were once much used in actuarial work, as we have mentioned at the very beginning of this section.

Suppose we are dealing with functions derived from a mortality table and we choose h so that $7h$ falls just within or just outside the limits of the table (i.e. $7h$ is, in fact, nearly 100), we can then write (10.8.4) as

$$\int_0^{\infty} y_x\,dx = h(0.28y_0 + 1.62y_h + 2.2y_{3h} + 1.62y_{5h} + 0.56y_{6h} + 1.62y_{7h}) \tag{10.8.5}$$

Hardy's formula (39a) is expressed in the above form.

[*Note:* A word of caution against thinking of quadrature formulae as means of computing areas under curves. These quadrature formulae are in fact methods of *computing the values of the definite integrals.* It is only when the integrands are ordinates of a curve, the quadrature formulae may be looked upon as means of evaluating the area under the curves. Usually the integrand may be any function of x, only known for equidistant values of x.]

10.9 Central difference quadrature formula

We may integrate Stirling's and Bessel's interpolation formulae and obtain rapidly converging quadrature formulae in terms of differences.

Suppose we are integrating Stirling's interpolation formula in the range $x_0 - h$ to $x_0 + h$ for x, which is the same as the range -1 to $+1$ for u, where $u = (x - x_0)/h$, $h\,du = dx$.

$$\int_{x_0-h}^{x_0+h} f(x)dx \simeq h \int_{-1}^{+1} \left\{ y_0 + u\frac{\Delta y_{-1} + \Delta y_0}{2} + \frac{u^2}{2!}\Delta^2 y_{-1} + \frac{u(u^2-1)}{3!}\frac{\Delta^3 y_{-2} + \Delta^3 y_{-1}}{2} + \frac{u^2(u^2-1)}{4!}\Delta^4 y_{-2} + \cdots \right\} du$$

$$= h\left[uy_0 + \frac{u^3}{3.2!}\Delta^2 y_{-1} + \left(\frac{u^5}{5.4!} - \frac{u^3}{3.4!}\right)\Delta^4 y_{-2} + \cdots \right]_{-1}^{+1}$$

$$= h\left[2y_0 + \frac{1}{3}\Delta^2 y_{-1} + \left(\frac{1}{60} - \frac{1}{36}\right)\Delta^4 y_{-2} + \cdots \right]$$

$$\int_{x_0-h}^{x_0+h} f(x)dx = h\int_{-1}^{+1} \phi(u)du = h\left[2y_0 + \frac{1}{3}\Delta^2 y_{-1} - \frac{1}{90}\Delta^4 y_{-2} + \cdots \right]$$

$$= 2h\left[y_0 + \frac{1}{6}\Delta^2 y_{-1} - \frac{1}{180}\Delta^4 y_{-2} + \cdots \right]$$

By advancing the subscripts of the y's by one unit we get the value of the integral from $x = x_0$ to $x = x_0 + 2h$. We may denote this integral by I_0^2 and write

$$I_0^2 = 2h\left[y_1 + \frac{1}{6}\Delta^2 y_0 - \frac{1}{180}\Delta^4 y_{-1} + \cdots \right]$$

Similarly, the integrals $I_2^4, I_4^6, \ldots, I_{n-2}^n$ may be written as

$$I_2^4 = 2h\left[y_3 + \frac{1}{6}\Delta^2 y_2 - \frac{1}{180}\Delta^4 y_1 + \cdots \right]$$

$$I_4^6 = 2h\left[y_5 + \frac{1}{6}\Delta^2 y_4 - \frac{1}{180}\Delta^4 y_3 + \cdots \right]\cdots$$

$$I_{n-2}^n = 2h\left[y_{n-1} + \frac{1}{6}\Delta^2 y_{n-2} - \frac{1}{180}\Delta^4 y_{n-3} + \cdots \right]$$

Adding all these separate integrals, we get

$$I_0^n = 2h\Big[(y_1 + y_3 + y_5 + \cdots + y_{n-1}) + \frac{1}{6}(\Delta^2 y_0 + \Delta^2 y_2 + \cdots + \Delta^2 y_{n-2}) - \frac{1}{180}(\Delta^4 y_1 + \Delta^4 y_3 + \cdots + \Delta^4 y_{n-3}) + \cdots\Big] \quad (10.9.1)$$

(we have ignored differences $\Delta^6 y$ and above)

where

$$I_0^n = \int_{x_0}^{x_0+nh} f(x)dx \text{ and } n \text{ is even.}$$

Again integrating Bessel's formula over the interval $x = x_0$ to $x = x_0 + h$, or $v = -1/2$ to $v = +1/2$, we have (noting that $v = (x - x_0)/h - 1/2$ and $hdv = dx$).

$$I_0^1 = \int_{x_0}^{x_0+h} f(x)dx = h\int_{-1/2}^{+1/2}\left[\frac{y_0 + y_1}{2} + v\Delta y_0 + \frac{v^2 - \frac{1}{4}}{2!}\cdot\frac{\Delta^2 y_{-1} + \Delta^2 y_0}{2}\right.$$

$$\left. + \frac{v\left(v^2 - \frac{1}{4}\right)}{3!}\Delta^3 y_{-1} + \frac{\left(v^2 - \frac{1}{4}\right)\left(v^2 - \frac{9}{4}\right)}{4!}\cdot\frac{\Delta^4 y_{-2} + \Delta^4 y_{-1}}{2}\right]dv$$

$$= h\left[v\cdot\frac{y_0 + y_1}{2} + \frac{\frac{v^3}{3} - \frac{v}{4}}{2!}\cdot\frac{\Delta^2 y_{-1} + \Delta^2 y_0}{2} + \frac{\frac{v^5}{5} - \frac{5}{2}\cdot\frac{v^3}{3} + \frac{9}{16}v}{4!}\cdot\frac{\Delta^4 y_{-2} + \Delta^4 y_{-1}}{2}\right]_{-1/2}^{+1/2}$$

$$= h\left[\frac{y_0 + y_1}{2} - \frac{1}{12}\cdot\frac{\Delta^2 y_{-1} + \Delta^2 y_0}{2} + \frac{11}{720}\cdot\frac{\Delta^4 y_{-2} + \Delta^4 y_{-1}}{2} + \cdots\right]$$

Advancing the subscripts by one unit at a time, we find the integrals over the succeeding intervals as

$$I_1^2 = h\left[\frac{y_1 + y_2}{2} - \frac{1}{12}\cdot\frac{\Delta^2 y_0 + \Delta^2 y_1}{2} + \frac{11}{720}\cdot\frac{\Delta^4 y_{-1} + \Delta^4 y_0}{2} + \cdots\right]$$

$$I_2^3 = h\left[\frac{y_2 + y_3}{2} - \frac{1}{12}\cdot\frac{\Delta^2 y_1 + \Delta^2 y_2}{2} + \frac{11}{720}\cdot\frac{\Delta^4 y_0 + \Delta^4 y_1}{2} + \cdots\right]\cdots$$

$$I_{n-1}^n = h\left[\frac{y_{n-1} + y_n}{2} - \frac{1}{12}\cdot\frac{\Delta^2 y_{n-2} + \Delta^2 y_{n-1}}{2} + \frac{11}{720}\cdot\frac{\Delta^4 y_{n-3} + \Delta^4 y_{n-2}}{2} + \cdots\right]$$

Adding these separate integrals,

$$I_0^n = h\left[\left(\frac{y_0}{2} + y_1 + y_2 + \cdots + y_{n-1} + \frac{y_n}{2}\right)\right.$$

$$- \frac{1}{12}\left(\frac{\Delta^2 y_{-1}}{2} + \Delta^2 y_0 + \Delta^2 y_1 + \cdots + \Delta^2 y_{n-2} + \frac{\Delta^2 y_{n-1}}{2}\right)$$

$$\left. + \frac{11}{720}\left(\frac{\Delta^4 y_{-2}}{2} + \Delta^4 y_{-1} + \Delta^4 y_0 + \cdots + \Delta^4 y_{n-3} + \frac{\Delta^4 y_{n-2}}{2}\right) + \cdots\right] \quad (10.9.2)$$

where n is either even or odd.

It will be observed that formulae (10.9.1) and (10.9.2) use only even differences, and that (10.9.2) uses all the even differences but (10.9.1) uses only half of them. However, formula (10.9.2) is rather cumbersome for practical use in the form as it stands in (10.9.2), but it can be transformed into a simpler and useful form as we shall see below.

From the definition, we have $\Delta^2 y_{-1} = \Delta y_0 - \Delta y_{-1}$

$$\Delta^2 y_0 = \Delta y_1 - \Delta y_0$$

$$\dots\dots\dots\dots\dots\dots$$

$$\Delta^2 y_{n-1} = \Delta y_n - \Delta y_{n-1}$$

$$\Delta^4 y_{-2} = \Delta^3 y_{-1} - \Delta^3 y_{-2}$$

$$\Delta^4 y_{-1} = \Delta^3 y_0 - \Delta^3 y_1$$

$$\dots\dots\dots\dots\dots\dots\dots\dots\dots$$

$$\Delta^4 y_{n-2} = \Delta^3 y_{n-1} - \Delta^3 y_{n-2}, \text{ etc.}$$

Substituting these in (10.9.2), and replacing the even differences by the corresponding odd differences of just lower order, we may observe that all differences except those at the beginning and end of the table cancel one another, and formula (10.9.2) reduces to

$$I_0^n = h\left[\left\{\frac{y_0}{2} + y_1 + y_2 + \cdots + y_{n-1} + \frac{y_n}{2}\right\} + \frac{1}{12}\left(\frac{\Delta y_{-1} + \Delta y_0}{2}\right)\right.$$

$$- \frac{11}{720}\left(\frac{\Delta^3 y_{-2} + \Delta^3 y_{-1}}{2}\right) + \cdots - \frac{1}{12}\left(\frac{\Delta y_{n-1} + \Delta y_n}{2}\right)$$

$$\left. + \frac{11}{720}\left(\frac{\Delta^3 y_{n-2} + \Delta^3 y_{n-1}}{2}\right) + \cdots\right]$$

$$= h\left[\left\{\frac{y_0}{2} + y_1 + y_2 + \cdots + y_{n-1} + \frac{y_n}{2}\right\} - \frac{1}{12}\left\{-\frac{\Delta y_{-1} + \Delta y_0}{2} + \frac{\Delta y_{n-1} + \Delta y_n}{2}\right\}\right.$$

$$\left. + \frac{11}{720}\left\{\frac{\Delta^3 y_{n-2} + \Delta^3 y_{n-1}}{2} - \frac{\Delta^3 y_{-2} + \Delta^3 y_{-1}}{2}\right\} + \cdots\right] \quad (10.9.3)$$

The results obtained by formulae (10.9.2) and (10.9.3) will be identical, but the computational labour involved in (10.9.3) is only a fraction of that required in using (10.9.2).

10.10 Euler-Maclaurin's summation formula

Euler-Maclaurin's summation formula states that

$$\sum_{r=0}^{n-1} f(a + rh)$$

can be written as

$$\sum_{r=0}^{n-1} f(a + rh) = \frac{1}{h}\int_a^{a+nh} f(x)dx - \frac{1}{2}\{f(a + nh) - f(a)\}$$

$$+ \frac{h}{12}\{f'(a + nh) - f'(a)\} - \frac{h^3}{720}\{f'''(a + nh) - f'''(a)\} \cdots$$

$$\therefore \frac{1}{h}\int_a^{a+nh} f(x)dx = \sum_{r=0}^{n} f(a + rh) - \frac{1}{2}\{f(a + nh) + f(a)\}$$

$$- \frac{h}{12}\{f'(a + nh) - f'(a)\} + \frac{h^3}{720}\{f'''(a + nh) - f'''(a)\} + \cdots$$

Proof. We assume that the function $f(x)$ is finite and continuous in the interval $(a, a + nh)$ and has continuous derivatives of all orders in the given interval. We may define a function

$F(x)$, such that $\Delta F(x) = f(x)$. Then

$$\sum_{x=a}^{x=a+\overline{n-1}\,h} f(x) = \Delta F(a) + \Delta F(a+h) + \cdots + \Delta F(a+\overline{n-1}) = F(a+nh) - F(a)$$

Further, $\Delta F(x) = f(x)$.

$$\therefore\ F(x) = \frac{1}{\Delta} f(x) = (e^{hD}-1)^{-1} f(x) \qquad [\because E \equiv e^{hD} \equiv 1+\Delta]$$

$$= \left[hD + \frac{h^2D^2}{2!} + \frac{h^3D^3}{3!} + \frac{h^4D^4}{4!} + \cdots\right]^{-1} f(x)$$

$$= \frac{1}{h}\cdot\frac{1}{D}\left[1 + \left(\frac{hD}{2!} + \frac{h^2D^2}{3!} + \frac{h^3D^3}{4!} + \cdots\right)\right]^{-1} f(x)$$

$$= \frac{1}{h}\cdot\frac{1}{D}\left[1 - \left(\frac{hD}{2!} + \frac{h^2D^2}{3!} + \frac{h^3D^3}{4!} + \cdots\right)\right.$$

$$+ \left(\frac{hD}{2!} + \frac{h^2D^2}{3!} + \frac{h^3D^3}{4!} + \cdots\right)^2 - \left(\frac{hD}{2!} + \frac{h^2D^2}{3!} + \frac{h^3D^3}{4!} + \cdots\right)^3$$

$$\left. + \left(\frac{hD}{2!} + \frac{h^2D^2}{3!} + \frac{h^3D^3}{4!} + \cdots\right)^4 - \cdots\right] f(x)$$

$$= \frac{1}{hD}\left[1 - \frac{h}{2!}D + h^2\left(\frac{1}{4} - \frac{1}{6}\right)D^2 + h^3\left(-\frac{1}{24} + \frac{1}{6} - \frac{1}{8}\right)D^3\right.$$

$$+ h^4\left(-\frac{1}{120} + \frac{1}{36} + \frac{1}{24} - \frac{1}{8} + \frac{1}{16}\right)D^4$$

$$\left. + h^5\left(-\frac{1}{720} + \frac{1}{120} + \frac{1}{72} - \frac{1}{32} + \frac{1}{12} - \frac{1}{24} - \frac{1}{32}\right)D^5 + \cdots\right] f(x)$$

$$= \frac{1}{h}D^{-1} f(x) - \frac{1}{2} f(x) + \frac{h}{12} Df(x) - \frac{h^3}{720} D^3 f(x) + \cdots$$

i.e. $F(x) = \dfrac{1}{h}\displaystyle\int f(x)dx - \dfrac{1}{2} f(x) + \dfrac{h}{12} f'(x) - \dfrac{h^3}{720} f'''(x) + \cdots$

$$\therefore\ F(a+nh) - F(a) = \frac{1}{h}\int_a^{a+nh} f(x)dx - \frac{1}{2}\{f(a+nh) - f(a)\}$$

$$+ \frac{h}{12}\{f'(a+nh) - f'(a)\} - \frac{h^3}{720}\{f'''(a+nh) - f'''(a)\} + \cdots$$

$$\therefore\ \sum_{r=0}^{n-1} f(a+rh) = F(a+nh) - F(a)$$

$$= \frac{1}{h}\int_a^{a+nh} f(x)dx - \frac{1}{2}\{f(a+nh) - f(a)\} + \frac{h}{12}\{f'(a+nh) - f'(a)\}$$

$$- \frac{h^3}{720}\{f'''(a+nh) - f'''(a)\} + \cdots \qquad \textbf{(10.10.1)}$$

$$\Rightarrow \sum_{r=0}^{r=n} f(a+rh) = \frac{1}{h}\int_a^{a+nh} f(x) + \frac{1}{2}\{f(a+nh)+f(a)\}$$

$$+ \frac{h}{12}\{f'(a+nh)-f'(a)\} - \frac{h^3}{720}\{f'''(a+nh)-f'''(a)\} + \cdots \quad (10.10.2)$$

$$\Rightarrow \frac{1}{2}f(a) + f(a+b) + \cdots + f(a+\overline{n-1}h) + \frac{1}{2}f(a+nh)$$

$$= \frac{1}{h}\int_a^{a+nh} f(x)dx + \frac{h}{12}\{f'(a+nh)-f'(a)\} - \frac{h^3}{720}\{f'''(a+nb)-f'''(a)\} + \cdots$$

$$(10.10.3)$$

This is the simple form of *Euler-Maclaurin's Summation Formula.*

10.11 Some applications of Euler-Maclaurin's summation formula

Example 10.11.1. Using Euler-Maclaurin's formula, obtain the sum $1^p + 2^p + 3^p + \cdots + n^p$.

Solution: Euler-Maclaurin's summation formula is given by

$$\frac{1}{h}\int_a^{a+nh} f(x)dx = \frac{1}{2}f(a) + f(a+h) + \cdots + f(a+\overline{n-1}h) + \frac{1}{2}f(a+nh)$$

$$- \frac{h}{12}\{f'(a+nh)-f'(a)\} + \frac{h^3}{720}\{f'''(a+nh)-f'''(a)\} - \cdots$$

Putting $a = 0$, $h = 1$, $f(x) = x^p$, we have

$$\int_0^n x^p dx = 1^p + 2^p + \cdots + (n-1)^p + \frac{1}{2}n^p - \frac{1}{12}\left[px^{p-1}\right]_0^n$$

$$+ \frac{1}{720}\left[p(p-1)(p-2)x^{p-3}\right]_0^n - \cdots$$

$$\Rightarrow 1^p + 2^p + \cdots + n^p = \int_0^n x^p dx + \frac{1}{2}n^p + \frac{1}{12}(pn^{p-1}) - \frac{1}{720}\{p(p-1)(p-2)n^{p-3}\} + \cdots$$

$$= \frac{n^{p+1}}{p+1} + \frac{n^p}{2} + \frac{p}{12}n^{p-1} - \frac{p(p-1)(p-2)}{720}n^{p-3} + \cdots$$

Example 10.11.2. Evaluate

$$\int_2^3 \log_e x\, dx$$

to five places of decimals.

Solution: Let us take $a = 2$, $h = 1/10$, $n = 10$.

$f(x) = \log_e x$

Euler-Maclaurin's formula gives

$$10\int_2^3 \log_e x\,dx = \frac{1}{2}\log_e 2 + \log_e 2.1 + \log 2.2 + \cdots + \log 2.9$$

$$+\frac{1}{2}\log_e 3 - \frac{1}{120}\left\{\frac{1}{3}-\frac{1}{2}\right\} + \frac{1}{720\times 10^3}\left\{\frac{2}{27}-\frac{2}{8}\right\}$$

$$= \frac{1}{2}\log_e 2 + \log_e 21 + \log_e 2.2 + \cdots + \log 2.9 + \frac{1}{2}\log_e 3$$

$$+\frac{1}{720} - \frac{1}{720000}\times\frac{19}{108}$$

$$= 9.094036404 + 0.00138889 = 9.095425293$$

$$\therefore \int_2^3 \log_e x\,dx = 0.90954 \text{ (correct to 5 decimal places).}$$

Actual result is

$$\left[x\log_e x - x\right]_2^3 = 3\log_e 3 - 2\log_e 2 - 1 = 0.909542505$$

Thus, the value of the integral correct to 5 decimal places by Euler-Maclaurin's summation formula is correct to the 5th decimal place.

EXAMPLE 10.11.3. Sum the series

$$\frac{1}{201^2}+\frac{1}{203^2}+\cdots+\frac{1}{299^2}$$

Solution: Let us write $a = 201$, $h = 2$, $n = 49$.

$$f(x) = \frac{1}{x^2}$$

By Euler-Maclaurin's formula

$$\frac{1}{2}\int_{201}^{299}\frac{1}{x^2}dx = \frac{1}{2}\cdot\frac{1}{201^2}+\frac{1}{202^2}+\cdots+\frac{1}{297^2}+\frac{1}{2}\cdot\frac{1}{299^2}-\frac{2}{12}\left[-\frac{2}{x^3}\right]_{201}^{299}$$

$$+\frac{8}{720}\left[-\frac{24}{x^5}\right]_{201}^{299}$$

$$\therefore \sum_{r=0}^{n} f(a+rh) = \frac{1}{2}\int_{201}^{299}\frac{1}{x^2}dx + \frac{1}{2}\left[\frac{1}{201^3}+\frac{1}{299^3}\right]+\frac{1}{16}\left[\frac{2}{201^2}-\frac{2}{299^2}\right]$$

$$-\frac{1}{90}\left[\frac{24}{201^5}-\frac{24}{299^5}\right].$$

$$\therefore \text{Sum} = \frac{1}{2}\left[-\frac{1}{x}\right]_{201}^{299} + \frac{1}{2}(0.00002475 + 0.00001118)$$

$$+\frac{1}{8}(0.000000123 - 0.000000037)$$

$$= 0.00163064 + 0.000017965$$

$$= 0.001648605 \text{ [the last part of the sum is negligible]}$$

EXAMPLE 10.11.4. Obtain the approximate integral

$$\int_0^1 \frac{dx}{1+x^2} = \frac{1}{4a}\left(1+\frac{1}{6a}\right)+\sum_{x=1}^{x=a}\frac{a}{a^2+x^2}$$

Solution: Let us take

$$f(x) = \frac{a}{a^2 + x^2}$$

and consider the integral

$$\int_0^a \frac{a}{a^2 + x^2}\,dx = \int_0^1 \frac{a^2\,dt}{a^2(1 + t^2)} \quad \text{(putting } x = at\text{)}$$

$$= \int_0^1 \frac{dt}{1 + t^2} = \int_0^1 \frac{dx}{1 + x^2} \qquad \text{(i)}$$

By Euler-Maclaurin's formula

$$\int_0^a \frac{a}{a^2 + x^2} = \sum_{x=0}^{x=a} \frac{a}{a^2 + x^2} - \frac{1}{2}\left(\frac{1}{a} + \frac{1}{2a}\right) - \frac{1}{12}\{f'(a) - f'(0)\}$$

$$= \sum_{x=0}^{x=a} \frac{a}{a^2 + x^2} + \frac{1}{a} - \frac{3}{4a} - \frac{1}{12}\left[\frac{-2ax}{(a^2 + x^2)^2}\right]_{x=0}^{x=a}$$

$$= \sum_{x=1}^{x=a} \frac{a}{a^2 + x^2} + \frac{1}{4a} + \frac{1}{24a^2}$$

$$= \sum_{x=1}^{x=a} \frac{a}{a^2 + x^2} + \frac{1}{4a}\left(1 + \frac{1}{6a}\right)$$

Hence

$$\int_0^1 \frac{dx}{1 + x^2} = \sum_{x=0}^{x=a} \frac{a}{a^2 + x^2} + \frac{1}{4a}\left(1 + \frac{1}{6a}\right) \quad \text{(Proved)}$$

10.12 Stirling's formula for approximation to factorials

If n is large,

$$n! \simeq e^{-n+(1/12n)-(1/360n^3)} \cdot n^{(n+1/2)} \cdot \sqrt{2\pi}$$

Proof. By Euler-Maclaurin's formula, we write

$$\int_1^n \log_e x\,dx = \frac{1}{2}\log_e 1 + \log_e 2 + \log_e 3 + \cdots + \log_e(n-1) + \frac{1}{2}\log_e n$$

$$- \frac{1}{12}\{f'(n) - f'(1)\} + \frac{1}{720}\{f'''(n) - f'''(1)\}\cdots$$

$$\Rightarrow [x\log_e x - x]_1^n = \log n! - \frac{1}{2}\log_e n - \frac{1}{12}\left(\frac{1}{n} - 1\right) + \frac{1}{720}\left(\frac{2}{n^3} - 2\right) - \cdots$$

$$\Rightarrow n\log_e n - n + 1 = \log n! - \frac{1}{2}\log_e n - \frac{1}{12n} + \frac{1}{360n^3} + \frac{1}{12} - \frac{1}{360}\cdots$$

$$\Rightarrow \log_e n! = \left(n + \frac{1}{2}\right)\log_e n - n - \frac{1}{12n} + \frac{1}{360n^3} + \cdots + c \qquad (10.12.1)$$

where c is a constant, independent of n.

We now evaluate the constant c, using Walli's formula in Trigonometry, as given below:

$$\frac{\pi}{2} = \lim_{n\to\infty}\left[\frac{2\cdot 2}{1\cdot 3}\cdot\frac{4\cdot 4}{3\cdot 5}\cdot\frac{6\cdot 6}{5\cdot 7}\cdots\frac{2n\cdot 2n}{(2n-1)(2n+1)}\right]$$

$$= \lim_{n\to\infty}\left[\frac{2^n\cdot n!\cdot 2^n\cdot n!\{2\cdot 4\cdot 6\cdots 2n\}^2}{(2n)!(2n)!(2n+1)}\right]$$

$$= \lim_{n\to\infty}\left[\frac{2^{4n}\cdot (n!)^4}{(2n)!^2(2n+1)}\right]$$

$$\therefore \log\frac{\pi}{2} = \lim_{n\to\infty}[4n\log 2 + 4\log n! - 2\log(2n)! - \log(2n+1)]$$

Substituting in the above, the value of $\log n!$ and $\log(2n)!$ as given in (10.12.1), we have

$$\log\frac{\pi}{2} = \lim_{n\to\infty}\left[4n\log 2 + 4\left\{\left(n+\frac{1}{2}\right)\log_e n - n - \frac{1}{12n} + \frac{1}{360n^3} + \cdots + c\right\}\right.$$
$$\left. - 2\left\{\left(2n+\frac{1}{2}\right)\log 2n - 2n - \frac{1}{24n} + \frac{1}{360\times 8n^3} + \cdots + c\right\} - \log(2n+1)\right]$$
$$= \lim_{n\to\infty}\left[(4n\log 2 + 4n\log n - 4n\log 2n) + 2\log_e n - \log 2n + (-4n+4n)\right.$$
$$\left. + \frac{3}{12n} + \frac{1}{90n^3} - \frac{1}{1440n^3} + \cdots + 2c - \log(2n+1)\right]$$
$$= \lim_{n\to\infty}\left[\log_e\frac{n^2}{2n(2n+1)} + 2c\right]$$
$$= \lim_{n\to\infty}\left[\log\frac{1}{2(2+1/n)} + 2c\right] = (-\log 4 + 2c)$$

$$\therefore \log\left(\frac{\pi}{2}\times 4\right) = 2c, \quad \text{or,} \quad c = \frac{1}{2}\log 2\pi = \log\sqrt{2\pi} \tag{10.12.2}$$

From (10.12.2),

$$n! \simeq e^c\cdot e^{-n-(1/12n)+(1/360n^3)}\cdot n^{(n+1/2)}$$
$$\simeq \sqrt{2\pi}\,n^{(n+1/2)}e^{-n-(1/12n)+(1/360n^3)}\ldots \tag{10.12.3}$$
$$\simeq \sqrt{2\pi}\,n^{(n+1/2)}e^{-n} \text{ (for large } n) \tag{10.12.4}$$

10.13 Newtọn-Cotes formula for numerical quadrature

We consider the integral

$$I = \int_a^b f(x)dx$$

The interval (a, b) is sub-divided into n equal parts, each of length h, by the points $a = x_0$, $x_0 + h$, $x_0 + 2h, \ldots, x_0 + nh = b$, where $n = (b-a)/h$, such that $x_i = x_0 + ih$, $i = 0(1)n$.

The corresponding entries $f(x_i)$, $i = 0(1), n$, are also available.

Newton-Cotes formula states

$$\int_{x_0}^{x_0+nh} f(x)dx = H_0 f(x_0) + H_1 f(x_1) + \cdots + H_n f(x_n) \tag{10.13.1}$$

where

$$H_i = \int_{x_0}^{x_0+nh} \frac{(x-x_0)(x-x_1)\cdots(x-x_{i-1})(x-x_{i+1})\cdots(x-x_n)}{(x_i-x_0)(x_i-x_1)\cdots(x_i-x_{i-1})(x_i-x_{i+1})\cdots(x_i-x_n)}\,dx$$

$$(i = 0, 1, 2, \ldots, n)$$

Proof. By Lagrange's formula (replacing $f(x)$ by the interpolating polynomial $\phi(x)$ of degree n in x), we obtain

$$\begin{aligned} f(x) \simeq \phi(x) &= \frac{(x-x_1)\cdots(x-x_n)}{(x_0-x_1)(x_0-x_2)\cdots(x_0-x_n)}f(x_0) \\ &+ \frac{(x-x_0)(x-x_2)\cdots(x-x_n)}{(x_1-x_0)(x_1-x_2)\cdots(x_1-x_n)}f(x_1) + \cdots \\ &+ \frac{(x-x_0)(x-x_1)\cdots(x-x_{i-1})(x-x_{i+1})\cdots(x-x_n)}{(x_i-x_0)(x_i-x_1)\cdots(x_i-x_{i-1})(x_i-x_{i+1})\cdots(x_i-x_n)}f(x_i) + \cdots \\ &+ \frac{(x-x_0)(x-x_1)\cdots(x-x_{n-1})}{(x_n-x_0)(x_n-x_1)\cdots(x_n-x_{n-1})}f(x_n) \end{aligned}$$

$$\begin{aligned} \therefore \int_{x_0}^{x_0+nh} f(x)dx &= \int_{x_0}^{x_0+nh} \sum_{i=0}^{n} \frac{(x-x_0)(x-x_1)\cdots(x-x_{i-1})(x-x_{i+1})\cdots(x-x_n)}{(x_i-x_0)(x_i-x_1)\cdots(x_i-x_{i-1})(x_i-x_{i+1})\cdots(x_i-x_n)}f(x_i) \\ &= \sum_{i=0}^{i=n} H_i f(x_i) \end{aligned}$$

where

$$H_i = \int_{x_0}^{x_0+nh} \frac{(x-x_0)(x-x_1)\cdots(x-x_{i-1})(x-x_{i+1})\cdots(x-x_n)}{(x_i-x_0)(x_i-x_1)\cdots(x_i-x_{i-1})(x_i-x_{i+1})\cdots(x_i-x_n)}\,dx$$

Writing $u = (x - x_0)/h$; $h\,du = dx$.

$$\begin{aligned} \therefore H_i &= h\int_0^n \frac{(uh)\{(u-1)h\}\cdots\{(u-i+1)h\}\{(u-i-1)h\}\cdots\{(u-n)h\}}{(ih)\{(i-1)h\}\{(i-2)h\}\cdots(1h)(-1h)(-2h)\cdots\{-(n-i)h\}}\,du \\ &= h\int_0^n \frac{h^n u(u-1)(u-2)\cdots(u-i+1)(u-i-1)\cdots(u-n)}{h^i i!(-1)^{n-i}h^{n-i}(n-i)!}\,du \\ &= \frac{(-1)^{n-i}}{i!(n-i)!}\cdot h\int_0^n \frac{u(u-1)(u-2)\cdots(u-n)}{(u-i)}\,du \qquad (10.13.2) \end{aligned}$$

The error in Newton-Cote's formula will be obtained by integrating the error term of Lagrange's interpolation formula over the range x_0 to $x_0 + nh$.

Now error term in Lagrange's formula:

$$\text{Error} = \frac{f^{(n+1)}(\xi)}{(n+1)!}(x-x_0)(x-x_1)\cdots(x-x_n)$$

where $x_0 < \xi < x_n$.

Hence error term in Newton-Cotes formula will be equal to

$$E = \int_{x_0}^{x_0+nh} \frac{f^{(n+1)}(\xi)}{(n+1)!}(x-x_0)(x-x_1)\cdots(x-x_n)dx$$

$$= h\int_0^n \frac{h^{(n+1)}f^{(n+1)}(\xi)}{(n+1)!}u(u-1)(u-2)\cdots(u-n)du \quad \text{(where } 0<\xi<n.\text{)}$$

$$\simeq h\int_0^n \frac{\Delta^{n+1}y_0}{(n+1)!}u(u-1)(u-2)\cdots(u-n)du \tag{10.13.3}$$

since

$$\frac{\Delta^{n+1}y_0}{h^{(n+1)}} \simeq f^{(n+1)}(\xi)$$

Corollary 10.13.1. To show $H_i = H_{n-i}$.

Proof. In (10.13.3), we put $n - v = u$. $\therefore$ $du = -dv$.

$$H_i = -h\int_n^0 \frac{(-1)^{n-i}}{i!(n-i)!}\cdot\frac{(n-v)(n-v-1)\cdots(-v)}{(n-v-i)}du$$

$$= h\int_0^n \frac{(-1)^{n-i}}{i!(n-i)!}\cdot\frac{(-v)(-\overline{v-1})\cdots(-\overline{v-n})}{-(\overline{v-n-i})}dv$$

$$= h\int_0^n \frac{(-1)^{n-i}}{i!(n-i)!}\cdot\frac{(-1)^{n+1}v(v-1)\cdots(v-n)}{(-1)(\overline{v-n-i})}dv$$

$$= h\int_0^n \frac{(-1)^{2n-i}}{i!(n-i)!}\cdot\frac{v(v-1)\cdots(v-n)}{(\overline{v-n-i})}dv$$

$$= h\int_0^n \frac{(-1)^{2i}(-1)^{-i}}{i!(n-i)!}\cdot\frac{v(v-1)\cdots(v-n)}{(\overline{v-n-i})}dv$$

$$= h\int_0^n \frac{(-1)^{i}}{(n-i)!i!}\cdot\frac{v(v-1)(v-2)\cdots(v-n)}{(\overline{v-n-i})}dv \equiv H_{n-i} \quad \text{(Proved)}$$

Thus, coefficients of Newton-Cotes formula are symmetric from both the ends.

Corollary 10.13.2. To show that $H_0 + H_1 + \cdots + H_n = nh$ (if interval of differencing is h).

Proof. From Lagrange's formula,

$$f(x) \simeq \sum_{r=0}^{r=n}\frac{F(x)}{(x-x_r)F'(x_r)}f(x_r)$$

where $F(x) = (x-x_0)(x-x_1)\cdots(x-x_n)$.

We have shown that

$$\sum_{r=0}^{r=n}\frac{F(x)}{(x-x_r)F'(x_r)} = 1$$

$$\therefore \int_{x_0}^{x_0+nh}\sum_{r=0}^{r=n}\frac{F(x)}{(x-x_r)F'(x_r)}dx = \int_{x_0}^{x_0+nh} 1\cdot dx = nh \tag{A}$$

Again

$$\int_{x_0}^{x_0+nh} \sum_{r=0}^{r=n} \frac{F(x)}{(x-x_r)F'(x_r)} dx = \sum_{r=0}^{r=n} h \int_0^n h^n \frac{u(u-1)(u-2)\cdots(u-n)}{(u-r)h^r r!(-1)^{n-r}\cdot(n-r)!} du$$

$$= \sum_{r=0}^{r=n} h \int_0^n \frac{(-1)^{n-r}}{r!(n-r)!} \cdot \frac{u(u-1)\cdot(u-n)}{(u-r)} du$$

$$= \sum_{r=0}^{r=n} H_r \qquad \text{(B)}$$

From (A) and (B), we obtain

$$\sum_{r=0}^{r=n} H_r = nh$$

10.14 Derivation of trapezoidal rule, Simpson's rules from Newton-Cotes formula

10.14.1 Trapezoidal rule

Let

$$\int_0^{x_0+h} f(x)dx = H_0 f(x_0) + H_1 f(x_1) \quad \text{(By Newton-Cotes formula)}$$

and $x_1 = x_0 + h$.

Here,

$$H_0 = \frac{(-1)^{1-0}}{0!1!} h \int_0^1 (u-1)du = -1h\left[\frac{u^2}{2} - u\right]_0^1 = \frac{h}{2}$$

$$H_1 = h\int_0^1 u\,du = \frac{h}{2}$$

$$\therefore I = \frac{h}{2}[f(x_0) + f(x_1)]$$

10.14.2 Simpson's one-third rule

Here $n = 2$.

$$\int_{x_0}^{x_0+2h} f(x)dx = H_0 f(x_0) + H_1 f(x_1) + H_2 f(x_2)$$

Also $H_0 = H_2$, $(H_i = H_{n-i})$.

$$\therefore H_0 = \frac{(-1)^{2-0}}{2!0!} h \int_0^2 \frac{u(u-1)(u-2)}{u} du = \frac{1}{2}h^2 \int_0^2 (u^2 - 3u + 2)du$$

$$= \frac{h}{2}\left[\frac{u^3}{3} - \frac{3u^2}{2} + 2u\right]_0^2 = \frac{h}{2}\left[\frac{8}{3} - 6 + 4\right] = \frac{h}{3} = H_2$$

$$H_1 = \frac{(-1)^{2-1}}{1!1!} h \int_0^2 u(u-2)du = -h\left[\frac{u^3}{3} - u^2\right]_0^2 = -h\left[\frac{8}{3} - 4\right] = \frac{4h}{3}$$

$$\therefore \int_{x_0}^{x_0+2h} f(x)dx = \frac{h}{3}[f(x_0) + 4f(x_1) + f(x_2)]$$

10.14.3 Simpson's three-eighths rule

Here $n = 3$.

$$\int_{x_0}^{x_0+3h} f(x)dx \simeq H_0 f(x_0) + H_1 f(x_1) + H_2 f(x_2) + H_3 f(x_3)$$

Also $H_0 = H_3$ and $H_1 = H_2$.

$$\therefore H_0 = \frac{(-1)^{3-0}}{0!3!} h \int_0^3 (u-1)(u-2)(u-3)du = -\frac{1}{6} h \int_0^3 (u^3 - 6u^2 + 11u - 6)du$$

$$= -\frac{h}{6}\left[\frac{u^4}{4} - 2u^3 + \frac{11}{2}u^2 - 64\right]_0^3 = -\frac{h}{6}\left[\frac{81}{4} - 54 + \frac{99}{2} - 18\right]$$

$$= -\frac{h}{6}\left[\frac{81+198}{4} - 72\right] = -\frac{h}{6}\left[\frac{279-288}{4}\right]$$

$$= -\frac{h}{6}\left(-\frac{9}{4}\right) = \frac{3h}{8} = H_3$$

$$H_1 = \frac{(-1)^{3-1}}{1!2!} h \int_0^3 u(u-2)(u-3)du = \frac{1}{2} h \left[\frac{u^4}{4} - \frac{5}{3}u^3 + 3u^2\right]_0^3$$

$$= \frac{h}{2}\left[\frac{81}{4} - 45 + 27\right] = \frac{h}{2}\cdot\frac{9}{4} = \frac{9h}{8} = H_2$$

$$\therefore \int_{x_0}^{x_0+3h} f(x)dx = \frac{3h}{8}[f(x_0) + 3f(x_1) + 3f(x_2) + f(x_3)]$$

which is the result of Simpson's Three-Eighths Rule.

10.15 Gauss's quadrature formula

Simpson's formula and Weddle's rule use ordinates which are equally spaced, but investigations were being made to see whether some other spacings of the ordinates could be found to yield more accurate result. It occurred to Gauss that if the ordinates be symmetrically placed about the mid-point of the range of integration, quadrature formula would give better result.

Let I denote the integral to be evaluated.

$$I = \int_a^b f(x)dx$$

We make the following substitution:

$$x = (b-a)u + \frac{a+b}{2}$$

As $x \to a$, $u \to -1/2$ and $x \to b$, $u \to +1/2$. Also $dx = (b-a)du$.

$$\therefore I = (b-a)\int_{-1/2}^{+1/2} \phi(u)du = (b-a)I_1, \quad \text{where } I_1 = \int_{-1/2}^{+1/2} \phi(u)du$$

where $\phi(u) = f\left\{\frac{a+b}{2} + (b-a)u\right\}$ and Gauss wrote

$$\int_{-1/2}^{+1/2} \phi(u)du = R_1\phi(u_1) + R_2\phi(u_2) + \cdots + R_n\phi(u_n) \tag{10.15.1}$$

where $u_1, u_2, \ldots, u_n$ and the points of sub-division of the interval $u = -1/2$ to $u = +1/2$.

The corresponding values of x are

$$x_1 = \frac{a+b}{2} + (b-a)u_1$$

$$x_2 = \frac{a+b}{2} + (b-a)u_2, \quad \text{etc.}$$

$$\therefore I = \int_a^b f(x)dx = (b-a)\int_{-1/2}^{+1/2} \phi(u)du$$

$$= (b-a)[R_1\phi(u_1) + R_2\phi(u_2) + \cdots + R_n\phi(u_n)] \tag{10.15.2}$$

Without giving the detailed derivation of Gauss's method, we shall only explain how the values of $u_1, u_2, \ldots, u_n$ are obtained and $R_1, R_2, \ldots, R_n$ found.

We shall assume that $\phi(u)$ can be expanded as a convergent power series in the interval $u = -1/2$ to $u = +1/2$. Thus, we write

$$\phi(u) = a_0 + a_1u + a_2u^2 + \cdots + a_mu^m + \cdots \tag{10.15.3}$$

We further assume that the integral can be expressed as a linear function of the ordinates of the form (10.15.2).

Integrating (10.15.3) between the limits $-1/2$ and $1/2$, we have

$$I = \int_{-1/2}^{+1/2} \phi(u)du = \int_{-1/2}^{+1/2} (a_0 + a_1u + a_2u^2 + \cdots + a_mu^m + \cdots)du$$

$$= \left[a_0u + a_1\frac{u^2}{2} + a_2\frac{u^3}{3} + a_3\frac{u^4}{4} + \cdots\right]_{-1/2}^{+1/2}$$

$$= 2\left[a_0\frac{1}{2} + a_2\frac{1}{24} + a_4\frac{1}{160} + \cdots\right]$$

$$= a_0 + \frac{1}{12}a_2 + \frac{1}{80}a_4 + \cdots \tag{10.15.4}$$

From (10.15.3), we have

$$\phi(u_1) = a_0 + a_1u_1 + a_2u_1^2 + a_3u_1^3 + a_4u_1^4 + \cdots$$

$$\phi(u_2) = a_0 + a_1u_2 + a_2u_2^2 + a_3u_2^3 + a_4u_2^4 + \cdots$$

....

$$\phi(u_n) = a_0 + a_1u_n + a_2u_n^2 + a_3u_n^3 + a_4u_n^4 + \cdots$$

Substituting in (10.15.3)

$$I_1 = \int_{-1/2}^{+1/2} \phi(u)du = R_1\phi(u_1) + R_2\phi(u_2) + R_3\phi(u_3) + \cdots$$
$$= R_1(a_0 + a_1 u_1 + a_2 u_1^2 + a_3 u_1^3 + \cdots)$$
$$+ R_2(a_0 + a_1 u_2 + a_2 u_2^2 + a_3 u_2^3 + \cdots) + \cdots$$
$$+ R_m(a_0 + a_1 u_m + a_2 u_m^2 + a_3 u_m^3 + \cdots) + \cdots \tag{10.15.5}$$

Equating coefficients of a_i $(i = 0, 1, 2, \ldots)$ from (10.15.4) and (10.15.4), we get

$$R_1 + R_2 + R_3 + \cdots = 1$$
$$R_1 u_1 + R_2 u_2 + R_3 u_3 + \cdots = 0$$
$$R_1 u_1^2 + R_2 u_2^2 + R_2 u_3^2 + \cdots = 1/12$$
$$R_1 u_1^3 + R_2 u_2^3 + R_2 u_3^2 + \cdots = 0$$
$$R_1 u_1^4 + R_2 u_2^4 + R_3 u_3^4 + \cdots = 1/80 \tag{10.15.6}$$
$$\cdots\cdots\cdots\cdots\cdots\cdots\cdots\cdots$$

We have obtained above $2n$ equations in $2n$ unknowns, namely $R_1, R_2, \ldots, R_n$ and $u_1, u_2, \ldots, u_n$ and it is theoretically possible to solve them simultaneously. The computational work is, however, prohibitive even for small values of n. We may, however, use a formula from higher algebra which would reduce the computational labour to a great extent.

It can be shown that if $\phi(u)$ be a polynomial of degree not higher than $(2n - 1)$, then $u_1, u_2, \ldots, u_n$ are the zeros of the Legendre polynomial $P_n(u)$ or the roots of $P_n(u) = 0$ and these roots can be found conveniently from the equation:

$$\frac{d^n}{du^n}\left[u^2 - \left(\frac{1}{2}\right)^2\right]^n = 0 \tag{10.15.7}$$

The n roots $u_1, u_2, \ldots, u_n$ of the nth degree equation (10.15.8) are all real and substituting them in (10.15.7) we may find the values of $R_1, R_2, \ldots, R_n$. We shall do this for $n = 5$.

The equation we propose to solve is

$$\frac{d^5}{du^5}\left[u^2 - \left(\frac{1}{2}\right)^2\right]^5 = 0,$$

or, $$\frac{d^5}{du^5}\left[u^{10} - 5 \cdot u^8 \cdot \frac{1}{4} + 10 \cdot u^6 \cdot \frac{1}{16} - 10 \cdot u^4 \cdot \frac{1}{64} + 5 \cdot u^2 \cdot \frac{1}{256} - \frac{1}{1024}\right] du = 0$$

$$\Rightarrow 10 \cdot 9 \cdot 8 \cdot 7 \cdot 6u^5 - \frac{5}{4} \cdot 8 \cdot 7 \cdot 6 \cdot 5 \cdot 4 \cdot u^3 + \frac{5}{8} \cdot 6 \cdot 5 \cdot 4 \cdot 3 \cdot 2u = 0$$

$$\Rightarrow 6 \cdot 5 \cdot u(1008u^4 - 280u^2 + 15) = 0$$

$\therefore$ either $u = 0$, or, $1008u^4 - 280u + 15 = 0$.

$$\Rightarrow u^2 = \frac{280 \pm \sqrt{78400 - 60480}}{2016} = \frac{280 \pm 133.8656}{2016} = 0.2053, \text{ or, } 0.0725$$

$\therefore u = 0, \pm\sqrt{0.2053}, \pm\sqrt{0.0725}$, or, $u = 0, \pm 0.4531, \pm 0.2693$

$\therefore u_0 = 0,\ u_{-1} = -0.2693,\ u_1 = +0.2693,\ u_{-2} = -0.4531,\ u_2 = +0.4531$

$$R_1 + R_2 + R_3 + R_4 + R_5 = 1$$

$$\therefore R_1(-0.4531) + R_2(-0.2693) + R_4(0.2693) + R_5(0.4531) = 0$$

$$R_1(0.2053) + R_2(0.0725) + R_4(0.0725) + R_5(0.2053) = 1/12$$

$$R_1(-0.09302) + R_2(-0.01952) + R_4(+0.01952) + R_5(+0.09302) = 0$$

$$R_1(0.04215) + R_2(0.005256) + R_4(0.005256) + R_5(0.04215) = 1/80$$

Solving, $R_3 = 0.2844$ $\quad u_1(= u_{-2}) = -0.4531$

$R_1 = R_5 = 0.2393$ $\quad u_2(= u_{-1}) = -0.2693$

$R_2 = R_4 = 0.1185$ $\quad u_3(= u_0) = 0$

$u_4(= u_1) = 0.2693$

$u_5(= u_2) = 0.4531$

Note: Gauss's formula should be used when only a few ordinates are obtainable and also when the importance of the result merits the great pains in computational work.

10.16 Change of scale in quadrature formula

Suppose the independent variable x is measured with origin at 0, and unit of measurement is 1 (as is usual).

Let us consider Simpson's One-Third Rule:

$$I = \int_0^2 y_x\, dx = \frac{1}{3}(y_0 + 4y_1 + y_2)$$

Let us take $u = hx$ (i.e. the unit of u is h).

$$I = \int_0^{2h} \frac{y_u\, du}{h} = \frac{1}{3}(y_0 + 4y_h + y_{2h}). \quad \int_0^{2h} y_u\, du = \frac{h}{3}(y_0 + 4y_h + y_{2h}) \tag{10.16.1}$$

We may state the rule for change of scale as: *If we make the scale h times, then in the quadrature formula, we are required to multiply the coefficients of the ordinates by h and each and the subscripts of y are also to be multiplied by h.*

10.17 Rule of change of origin

If we change the origin to h, the formula [10.16(a)] reduces to

$$\int_{-h}^{h} y_u\, du = \frac{h}{3}(y_{-h} + 4y_0 + y_h) \tag{10.17.1}$$

This means that the coefficients of the ordinates will remain unaltered but the subscripts should be lowered by h in each case.

Example 10.17.1. If $u_x = a + bx + cx^2$, prove that

$$\int_1^3 u_x\, dx = 2u_2 + \frac{1}{12}(u_0 - 2u_2 + u_4)$$

and hence find an approximate value for

$$\int_{-1/2}^{+1/2} e^{-x^2/10}\, dx$$

Solution: We have to prove

$$\int_1^3 u_x\, dx = 2u_2 + \frac{1}{12}(u_0 - 2u_2 + u_4)$$

Shifting the origin to 2, we have to prove

$$\int_{-1}^1 u_x = 2u_0 + \frac{1}{12}(u_{-2} - 2u_0 + u_2)$$

Since $u_x = a + bx + cx^2$, we have $u_0 = a$, $u_{-2} = a - 2b + 4c$, $u_2 = a + 2b + 4c$.

$$\therefore \text{ RHS } = 2a + \frac{1}{12}\{2(a + 4c) - 2a\} = 2a + \frac{2c}{3} = 2\left(a + \frac{c}{3}\right)$$

Again

$$\text{LHS} = \int_{-1}^1 (a + bx + cx^2)dx = \left[ax + \frac{bx^2}{2} + \frac{cx^3}{3}\right]_{-1}^{+1} = 2\left(a + \frac{c}{3}\right)$$

$\therefore$ LHS = RHS (Proved).

Again let $u_x = e^{-x^2/10}$.

$$\therefore \int_{-1}^{+1} u_x\, dx = 2u_0 + \frac{1}{12}(u_{-2} - 2u_0 + u_2)$$

Changing the scale,

$$\int_{-1/2}^{+1/2} u_x\, dx = 2\frac{1}{2}u_0 + \frac{1}{12}\cdot\frac{1}{2}(u_{-1} - 2u_0 + u_1) = u_0 + \frac{1}{24}(u_1 - 2u_0 + u_1)$$

$$= 1 + \frac{1}{24}(e^{-1/10} - 2 + e^{-1/10}) = 1 + \frac{1}{12}(e^{-1/10} - 1) = 1 - \frac{1}{12}(1 - e^{-1/10})$$

$$= 1 - \frac{1}{12}(1 - 0.9048) = 1 - \frac{1}{12}(0.0952) = 1 - 0.0079 = 0.9921$$

Example 10.17.2. Prove that

$$\int_{-1/2}^{3/2} f(x)dx = \frac{1}{24}\{27f(0) + 17f(1) + 5f(2) - f(3)\}$$

Solution: Since the integral is expressed as a linear function of four ordinates $f(0)$, $f(1)$, $f(2)$, $f(3)$, we may assume that $f(x)$ is a polynomial of degree 3 in x,

Let $f(x) = a + bx + cx^2 + dx^3$.

Shifting the origin to −1/2, we have to show

$$\int_{-1}^{+1} f(x) = \frac{1}{24}\left\{27f\left(-\frac{1}{2}\right) + 17f\,(12) + 5f\left(\frac{3}{2}\right) - f\left(\frac{5}{2}\right)\right\} \qquad (10.17.2)$$

Doubling the scale, we have to show

$$\int_{-2}^{+2} f(x)dx = \frac{1}{12}\{27f(-1) + 17f(1) + 5f(3) - f(5)\} \qquad (10.17.3)$$

Now LHS of (2) is

$$\int_{-2}^{+2} (a + bx + cx^2 + dx^3)dx = \left[ax + \frac{bx^2}{2} + \frac{cx^3}{3} + \frac{dx^4}{4}\right]_{-2}^{+2} = 4\left(a + \frac{4c}{3}\right)$$

Again RHS of (2) is

$$\frac{1}{12}\{27f(1) + 17f(1) + 5f(3) - f(5)\}$$

$$= \frac{1}{12}\{27(a - b + c - d) + 17(a + b + c + d) + 5(a + 3b + 9c + 27d) - (a + 5b + 25c + 125d)\}$$

$$= \frac{1}{12}(48a + 64c) = 4a + \frac{16c}{3} = 4\left(a + \frac{4c}{3}\right)$$

$\therefore$ LHS = RHS (Proved).

EXERCISES

1. Derive Newton-Cotes Quadrature formula and deduce Simpson's one-third rule and Simpson's three-eighths rule as special cases. Also give the error term of the above quadrature formulae.
2. State and prove Euler-Maclaurin's summation formula and hence obtain the sum of the fourth power of the first n natural integers.
3. Obtain Gauss's Quadrature formula and discuss the merits and demerits, if any, of this quadrature formula.
4. Describe the effect of change of origin and scale on Simpson's rule and Weddle's rule.
5. Taking 11 equidistant ordinates at $x = 0, \pi/20, 2\pi/20, 3\pi/20, 4\pi/20, 5\pi/20, 6\pi/20, 7\pi/20, 8\pi/20, 9\pi/20, 10\pi/20$, obtain the approximate value of the integral $\int_0^{\pi/2} \sin x\, dx$ by:
   (a) Trapezoidal rule, [*Ans:* 0.997943.]
   (b) Simpson's one-third rule. [*Ans:* 1.00000334.]
   Also find the errors in both the cases.
6. Using six intervals of equal length, obtain the approximate value of
   $$\int_0^1 \frac{1}{1+x}\,dx$$
   by using (a) Simpson's one-third and (b) Weddle's rules and hence give an approximate value of $\log_e 2$ by these formulae and give the error in each case.
   [*Ans:* (a) 0.6931698; (b) 0.6931493, True value = 0.6931472.]
7. Evaluate the approximate value of the following integrals by using Simpson's one-third rule (take 7 equispaced ordinates):
   (a) $\int_5^{11} \log_{10} x\, dx$ [*Ans:* 5.3546696.]
   (b) $\int_1^4 \frac{dx}{2x+1}$ [*Ans:* 0.5494709.]
   (c) $\int_0^1 \sqrt{8 + x + x^2}\,dx$
   correct to six decimals, using 11 ordinates (*equispaced*). [*Ans:* 2.97049.]
   (d) $\int_5^{6.2} \log_e x\, dx.$ [*Ans:* 2.0650160.]

8. Using Euler-Maclaurin's summation formula, find the sum of the following series

(a) $\frac{1}{51^2} + \frac{1}{53^2} + \frac{1}{55^2} + \cdots + \frac{1}{99^2}$ [*Ans:* 0.004998833.]

(b) $\frac{1}{201^2} + \frac{1}{203^2} + \cdots + \frac{1}{249^2}$ [*Ans:* 0.000833290.]

(c) $\frac{1}{21^3} + \frac{1}{23^3} + \cdots + \frac{1}{35^3}$ [*Ans:* 0.000427345.]

9. Using Euler-Maclaurin's summation formula, obtain the sum of the cubes of first n natural numbers. [*Ans:* $n^2(n+1)^2/4$.]

10. Evaluate the integral

$$\int_0^{\pi/2} \sqrt{1 - 0.162 \sin^2 \phi}\, d\phi$$

by Simpson's one-third rule and Weddle's rule.

[*Ans:* $I_{1/3} = 1.505103$, $I_w = 1.505103$.]

11. By computing the integral

$$\int_0^1 \frac{1}{1+x^2} dx$$

by (a) Simpson's one-third rule and (b) Weddle's rule, compute the value of π correct to six decimal places and obtain the error in each case.

[*Ans:* (a) $I_{1/3} = 0.7853979 \simeq \pi/4$, $E_{1/3} = 0.0000002634$; (b) $I_w = 0.7853996 \simeq \pi/4$, $E_w = -0.0000014366$.]

12. If the third differences are constant, prove that

(a) $\int_{-1}^{+1} f(x)dx = \frac{2}{3}\left[f(0) + f\left(\frac{1}{\sqrt{2}}\right) + f\left(-\frac{1}{\sqrt{2}}\right)\right]$

(b) $\int_0^2 f(x)dx = \frac{1}{24}\left[f\left(-\frac{1}{2}\right) + 23f\left(\frac{1}{2}\right) + 23f\left(\frac{3}{2}\right) + f\left(\frac{5}{2}\right)\right]$

13. By taking six equal sub-intervals, obtain the value of the following integral

$$\int_1^3 \frac{dx}{\sqrt{3 + 2x - x^2}}$$

by Simpson's one-third rule and hence obtain the value of π (*approximately*).

[*Ans:* $I_{1/3} = 3.1101262 \simeq \pi$.]

14. Taking 10 equal sub-intervals, obtain by Simpson's rule (one-third) the value of

$$\int_0^1 \frac{\log(1+x^2)}{1+x^2} dx$$

[*Ans:* 0.1730.]

15. Obtain the approximate quadrature formula:

(a) $\int_{1.5}^{2.5} y_x\, dx = \frac{1}{24}(y_1 + 22y_2 + y_3)$

(b) $\int_0^n y_x\, dx = n\left[\frac{3}{8}y_0 + \frac{1}{24}(19y_n - 5y_{2n} + y_{3n})\right]$

16. If third differences are constant, prove that

(a) $\int_{-1/2}^{+1/2} f(x)dx = \frac{1}{2}\left[\left\{f\left(-\frac{1}{2}\right) + f\left(\frac{1}{2}\right)\right\} + \frac{1}{24}\left\{\Delta f\left(-\frac{3}{2}\right) - \Delta f\left(\frac{1}{2}\right)\right\}\right]$

(b) $\int_0^2 f(x)dx = \frac{8}{9}f(1) + \frac{5}{2}\left[f\left(1 + \sqrt{0.6}\right) + f(1 - 0.6)\right]$

17. Obtain the approximate quadrature formula:

(a) $\int_0^1 u_x\,dx = \frac{1}{12}[13(u_1 + u_{-1}) - (u_3 + u_{-3})]$

(b) $\int_0^1 y_x\,dx = \frac{1}{12}(5y_0 + 8y_1 - y_2)$

(c) $\int_{-1/2}^{1/2} y_x\,dx = \frac{1}{24}\left[13f\left(-\frac{1}{2}\right) + 13f\left(\frac{1}{2}\right) - f\left(\frac{3}{2}\right) - f\left(-\frac{3}{2}\right)\right]$

(d) $\int_1^2 y_x\,dx = \frac{1}{24}(-y_0 + 13y_1 + 13y_2 - y_3)$

18. Prove that

$$\int_{-3}^{+3} y_x\,dx = \frac{3}{10}[11(y_{-2} + y_2) - 14(y_{-1} + y_1) + 26y_0]$$

19. Use Gauss's quadrature formula to evaluate

$\int_2^5 \frac{1}{\sqrt{1+x}}$ by taking:

(a) 3 ordinates and

(b) 5 ordinates,

symmetrically selected about the mid-point of the range.

20. Use Simpson's three-eighths rule to evaluate

$$\int_0^{0.3} (1 - 8x^3)^{1/2}\,dx$$

[*Ans:* 0.29159.]

21. Evaluate

$$\int_0^1 \sqrt{\sin x + \cos x}\,dx$$

correct to 3 decimal places using six equal intervals. [*Ans:* 1.139.]

22. If $f(x)$ is a polynomial of degree 3 in x and

$$u_{-1} = \int_{-3}^{-1} f(x)dx, \quad u_0 = \int_{-1}^{+1} f(x)dx, \quad u_1 = \int_1^3 f(x)dx$$

then prove

$$f(0) = \frac{1}{2}\left[u_0 - \frac{1}{24}\Delta^2 u_{-1}\right]$$

23. If $y_x = a + bx + cx^2 + dx^3 + ex^4 + fx^5$, then establish Hardy's formula given by

$$\int_0^6 y_x\,dx = 0.28(y_0 + y_6) + 1.62(y_1 + y_5) + 2.2y_3$$

24. A river is 80 feet wide. The depth d (*in feet*) of the river at a distance x from one bank is given by the following table.

x	0	10	20	30	40	50	60	70	80
d	0	4	7	9	12	15	14	8	3

Find approximately the area of the cross section of the river. [*Ans:* 710 sq. ft]

25. If $H_0, H_1, \ldots, H_n$ are Cotes coefficients, show that

(a) $H_0 + H_1 + H_2 + \cdots + H_n = nh$

(b) $H_r = H_{n-r}$.

26. Compute the values of the function $y = 1/(1 + x^2)$ from $x = -0.3$ to $x = 1.3$ at an interval of 0.1 and then form a table of differences. Obtain from it the value of

$$I = \int_0^1 \frac{dx}{1 + x^2}$$

by an appropriate central difference quadrature formula and evaluate the value of π.
[*Ans:* 0.78539816, π = 4× 0.78539816 = 3.14159264 (True value of π to nine figure is π = 3.14159265.]

27. Compute the value of the function $y = 1/x$ from $x = 0.7$ to $x = 2.3$ at an interval of 0.1 and form a table of differences. Use this table to obtain

$$\int_1^2 \frac{dx}{x}$$

by using any suitable central difference quadrature formula. [*Ans:* 0.693147185.]

Chapter 11

Summation of Series

Summary

11.1 Introduction

An important application of finite difference algebra is the problem of finding a formula giving the sum to n terms of the given series. We shall presently see that the summation problems discussed in this chapter boil down to the problem of finite integration.

11.2 To find the sum to n terms of the series whose general term is the first difference of a function

Suppose $u_1, u_2, u_3, \cdots$ be the series whose sum up to n terms is required. Let u_x be the general term of the series and suppose

$$u_x = \Delta f(x), \quad \text{or,} \quad f(x) = \frac{1}{\Delta} u_x$$

We are to find the sum

$$\sum_{x=1}^{n} u_x$$

If the interval of differencing be h, we have

$u_a = \Delta f(a) = f(a+h) - f(a)$

$u_{a+h} = \Delta f(a+h) = f(a+2h) - f(a+h)$

....

$u_{a+\overline{n-1}h} = \Delta f(a+\overline{n-1}h) = f(a+nh) - f(a+\overline{n-1}h)$

∴ adding

$$\sum_{x=a}^{a+\overline{n-1}h} u_x = f(a+nh) - f(a) \tag{11.2.1}$$

By a suitable change of origin and scale, we can take $h = 1$, $a = 1$ and we get from (11.2.1)

$$\sum_{x=1}^{n} u_x = f(n+1) - f(1) = \left[f(x)\right]_1^{n+1} \tag{11.2.2}$$

This leads to a sufficient condition for finding the sum $\sum u_x$. To find the sum, our job is to find a function $f(x)$ such that $\Delta f(x) = u_x$.

We shall call $f(x)(= (1/\Delta)u_x)$ the indefinite integral of u_x. When definite limits are taken, we have the definite integral of u_x.

Thus,

$$\frac{1}{\Delta}u_x = f(x) + c$$

is an indefinite integral.

Note 1: Though it is possible to find the difference of a function, there are a few functions which are explicit functions of x. Some such important functions are given below in which the interval of differencing is h.

1. $\Delta\left[\dfrac{a^x}{a^h - 1}\right] = a^x$

2. $\Delta\left[\dfrac{(ax+b)^m}{am}\right] = (ax+b)^{m-1}$

3. $\Delta\left[\dfrac{-(ax+b)^{-m}}{am}\right] = (ax+b)^{-m-1}$

(2 and 3: m is a positive integer)

4. $\Delta\left[\binom{x}{m}\right] = \binom{x}{m-1}$ (interval of differencing is unity)

5. $\Delta\left[\dfrac{-\cos\left(ax+b-\frac{1}{2}a\right)}{2\sin\frac{a}{2}}\right] = \sin(ax+b)$

6. $\Delta\left[\dfrac{\sin\left(ax+b-\frac{a}{2}\right)}{2\sin\frac{a}{2}}\right] = \cos(ax+b)$;

(5 and 6: $a \neq 0$)

7. $\Delta\left[\dfrac{a^{x+1}\sin b(x-1)-a^x\sin bx}{a^2-2a\cos b+1}\right]=a^x\sin bx$

8. $\Delta\left[\dfrac{\cosh\left(ax+b-\dfrac{a}{2}\right)}{2\sinh\dfrac{a}{2}}\right]=\sinh(ax+b),\ a\neq 0$

Note 2: The operator Δ converts exponential function into exponential function, polynomials into polynomials, trigonometrical functions into trigonometrical functions.

Note 3: The formula (11.2.2) can be used to find the sum to n terms of the series whose nth term can be expressed as (a) ka^x (b) kx^{-m-1} (c) $k\binom{x}{m-1}$ (d) *a* polynomial in x.

EXAMPLE 11.2.1. Use the method of finite differences to find the sum to n terms of the series whose rth term is (a) $(r+4)(r+5)(r+6)$ (b) $r(r+3)(r+6)$ (c) $(3r+1)(3r+4)(3r+7)$.

Solution: (a) $u_r=(r+4)(r+5)(r+6)=(r+6)^{(3)}$.

$$\therefore \sum_{r=1}^{n}u_r=\left[\frac{1}{\Delta}u_r\right]_{r=1}^{n+1}=\left[\frac{1}{4}(r+6)^{(4)}\right]_{r=1}^{n+1}=\frac{1}{4}\left[(n+7)^{(4)}-7^{(4)}\right],$$

i.e. $$\sum_{r=1}^{n}u_r=\frac{1}{4}[(n+7)(n+6)(n+5)(n+4)-7.6.5.4]$$

$$=\frac{1}{4}(n+4)(n+5)(n+6)(n+7)-210$$

(b) $u_r=r(r+3)(r+6)=r^3+9r^2+18r=r^{(3)}+12r^{(2)}+28r^{(1)}$.

$$\therefore \sum_{r=1}^{n}u_r=\left[\frac{1}{\Delta}u_r\right]_1^{n+1}=\left[\frac{1}{\Delta}r^{(3)}+12\cdot\frac{1}{\Delta}r^{(2)}+28\cdot\frac{1}{\Delta}r^{(1)}\right]_1^{n+1}$$

$$=\frac{(n+1)^{(4)}}{4}+12\cdot\frac{1}{3}(n+1)^{(3)}+\frac{28}{2}(n+1)^{(2)}-\left[\frac{1}{4}\cdot 1^{(4)}+12\cdot\frac{1}{3}\cdot 1^{(3)}+\frac{28}{2}\cdot 1^{(2)}\right]$$

$$=\frac{1}{4}n(n+1)(n-1)(n-2)+4n(n+1)(n-1)+\frac{1}{14}n(n+1)-0$$

$$=\frac{1}{28}n(n+1)\{7(n-1)(n-2)+112(n-1)+2\}$$

$$=\frac{n(n-1)}{28}\{7n^2-21n+14+112n-112+2\}$$

$$=\frac{n(n-1)}{28}\{7n^2+91n-96\}$$

(c) $u_r=(3r+1)(3r+4)(3r+7)=27r^3+108r^2+117r+28=27r^{(3)}+189r^{(2)}+252r^{(1)}+28.$

$$\therefore \sum_{r=1}^{n} u_r = \left[27 \cdot \frac{1}{\Delta} r^{(3)} + 189 \cdot \frac{1}{\Delta} r^{(2)} + 252 \cdot \frac{1}{\Delta} r^{(1)} + \frac{1}{\Delta} 28\right]_1^{n+1}$$

$$= \left[\frac{27}{4} r^{(4)} + \frac{189}{3} r^{(3)} + \frac{252}{2} r^{(2)} + 28 r^{(1)}\right]_1^{n+1}$$

$$= \frac{27}{4}(n+1)^{(4)} + 63(n+1)^{(3)} + 126(n+1)^{(2)} + 28(n+1)^{(1)}$$

$$- \left[\frac{27}{4} \cdot 1^{(4)} + 63 \cdot 1^{(3)} + 126 \cdot 1^{(2)} + 28 \cdot 1^{(1)}\right]$$

$$= \frac{27}{4}(n+1) \cdot n(n-1)(n-2) + 63(n+1) \cdot n(n-1)$$

$$+ 126n(n+1) + 28(n+1) - 28$$

$$= \frac{n+1}{4}\{27n(n-1)(n-2) + 252n(n-1) + 504n - 112\}$$

Example 11.2.2. Find $\sum_{x=1}^{n} u_x$, where

(a) $u_x = \dfrac{1}{(x+1)(x+2)(x+3)}$

(b) $u_x = \dfrac{1}{(x-1)(x)(x+1)}$

(c) $u_x = \left|\dfrac{x+3}{x(x+1)(x+2)}\right|$.

Solution: (a) Here $u_x = \dfrac{1}{(x+1)(x+2)(x+3)} = x^{(-3)}$.

$$\therefore \sum_{x=1}^{n} u_x = \left[\frac{1}{\Delta} x^{(-3)}\right]_1^{n+1} = \left[-\frac{1}{2} x^{(-2)}\right]_1^{n+1} = \left[-\frac{1}{2}(n+1)^{(-2)} + \frac{1}{2} 1^{(-2)}\right]$$

$$= -\frac{1}{2} \cdot \frac{1}{(n+2)(n+3)} + \frac{1}{2} \cdot \frac{1}{2.3} = \frac{1}{2}\left[\frac{1}{6} - \frac{1}{(n+2)(n+3)}\right]$$

(b) Here $u_x = (x-2)^{(-3)}$.

$$\therefore \sum_{x=1}^{n} u_x = \left[\frac{1}{\Delta}(x-2)^{(-3)}\right]_1^{n+1} = \left[-\frac{1}{2}(x-2)^{(-2)}\right]_1^{n+1}$$

$$= -\frac{1}{2}(n-1)^{(-2)} + \frac{1}{2}(-1)^{(-2)} = -\frac{1}{2} \cdot \frac{1}{n(n+1)} + 0 = -\frac{1}{2} \cdot \frac{1}{n(n+1)}$$

(c) Here $u_x = \dfrac{x+3}{x(x+1)(x+2)} = \dfrac{1}{(x+1)(x+2)} + \dfrac{3}{x(x+1)(x+2)}$

$$= (x)^{(-2)} + 3(x-1)^{(-3)}.$$

$$\therefore \sum_{x=1}^{n} u_x = \left[\frac{1}{\Delta} x^{(-2)} + 3\frac{1}{\Delta}(x-1)^{(-3)}\right]_1^{n+1} = \left[-x^{(-1)} - \frac{3}{2}(x-1)^{(-2)}\right]_1^{n+1}$$

$$= -(n+1)^{(-1)} - \frac{3}{2} n^{(-2)} - \left[-1^{(-1)} - \frac{3}{2} 0^{(-2)}\right]$$

$$= -\frac{1}{(n+2)} - \frac{3}{2} \cdot \frac{1}{(n+1)(n+2)} - \left[-\frac{1}{2} - \frac{3}{2} \cdot \frac{1}{1.2}\right]$$

$$= -\frac{1}{n+2}\left[1 + \frac{3}{2} \cdot \frac{1}{n+1}\right] + \frac{5}{4}$$

Example 11.2.3. Sum the following series to n terms: 2, 22, 60, 122, 214, 342, ...

Solution: The difference table is

u_x	Δu_x	$\Delta^2 u_x$	$\Delta^3 u_x$
2			
	20		
22		18	
	38		6
60		24	
	62		6
122		30	
	92		6
214		36	
	128		
342			

Since the third differences are constant, we may write

$$u_x = u_0 + \binom{x}{1}\Delta u_0 + \binom{x}{2}\Delta^2 u_0 + \binom{x}{3}\Delta^3 u_0 = 2 + x^{(1)} \cdot 20 + x^{(2)} \cdot 18 + x^{(3)} \cdot 6$$

$$\therefore \sum_{x=0}^{n-1} u_x = \sum_{x=0}^{n-1}[2 + 20x^{(1)} + 18x^{(2)} + 6x^{(3)}] = \left[\frac{1}{\Delta}\{2 + 20x^{(1)} + 18x^{(2)} + 6x^{(3)}\}\right]_0^n$$

$$= \left[2n + 10n^{(2)} + 6n^{(3)} + \frac{3}{2}n^{(4)}\right]$$

$$\Rightarrow \sum_{x=0}^{n-1} u_x = 2n + 10n(n-1) + 6n(n-1)(n-2) + \frac{3}{2}n(n-1)(n-2)(n-3)$$

11.3 To find the sum of the series with general term $u_x = f(x) + ar^x$

In this section we consider finding the sum of the series in which the general term is of the form:

$$u_x = f(x) + ar^x$$

where $f(x)$ is a rational integral function of x of degree n.

From the given series, we form the successive differences. Any such difference is the sum of two parts, one arising from $f(x)$ and the other from the term ar^x. Since $f(x)$ is a polynomial of degree n in x, the $(n + 1)$th order difference of the part arising from $f(x)$, will be zero. The contribution of the first part in the successive higher order differences will also be zero, and therefore differences of the order $(n + 1)$ and above will give a series in Geometric Progression (GP) with common ratio r.

For example, let $f(x) = a + bx + cx^2$. Then

$$u_x = a + bx + cx^2 + ar^x$$
$$\Delta u_x = b(x + 1) + c(x + 1)^2 - bx - cx^2 + (ar^{x+1} - ar^x)$$
$$= b + 2cx + c + ar^x(r - 1) = (b + c) + 2cx + (r - 1)ar^x$$
$$\Delta^2 u_x = 2c(x + 1) - 2cx + (r - 1)(ar^{x+1} - ar^x) = 2c + a(r - 1)^2 r^x$$
$$\Delta^3 u_x = (r - 1)^3 \cdot ar^x \cdots$$
$$\Delta^n u_x = (r - 1)^n \cdot ar^x$$

so that the third and all the succeeding differences are in GP with common ratio r.

Hence if the first few terms of a series be given, and if the nth differences of the series form a GP with common ratio r, we may assume that the general term u_x of the series may be written as $u_x = f(x) + ar^x$, where $f(x)$ is a rational integral function of x of degree $(n - 1)$.

To find

$$\sum_{x=1}^{n} u_x$$

it will be convenient to express $f(x)$ in factorial notation and use the following results

$$\sum_{1}^{n} x^{(r)} = \left[\frac{x^{(r+1)}}{r+1}\right]_1^{n+1} \quad \text{and} \quad \sum_{x=1}^{n} a^x = \left[\frac{a^x}{a-1}\right]_1^{n+1}$$

Example 11.3.1. Find the general term and the sum to n terms of the series: 1, 2, 4, 8, 18, 50, 168, ...

The difference table is

u_x	Δu_x	$\Delta^2 u_x$	$\Delta^3 u_x$
1			
	1		
2		1	
	2		1
4		2	
	4		4
8		6	
	10		16
18		22	
	32		64
50		86	
	118		
168			

Since the third differences are in GP with common ratio 4, we assume that

$$u_x = a\cdot 4^x + bx^{(2)} + cx^{(1)} + d$$

$$\Delta u_x = a(4-1)\cdot 4^x + 2bx^{(1)} + c = 3a4^x + 2bx^{(1)} + c$$

$$\Delta^2 u_x = 3^2 a4^x + 2b = 9a\cdot 4^x + 2b$$

$$\Delta^3 u_x = 27a\cdot 4^x$$

Putting $x = 1$ in these results, we have

$$\begin{aligned} 4a \quad\quad +c+d &= 1\\ 12a+2b+c \quad &= 1\\ 36a+2b \quad\quad &= 1\\ 108a \quad\quad\quad &= 1 \Rightarrow a = \frac{1}{108}\\ 2b = 1-\frac{1}{3} = \frac{2}{3} \quad &\Rightarrow b = \frac{1}{3}\\ c = 1-\frac{2}{3}-\frac{1}{9} &= \frac{2}{9}\\ d = 1-\frac{2}{9}-\frac{1}{27} &= \frac{20}{27}\end{aligned}$$

$\therefore$ general term of the series

$$u_x = \frac{1}{108}\cdot 4^x + \frac{1}{3}x^{(2)} + \frac{2}{9}x^{(1)} + \frac{20}{27}$$

$$\begin{aligned}\therefore \sum_{x=1}^{n} u_x &= \left[\frac{1}{\Delta}\left\{\frac{1}{108}4^x + \frac{1}{3}x^{(2)} + \frac{2}{9}x^{(1)} + \frac{20}{27}\right\}\right]_1^{n+1}\\
&= \left[\frac{1}{108}\cdot\frac{4^x}{4-1} + \frac{1}{9}x^{(3)} + \frac{1}{9}x^{(2)} + \frac{20}{27}x^{(1)}\right]_1^{n+1}\\
&= \frac{1}{324}\left\{(n+1)^x - 1\right\} + \frac{1}{9}\left\{(n+1)^{(3)} - 1^{(3)}\right\} + \frac{1}{9}\left\{(n+1)^{(2)} - 1^{(2)}\right\}\\
&\quad + \frac{20}{27}\left\{(n+1)^{(1)} - 1^{(1)}\right\}\\
&= \frac{(n+1)^x - 1}{324} + \frac{1}{9}(n+1)^{(3)} + \frac{1}{9}(n+1)^{(2)} + \frac{20}{27}(n+1)^{(1)} - \frac{20}{27}.\end{aligned}$$

11.4 Summation by parts

We consider the series in which the general term is the product of two functions, that is, the general term is of the form $u_x = f_1(x)\cdot f_2(x)$ and the value of each of

$\sum f_1(x)$ and $\sum f_2(x)$ is known

We may then derive a method known as Summation by Parts. We have

$$\begin{aligned}\Delta[f_1(x)f_2(x)] &= f_1(x+1)f_2(x+1) - f_1(x)f_2(x)\\
&= f_1(x+1)\{f_2(x+1) - f_2(x)\} + f_2(x)\{f_1(x+1) - f_1(x)\}\\
&= f_1(x+1)\Delta f_2(x) + f_2(x)\Delta f_1(x),\end{aligned}$$

or, $f_2(x)\Delta f_1(x) = \Delta[f_1(x)f_2(x)] - f_1(x+1)\Delta f_2(x)$

$$\therefore \sum_{1}^{n} f_2(x)\Delta f_1(x) = \left[f_1(x)f_2(x)\right]_1^{n+1} - \sum_{1}^{n} f_1(x+1)\Delta f_2(x) \tag{11.4.1}$$

It follows therefore that when the function $f_1(x)f_2(x)$ can be put in the form $f_2(x)\Delta f_1(x)$, the summation can be performed at once if

$$\sum\{f_1(x+1)\Delta f_2(x)\}$$

can be evaluated.

EXAMPLE 11.4.1. Find the sum to n terms of the series whose xth term is $2^x(x^3+x)$.

Solution: We denote 2^x by $f_1(x)$ and x^3+x by $f_2(x)$.

Now

$$\Delta[f_1(x)f_2(x)] = f_1(x+1)\Delta f_2(x) + f_2(x)\Delta f_1(x)$$
$$\Rightarrow f_2(x)\Delta f_1(x) = \Delta[f_1(x)f_2(x)] - f_1(x+1)\Delta f_2(x)$$

From (11.4.1), we have

$$\therefore \sum_{1}^{n} f_2(x)\Delta f_1(x) = \left[f_1(x)f_2(x)\right]_1^{n+1} - \sum_{1}^{n}\left[2^{x+1}3x^{(2)} + 3x^{(1)} + 2\right]$$

$$\sum_{1}^{n} 2^x(x^3+x) = \sum_{1}^{n}\left\{x^{(3)} + 3x^{(2)} + 2x^{(1)}\right\}\Delta 2^x$$

$$= \left[2^x\left\{x^{(3)} + 3x^{(2)} + 2x^{(1)}\right\}\right]_1^{n+1} - \sum_{1}^{n} 2^{x+1}\left\{3x^{(2)} + 6x^{(1)} + 2\right\}$$

$$= 2^{n+1}\{(n+1)n(n-1) + 3n(n+1) + 2(n+1)\}$$

$$- 2(2) - \left[2^{n+1}\left\{3x^{(2)} + 6x^{(1)} + 2\right\}\right]_1^{n+1} + \sum_{1}^{n} 2^{x+2}\left\{6x^{(1)} + 6\right\}$$

$$= 2^{n+1}(n^3 - n + 3n^2 + 3n + 2n + 2) - 4$$

$$- 2^{n+2}\{3n(n+1) + 6(n+1) + 2\} + 2^2(6+2)$$

$$+ \left[2^{x+2}\left\{6x^{(1)} + 6\right\}\right]_1^{n+1} - 6\sum_{1}^{n} 2^{x+3}$$

$$= 2^{n+1}(n^3 + 3n^2 + 4n + 2 - 6n^2 - 6n - 12n - 12 - 4) - 4 + 32$$

$$+ 2^{n+3}\{6(n+1) + 6\} - 2^3(6+6) - 6\cdot 2^4(2^n - 1)$$

$$= 2^{n+1}(n^3 - 3n^2 - 14n - 14) + 28 + 2^{n+1}(24n + 24 + 24) - 96 + 96 - 48\cdot 2^{n+1}$$

$$= 2^{n+1}(n^3 - 3n^2 + 10n - 14) + 28$$

11.5 The relation between the operators Σ and Δ

We may see that if $\Delta f(x) = \phi(x)$, then

$$f(x) = \sum \phi(x)$$

where the summation is taken between certain limits.

Omitting these limits for the present, we may say, with certain reservations, that summation is the inverse process of differencing. Hence

$$\phi(x) = \Delta f(x) = \Delta\Sigma\phi(x)$$

∴ symbolically, we may write

$$1 \equiv \Delta\Sigma \Rightarrow \Sigma \equiv \frac{1}{\Delta} \equiv \Delta^{-1}$$

11.5.1 The operation ΔΣ is not commutative

We shall show that

$$\Delta\Sigma \equiv 1 \not\equiv \Sigma\Delta$$

We know that

$$\Delta f(x) = \Delta\{f(x) + c\}$$

where c is some arbitrary non-zero constant.

Thus, if

$$\Delta\{f(x) + c\} = \phi(x)$$

then

$$f(x) + c = \Delta^{-1}\phi(x) = \Sigma\phi(x) = \Sigma\Delta\phi(x). \therefore \Sigma\Delta \not\equiv 1$$

$$\therefore \Sigma\Delta \not\equiv \Delta\Sigma$$

The symbol Σ may be used in place of the inverse symbol Δ^{-1}, provided we remember that Σ (or Δ^{-1}) and Δ are not commutative.

Thus, the process of summation in finite differences is similar to the corresponding process in integral calculus. Σu_x is therefore said to be the indefinite finite integral of u_x.

Corollary 11.5.1. To prove that

$$\sum_1^n u_x = \left[\Delta^{-1}u_x\right]_1^{n+1}$$

where $\Delta^{-1} \equiv 1/\Delta$.

Proof.

$$\sum_1^n u_x = u_1 + u_2 + u_3 + \cdots + u_n = (1 + E + E^2 + \cdots + E^{n-1})u_1$$

$$= \frac{E^n - 1}{E - 1} \cdot u_1 = \frac{(E^n - 1)u_1}{\Delta} = \frac{u_{n+1} - u_1}{\Delta} = \Delta^{-1}(u_{n+1} - u_1)$$

$$= \Delta^{-1}u_{n+1} - \Delta^{-1}u_1 = \left[\Delta^{-1}u_x\right]_1^{n+1}$$

Corollary 11.5.2. If

$$\psi(x) \equiv \frac{d}{dx}\log\Gamma(x) = \frac{\Gamma'(x)}{\Gamma(x)}$$

prove that

$$\Delta^{-1}\left[(x - b)^{-n}\right] = \frac{(-1)^{1-n}}{(n-1)!}\psi^{n-1}(x - b)$$

Proof. We have

$\Gamma(x+1) = x\Gamma(x)$

$\therefore \log\Gamma(x+1) = \log x + \log\Gamma(x)$

Differentiating both sides w.r.t. x,

$$\frac{\Gamma'(x+1)}{\Gamma(x+1)} = \frac{1}{x} + \frac{\Gamma'(x)}{\Gamma(x)}. \quad \therefore \psi(x+1) = \frac{1}{x} + \psi(x) \tag{11.5.1}$$

$$\Rightarrow \psi(x+1) - \psi(x) = \frac{1}{x}. \quad \therefore \Delta\psi(x) = \frac{1}{x} \tag{11.5.2}$$

Differentiating (11.5.1), w.r.t. x,

$$\psi'(x+1) = -\frac{1}{x^2} + \psi'(x). \quad \therefore \psi'(x+1) - \psi'(x) = -\frac{1}{x^2}$$

$$\Rightarrow \Delta\psi'(x) = -\frac{1}{x^2}. \quad \therefore \Delta^{-1}\left(\frac{1}{x^2}\right) = -\psi'(x) \tag{11.5.3}$$

Similarly,

$$\Delta^{-1}\left(-\frac{2}{x^3}\right) = -\psi''(x) \Rightarrow \Delta^{-1}\left(\frac{1}{x^3}\right) = \frac{1}{2}\psi''(x) = \frac{(-1)^2}{2!}\psi''(x)$$

$$\Delta^{-1}\left(-\frac{3}{x^4}\right) = \frac{1}{2}\psi'''(x) \Rightarrow \Delta^{-1}\left(\frac{1}{x^4}\right) = \frac{(-1)^3}{3!}\psi'''(x)$$

Noting the symmetry at each stage, we may write

$$\Delta^{-1}\left(\frac{1}{x^n}\right) = \frac{(-1)^{n-1}}{(n-1)!}\psi^{n-1}(x)$$

$$\Rightarrow \Delta^{-1}\{(x-b)^{-n}\} = \frac{(-1)^{n-1}}{(n-1)!}\psi^{n-1}(x-b) = \frac{(-1)^{n-1}}{(n-1)!}\psi^{n-1}(x-b)$$

11.6 An important result

Theorem 11.6.1. If $\phi(E)$ is a polynomial in E and $f(x)$ is a function of x, then

$$\phi(E)\{a^x f(x)\} = a^x\phi(aE)\{f(x)\}$$

Proof. Let

$$\phi(E) \equiv b_0 + b_1E + b_2E^2 + \cdots + b_rE^r + \cdots \quad (b_i's \text{ being constant})$$

Now

$$\begin{aligned} b_rE^r\{a^x f(x)\} &= b_r a^x \cdot a^r f(x+r) = a^x \cdot \{b_r a^r f(x+r)\} \\ &= a^x \cdot \{b_r a^r E^r f(x)\} = a^x\{b_r(aE)^r f(x)\} \\ &= a^r b_r (aE)^r f(x) \end{aligned}$$

$$\therefore \phi(E)\{a^x f(x)\} = \sum_{r=0}^{\infty}[b_rE^r\{a^x f(x)\}]$$

$$\begin{aligned} \therefore \phi(E)a^x f(x) &= \sum_{r=0}^{\infty} a^x b_r(aE)^r f(x) = a^x \sum_{r=0}^{\infty} b_r(aE)^r f(x) \\ &= a^x \cdot \phi(aE)f(x) \text{ (Proved)} \end{aligned}$$

Example 11.6.1. Use the result in 11.6 to evaluate

$$\sum_{x=0}^{\infty} a^x f(x)$$

Solution: We have

$$\sum_{x=0}^{\infty} a^x f(x) = \Delta^{-1}\{a^x f(x)\} = (E-1)^{-1} a^x f(x)$$

writing $(E-1)^{-1} = \phi(E)$.

We have the sum

$$= \phi(E)\{a^x f(x)\} = a^x \phi(aE) f(x) = a^x (aE-1)^{-1} f(x)$$
$$= a^x\{a(1+\Delta)-1\}^{-1} f(x) = a^x\{(a-1)+a\Delta\}^{-1} f(x)$$
$$= \frac{a^x}{a-1}\left\{1+\frac{a}{a-1}\Delta\right\}^{-1} f(x)$$
$$= \frac{a^x}{a-1}\left\{1-\frac{a\Delta}{a-1}+\frac{a^2}{(a-1)^2}\Delta^2 \cdots\right\} f(x)$$
$$= \frac{a^x}{a-1}\left[1-\frac{a\Delta}{a-1}+\frac{a^2}{(a-1)^2}\Delta^2-\frac{a^3}{(a-1)^3}\Delta^3+\cdots\right] f(x) + \text{constant}$$
$$= \frac{a^x}{a-1} f(x) - \frac{a^{x+1}}{(a-2)^2}\Delta f(x) + \frac{a^{x+2}}{(a-1)^3}\Delta^2 f(x) + \cdots + \text{constant}$$

Example 11.6.2. Find the general term of the following series and also the sum to n terms of the series $2.2 + 6.4 + 12.8 + 20.16 + 30.32 + \cdots$.

Solution: Let us first find the general term of the series 2, 6, 12, 20, 30, $\cdots$.

x	Δx	$\Delta^2 x$
2		
	4	
6		2
	6	
12		2
	8	
20		2
	10	
30		

Since second difference is constant the general term of the series:

$2, 6, 12, 20, 30, \cdots$

is of the form $= b_0 + b_1x^{(1)} + b_2x^{(2)}$, where $b_0 = 0$, $b_1 = 2$, $b_2 = 1$. Thus, the general term of the given series

$$u_x = 2^x\left\{b_0 + b_1x^{(1)} + b_2x^{(2)}\right\}$$

$$\therefore \sum_1^n u_x = \sum_{x=1}^n 2^x\left\{2x^{(1)} + x^{(2)}\right\}$$

$$= \sum\left\{x^{(2)} + 2x^{(1)}\right\}\Delta 2^x$$

$$= \left[\left\{x^{(2)} + 2x^{(1)}\right\}2x\right]_1^{n+1} - \sum_1^n 2^{x+1}\left\{2x^{(1)} + 2\right\}$$

$$= \{n(n-1) + 2(n+1)\}\,2^{n+1} - 2(2) - \sum_1^n\left\{2x^{(1)} + 2\right\}\Delta 2^{x+1}$$

$$= (n+1)(n+2)\cdot 2^{n+1} - 4 - \left[2^{x+1}\left\{2x^{(1)} + 2\right\}\right]_1^{n+1} + \sum_{x=1}^n 2^{x+2}\cdot 2$$

$$= (n+1)(n+2)\cdot 2^{n+1} - 4 - 2^{n+2}\{2(n+1) + 2\} + 2^2(2+2) + 2\cdot\frac{2^3(2^n - 1)}{2-1}$$

$$= (n+1)(n+2)\cdot 2^{n+1} - 2^{n+1}(2n+4) - 4 + 16 + 2^4(2^n - 1)$$

$$= 2^{n+1}(n^2 + 3n + 2 - 4n - 8 + 8) - 4 + 16 - 16$$

$$= 2^{n+1}(n^2 - n + 2) - 4$$

Alternatively,

$$u_x = 2^x(x^2 + x) = 2^x\left\{x^{(2)} + 2x^{(1)}\right\}$$

Denoting $x^{(2)} + 2x^{(1)}$ by $f(x)$, we have

$$\Delta f(x) = 2x^{(1)} + 2$$

and $\Delta^2 f(x) = 2$

$$\sum_1^n u_x = \left[\Delta^{-1}\left\{2^x(x^{(2)} + 2x^{(1)})\right\}\right]_1^{n+1} = \left[(E-1)^{-1}\left\{2^x(x^{(2)} + 2x^{(1)})\right\}\right]_1^{n+1}$$

$$= \left[2^x(2E-1)^{-1}\left\{x^{(2)} + 2x^{(1)}\right\}\right]_1^{n+1} = \left[2^x(1+2\Delta)^{-1}\left\{x^{(2)} + 2x^{(1)}\right\}\right]_1^{n+1}$$

$$= \left[2^x(1 - 2\Delta + 4\Delta^2)f(x)\right]_1^{n+1} = \left[2^x\left\{x^{(2)} + 2x^{(1)} - 4x^{(1)} - 4 + 8\right\}\right]_1^{n+1}$$

$$= 2^{n+1}\{n(n+1) - 2(n+1) + 4\} - 2(-2+4) = 2^{n+1}(n^2 - n + 2) - 4$$

EXAMPLE 11.6.3. Apply relation between $\sum$ and Δ to evaluate

$$\sum_1^n 3^x(x^3 + x^2 + x + 1)$$

Solution: Denoting $x^3 + x^2 + x + 1$ by $f(x)$, we may express

$$f(x) = x^{(3)} + 4x^{(2)} + 3x^{(1)} + 1$$

$$\therefore \sum_{1}^{n} 3^x(x^3 + x^2 + x + 1) = \left[\Delta^{-1}\left\{3^x(x^{(3)} + 4x^{(2)} + 3x^{(1)} + 1)\right\}\right]_1^{n+1}$$

$$= \left[(E-1)^{-1}\left\{3^x(x^{(3)} + 4x^{(2)} + 3x^{(1)} + 1)\right\}\right]_1^{n+1}$$

$$= \left[3^x(3E-1)^{-1}\left\{x^{(3)} + 4x^{(2)} + 3x^{(1)} + 1\right\}\right]_1^{n+1}$$

$$= \left[3^x(2+3\Delta)^{-1}\left\{x^{(3)} + 4x^{(2)} + 3x^{(1)} + 1\right\}\right]_1^{n+1}$$

$$= \left[\frac{3^x}{2}\left(1 - \frac{3}{2}\Delta + \frac{9}{4}\Delta^2 - \frac{27}{8}\Delta^3\right)f(x)\right]_1^{n+1}$$

Now, $\Delta f(x) = 3x^{(2)} + 8^{(1)} + 3$

$\Delta^2 f(x) = 6x^{(1)} + 8$

$\Delta^3 f(x) = 6$

$$\therefore \sum 3^x(x^3 + x^2 + x + 1)$$

$$= \left[\frac{3^x}{2}\left\{(x^{(3)} + 4x^{(2)} + 3x^{(1)} + 1) - \frac{3}{2}\left\{3x^{(2)} + 8x^{(1)} + 3\right\} + \frac{9}{4}\left\{6x^{(1)} + 8\right\} - \frac{27}{8}\cdot 6\right\}\right]_1^{n+1}$$

$$= \left[\frac{3^x}{2}\left\{x^{(3)} + \left(4 - \frac{9}{2}\right)x^{(2)} + \frac{9}{2}x^{(1)} - \frac{23}{4}\right\}\right]_1^{n+1}$$

$$= \left[\frac{3^{n+1}}{2}\left\{n(n^2-1) - \frac{1}{2}n(n+1) + \frac{9}{2}(n+1) - \frac{23}{4}\right\}\right] - \left[\frac{3}{2}\left\{\frac{9}{2} - \frac{23}{4}\right\}\right]$$

$$= \frac{3^{n+1}}{8}[4n^3 - 4n - 2n^2 - 2n + 18n + 18 - 23] + \frac{15}{8}$$

$$= \frac{3^{n+1}}{8}[4n^3 - 2n^2 + 12n - 5] + \frac{15}{8}$$

Corollary 11.6.1. Evaluate $\Sigma 3^x(x^3 + x^2 + x + 1)$.

$$\sum 3^x(x^3 + x^2 + x + 1) = \left[\Delta^{-1}\left\{3^x(x^3 + x^2 + x + 1)\right\}\right] + \text{const.}$$

$$= \left[3^x(3E-1)^{-1}\left\{x^{(3)} + 4x^{(2)} + 3x^{(1)} + 1\right\}\right] + \text{const.}$$

$$= \left[3^x(2+3\Delta)^{-1}f(x)\right] + \text{const.}$$

$$= \frac{3^x}{2}\left(1 + \frac{3}{2}\Delta\right)^{-1} f(x) + \text{const.}$$

$$= \frac{3^x}{2}\left(1 - \frac{3}{2}\Delta + \frac{9}{4}\Delta^2 - \frac{27}{8}\Delta^3\right)f(x) + \text{const.}$$

$$= \frac{3^x}{2}\left[(x^3 + x^2 + x + 1) - \frac{3}{2}\left\{3x(x-1) + 8x + 3\right\} + \frac{9}{4}(6x + 8) - \frac{27}{8}\cdot 6\right] + \text{const.}$$

$$= \frac{3^x}{2}\left[(x^3 + x^2 + x + 1) - \frac{3}{2}\left\{(3x^2 - 3x) + 8x + 3\right\} + \frac{27}{2}x + 18 - \frac{81}{4}\right] + \text{const.}$$

$$= \frac{3^x}{2}\left[x^3 + \left(1 - \frac{9}{2}\right)x^2 + \left(1 + \frac{9}{2} - 12 + \frac{27}{2}\right)x + \left(1 - \frac{9}{2} - \frac{81}{4} + 18\right)\right] + \text{const.}$$

$$= \frac{3^x}{2}\left[x^3 - \frac{7}{2}x^2 + 7x - \frac{23}{4}\right] + \text{const.} = \frac{3^x}{8}[4x^3 - 14x^2 + 28x - 23] + \text{const.}$$

EXERCISES

1. Sum the following series:

   (a) 2, 12, 36, 80, 150, 252, $\cdots$ to n terms. [*Ans:* $\frac{n}{6}(n+1)(n+2)(3n+1)$.]

   (b) 10, 9, 7, 4, 0, −5, $\cdots$ to 30 terms. [*Ans:* −4195.]

   (c) 1, 2, 4, 8, 17, 40, 104, $\cdots$ to n terms. [*Ans:* $\frac{1}{8}\{\frac{1}{2}(3^n-1)+5n+\frac{n}{3}(n+1)(2n+1)\}$.]

   (d) 10, 14, 10, 6, $\cdots$ to n terms. [*Ans:* $\frac{1}{3}(n^4-10n^3+29n^2+10n)$.]

   (e) 125, 343, 729, 1331, 2197, $\cdots$ to n terms. [*Ans:* $2n^4+16n^3+47n^2+60n$.]

   (f) 0, 10, 33, 77, 150, $\cdots$ to n terms. [*Ans:* $\frac{n}{6}(n-1)(2n^2+3n+6)$.]

   (g) 5, 10, 17, 28, 47, 82, $\cdots$ to n terms. [*Ans:* $2^{n+1}-2+\frac{3}{2}n(n+1)$.]

   (h) 1, 0, 1, 8, 29, 80, 193, $\cdots$ to n terms. [*Ans:* $2^{n+2}-4-\frac{n}{6}(n+1)(2n+7)$.]

2. Find sum to n terms of the series whose xth terms is:

   (a) $x(x+1)(x+2)$; [*Ans:* $\frac{n(n+1)}{4}(n^2+5n+6)$.]

   (b) $x(x-1)(x-3)$; [*Ans:* $\frac{n}{12}(n^2-1)(3n-10)$.]

   (c) $(3x-1)(3x+2)(3x+5)(3x+8)$;
   [*Ans:* $\frac{1}{15}\{(3n-1)(3n+2)(3n+5)(3n+8)(3n+11)+880\}$.]

   (d) $2^x(x^3+x)$; [*Ans:* $2^{n+1}(n^3-3n^2+10n-14)+28$.]

   (e) $4^x(x^2+x+1)$. [*Ans:* $\frac{1}{27}\{4^{n+1}(9n^2+3n+11)-44\}$.]

3. Sum to n terms of the series whose xth term is:

   (a) $\dfrac{1}{(x+1)(x+2)(x+3)}$; [*Ans:* $\frac{1}{12}-\frac{1}{2(n+2)(n+3)}$.]

   (b) $\dfrac{1}{(2x-1)(2x+1)(2x+3)}$; [*Ans:* $\frac{1}{10}-\frac{1}{2(2n+1)(2n+3)}$.]

   (c) $\dfrac{1}{4x^2-1}$; [*Ans:* $\frac{n}{2n+1}$.]

   (d) $\dfrac{x+3}{x(x+1)(x+2)}$. [*Ans:* $\frac{n(5n+11)}{4(n+1)(n+2)}$.]

4. Sum the following series to n terms:

   (a) 1.4 + 4.7 + 7.10 + 10.13 + $\cdots$; [*Ans:* $2+(n+1)(3n-2)$.]

   (b) 2.4.6 + 4.6.8 + 6.8.10 + $\cdots$; [*Ans:* $2n(n+1)(n+2)(n+3)$.]

   (c) $\dfrac{1}{1.4}+\dfrac{1}{4.7}+\dfrac{1}{7.10}+\cdots$; [*Ans:* $\frac{n}{3n+1}$.]

   (d) $\dfrac{4}{1.2.3}+\dfrac{5}{2.3.4}+\dfrac{6}{3.4.5}+\cdots$. [*Ans:* $\frac{5}{2}-\frac{2n+5}{2(n+1)(n+2)}$.]

5. Find the general term and the sum to n terms of the series:

   (a) 10, 23, 60, 169, 494, 1467, $\cdots$; [*Ans:* $u_x(3+x)+2\cdot 3^x$, $S_n=3^{n+1}+\frac{1}{2}(n^2+7n-6)$.]

   (b) $\dfrac{5}{1.2}\cdot\dfrac{1}{3}+\dfrac{7}{2.3}\cdot\dfrac{1}{3^2}+\dfrac{9}{3.4}\cdot\dfrac{1}{3^3}+\dfrac{11}{4.5}\cdot\dfrac{1}{3^4}+\cdots$.
   [*Ans:* $u_x=\frac{2x+3}{x(x+1)}\cdot\frac{1}{3^x}$, $S_n=1-\frac{1}{n+1}\cdot\frac{1}{3^n}$.]

6. Find the sum to n terms $a+2a^2+3a^3+4a^4+\cdots$.
   [*Ans:* $(n+1)\frac{a^{n+1}}{a-1}-\frac{a}{a-1}-\frac{a^{n+2}}{(a-1)^2}+\frac{a^2}{(a-1)^2}$.]

7. Evaluate $\Delta^{-1}\left\{\dfrac{x+4}{(x+2)(x+3)(x+5)(x+6)}\right\}$.

[*Ans:* $\frac{31}{18}-\frac{1}{4}\{(n^2-1)n^{(-5)}+n^{(-4)}\}-\frac{3}{2}n(n+1)^{(-3)}-\frac{7}{4}(n+1)^{(-3)}-\frac{1}{4}(n+3)n^{(-2)}-n^{(-1)}$.]

8. (a) Evaluate $\Delta^{-1}\left\{\dfrac{(x+1)^2}{x(x+2)(x+3)(x+4)}\right\}$. [*Ans:* $C-\dfrac{12x^2+36x^2+28x+3}{12x(x+1)(x+2)(x+3)}$.]

(b) Find $\sum_{1}^{n} u_x$, where $u_x = x(x+2)(x+5)$. [*Ans:* $\frac{n}{12}(n+1)(3n^2+31n+74)$.]

9. Evaluate $\sum_{1}^{n}\left\{\dfrac{x^2 4^x}{(x+1)(x+2)}\right\}$. [*Ans:* $\frac{2}{3}+\frac{4^{n+1}}{3}\left(\frac{n-1}{n+2}\right)$.]

10. Obtain the indefinite integral of $(2x-1)3^x$. [*Ans:* $3^x\cdot(x-2)+$ constant.]

11. Sum the following series: $x^3+(x+1)^3+(x+2)^3+\cdots+(x+n-1)^3$.

[*Hints:* $\text{Sum} = (1+E+E^2+\cdots+E^{n-1}r)x^3 = \dfrac{E^n-1}{E-1}x^3 = (E^n-1)\Delta^{-1}x^3$

$$= \Delta^{-1}(E^n-1)x^3 = \Delta^{-1}(x+n)^3 - \Delta^{-1}x^3$$

$$= \Delta^{-1}\left\{(x+n)^{(3)}+3(x+n)^{(2)}+(x+n)^{(1)}\right\} - \Delta^{-1}\left\{x^{(3)}+3x^{(2)}+x^{(1)}\right\}$$

$$= \frac{1}{4}(x+n)^{(4)}+(x+n)^{(3)}+\frac{1}{2}(x+n)^{(2)}-\left\{\frac{1}{4}x^{(4)}+x^{(3)}+\frac{1}{2}x^{(2)}\right\}$$

$$= \frac{1}{4}(x+n)^2(x+n-1)^2-\frac{1}{4}x^2(x-1)^2 \text{ (on simplification)}]$$

12. Evaluate

$$\frac{1}{x+1}-\frac{n}{(x+1)(x+2)}+\frac{n(n-1)}{(x+1)(x+2)(x+3)}-\cdots+\frac{(-1)^n\cdot n!}{(x+1)(x+2)\cdots(x+n+1)}$$

[*Hints:* The series $= x^{(-1)}-nx^{(-2)}+n(n-1)x^{(-3)}+\cdots+(-1)^n\cdot n!x^{(-n-1)}$

$$= x^{(-1)}+\binom{n}{1}\Delta x^{(-1)}+\binom{n}{2}\Delta^2x^{(-1)}+\cdots+\binom{n}{n}\Delta^n x^{(-1)}$$

$$= (1+\Delta)^n\cdot x^{(-1)} = E^n x^{(-1)} = (x+n)^{(-1)} = \frac{1}{x+n+1}]$$

Chapter 12

Solution of Difference Equations

Summary

12.1 Introduction

A difference equation is an equation involving an independent variable, dependent variable and successive differences of the dependent variable. We have already seen that the successive differences of the dependent variable can be expressed in terms of the successive values of the dependent variable. For example,

$\Delta^3 y_x = y_{x+3} + 3y_{x+2} + 3y_{x+1} + y_x$

Therefore, a difference equation may also be defined as an equation which expresses a relation between an independent variable and the successive values of the dependent variable. We may give below some examples of difference equations:

1. $y_{x+2} - 2y_{x+1} - 3y_x = 0$,
2. $y_{x+3} - 3y_{x+2} = 2^x$,
3. $\Delta^3 y_x - 3\Delta^2 y_x + \Delta y_x + y_x = \cos 2x$,

where Δ is our usual finite difference operator given by $\Delta y_x = y_{x+1} - y_x$.

The first difference equation involves successive values of the dependent variable, while the third difference equation involves successive differences of the dependent variable. It may, however, appear more convenient to work with difference equations involving successive values of the dependent variable.

A difference **equation not involving successive** values of y_x greater than y_{x+n} is said to be of order *n*. **Hence difference equation (1)** may be said to be of order 2. Difference equation (2) is **however, of order 1, since by putting** $u_x = y_{x+2}$, equation (2) reduces to the form $u_{x+1} - 3u_x = 2^x$.

A solution of a difference equation is a relation between the independent variable and the dependent variable, which satisfies the equation. Thus,

$y_x = C_1 \cdot 3^x + C_2 \cdot (-1)^x$

is a solution of the difference equation (1), since

$$\begin{aligned}y_{x+2} - 2y_{x+1} - 3y_x &= C_1 \cdot 3^{x+2} + C_2(-1)^{x+2} - 2\{C_1 \cdot 3^{x+1} + C_2(-1)^{x+1}\}\\ &\quad - 3C_1 \cdot 3^x - 3C_2(-1)^x\\ &= 9C_1 \cdot 3^x + C_2(-1)^x - 6C_1 \cdot 3^x + 2C_2(-1)^x - 3C_1 \cdot 3^x\\ &\quad - 3C_2(-1)^x = 0\end{aligned}$$

holds for the given solution.

We may also see that the difference equation of order 2, involves two arbitrary constants C_1 and C_2. The solution of equation of order *n* will generally contain *n* arbitrary constants $C_1, C_2, \ldots, C_n$. A solution containing arbitrary constants equal in number to the order of the difference equation will be called the *general solution.* Any other solution by assigning particular values to one or more of the arbitrary constants is called a *particular solution.* Thus,

$y_x = 3^x + 2(-1)^x$

is a particular solution of the difference equation (1).

An equation of the form:

$y_{x+n} + p_1 y_{x+n-1} + p_2 y_{x+n-2} + \cdots + p_n y_x = P(x)$

where p_i's and $P(x)$ are functions of x or constants, is called a *linear difference equation* of order *n*.

If $P(x) = 0$, the equation is called a *linear homogeneous equation;* otherwise it is called a complete equation.

12.2 Formation of difference equations

Let there be a functional relationship between the independent variable x, dependent variable y and a number of constants $C_1, C_2, \ldots, C_n$ of the form:

$$P(y, x, C_1, C_2, \ldots, C_n) = 0 \qquad (12.2.1)$$

We shall eliminate the constants from the above equation with the help of certain operations, which will ultimately give an equation in x, y, and various orders of differences of y. To illustrate this point, we may start by taking one constant C_1 (say). Then

$$P(y, x, C_1) = 0 \qquad (12.2.2)$$

Operating Δ on (12.2.2), we get

$$\Delta P(y, x, C_1) = 0 \qquad (12.2.3)$$

Eliminating C_1 from (12.2.2) and (12.2.3), we have a relation of the form

$$F(y, x, \Delta y) = 0 \qquad (12.2.4)$$

where $\Delta y = \Delta f(x) = f(x+1) - f(x)$.

Putting this value of Δy in (12.2.4), we get

$$F_1[f(x+1), f(x), x] = 0 \tag{12.2.5}$$

It is thus evident from (12.2.4) and (12.2.5) that if there be one constant in the relation (12.2.1), then it can be replaced by $\Delta f(x)$ or $f(x+1)$ to obtain the difference equation. Proceeding in this way, we may show that if (12.2.1) contains two constants C_1 and C_2 (*say*), then they can be replaced by Δy and $\Delta^2 y$ or $f(x+1)$ and $f(x+2)$. In the same way we can show that if (12.2.1) contains n constants $C_1, C_2, \ldots, C_n$ (say), then operating Δ, $\Delta^2, \ldots, \Delta^n$ in succession, we may ultimately get a relation of the form:

$$F[x, f(x), \Delta f(x), \Delta^2 f(x), \ldots, \Delta^n f(x)] \tag{12.2.6}$$

$$\text{or, } F_1[x, f(x), f(x+1), f(x+2), \ldots, f(x+n)] \tag{12.2.7}$$

The relation given in (12.2.6) and (12.2.7) is known as *difference equation.*

Note: The constants in the equation (12.2.1) are not necessarily absolute constants. They may well be periodic function of x, such that they remain constant with regard to the operation Δ.

Example 12.2.1. Given

$y_x \equiv f(x) = C_1 3^x + C_2(-2)^x$;

find the corresponding difference equation.

Solution: Since there are two constants C_1, C_2 in the equation, we shall use two operations Δy_x and $\Delta^2 y_x$.

Now

$$\begin{aligned}\Delta f(x) &= C_1 3^{x+1} + C_2(-2)^{x+1} - C_1 3^x - C_2(-2)^x \\ &= C_1 3^x(3-1) + C_2(-2)^x(-2-1) = 2C_1 3^x - 3C_2(-2)^x\end{aligned}$$

$$\therefore\ 2f(x) - \Delta f(x) = 5C_2(-2)^x \Rightarrow \frac{2}{5}f(x) - \frac{1}{5}\Delta f(x) = C_2(-2)^x \tag{i}$$

Again

$$\begin{aligned}\Delta^2 f(x) &= C_1 3^{x+2} C_2(-2)^{x+2} - 2C_1 3^{x+1} - 2C_2(-2)^{x+1} + C_1 3^x + C_2(-2)^x \\ &= 4C_1 3^x + 9C_2(-2)^x\end{aligned} \tag{ii}$$

From (i) and (ii),

$$\begin{aligned}C_1 \cdot 3^x &= \frac{1}{4}\left[\Delta^2 f(x) - 9\left\{\frac{2}{5}f(x) - \frac{1}{5}\Delta f(x)\right\}\right] \\ &= \frac{1}{4}\Delta^2 f(x) - \frac{9}{10}f(x) + \frac{9}{20}\Delta f(x)\end{aligned} \tag{iii}$$

Hence

$$\begin{aligned}y_x &= \frac{1}{4}\Delta^2 f(x) - \frac{9}{10}f(x) + \frac{9}{20}\Delta f(x) + \frac{2}{5}f(x) - \frac{1}{5}\Delta f(x) \\ &= -\frac{1}{2}y_x + \frac{1}{4}\Delta y_x + \frac{1}{4}\Delta^2 y_x\end{aligned} \tag{iv}$$

$$\therefore f(x) = \frac{1}{4}[\Delta^2 y_x + \Delta y_x - 2y_x]$$

$$\Rightarrow 4y_x = y_{x+2} - 2y_{x+1} + y_x + y_{x+1} - y_x - 2y_x$$

$$\Rightarrow y_{x+2} - y_{x+1} - 6y_x = 0. \quad \text{(v)}$$

EXAMPLE 12.2.2. Find the corresponding difference equation given

$f(x) = C \cdot 2^x + x \cdot 3^{x-1}$

Solution: Here we have one constant C in the relation. Therefore, operating Δ on $f(x)$, we obtain

$$\begin{aligned}\Delta f(x) &= C \cdot 2^{x+1} + (x+1) \cdot 3^x - C \cdot 2^x - x \cdot 3^{x-1} \\ &= C \cdot 2^x(2-1) + 3(x+1) \cdot 3^{x-1} - x \cdot 3^{x-1} \\ &= C \cdot 2^x + (2x+3) \cdot 3^{x-1}\end{aligned}$$

$$\therefore \ C \cdot 2^x = \Delta f(x) - (2x+3) \cdot 3^{x-1} \quad \text{(i)}$$

Eliminating C between (i) and the given equation

$$\begin{aligned}y_x - x \cdot 3^{x-1} &= \Delta y_x - (2x+3)3^{x-1} \\ &= y_{x+1} - y_x - (2x+3) \cdot 3^{x-1}\end{aligned}$$

$$\therefore y_{x+1} - 2y_x - (x+3) \cdot 3^{x-1} = 0$$

$$\Rightarrow y_{x+1} - 2y_x = (x+3) \cdot 3^{x-1}$$

which is the required difference equation.

12.3 Solution of simple difference equations

Many difference equations can be solved either by (i) a simple substitution or (ii) on multiplication by a suitable function, known as the integrating factor, and then reducing the equation to an exact difference form:

EXAMPLE 12.3.1. Solve: $u_{x+1} - bu_x = 0;\ b \neq 1.$

Solution: Let $u_x = kt^x$ be the trial solution, ($t^x \neq 0$). Then

$$u_{x+1} - bu_x = kt^{x+1} - bkt^x = 0$$

$$\Rightarrow t^x k(t-b) = 0$$

$$\Rightarrow t = b \quad \therefore u_x = kb^x$$

is the general solution, where k is a periodic function of period 1.

Alternatively, multiplying both sides of the given equation by b^{-x-1}, we have

$$b^{-(x+1)}u_{x+1} - b^{-x}u_x = 0,$$

or, $E\{b^{-x}u_x\} - b^{-x}u_x = 0,$

or, $\Delta\{b^{-x}u_x\} = 0$

$\therefore b^{-x}u_x = k$ (constant) $\quad \therefore u_x = kb^x$

where k is a periodic function of period 1 in the general solution.

Note: If we write $k = 2$, then $u_k = 2b^x$ is also a solution of (1). Thus, for different values of k, (1) may have infinite number of solutions. $u_x = 2b^x$ is a *particular* solution of (1) and $u_x = kb^x$ is the general solution of (1) for $k = 1, 2, 3, \cdots$.

12.4 Solution of the equation $y_{x+1} = ay_x + b$, (Markoff's form), when a, b are constants ($a \neq 1$)

We have the equation

$$y_{x+1} - ay_x = b$$

To find the complimentary function, we consider the homogeneous equation

$$y_{x+1} - ay_x = 0$$

Let $y_x = kt^x$ be a trial solution $k \neq 0$. Then

$$kt^{x+1} - akt^x = 0;$$

$$kt^x(t - a) = 0$$

$$\therefore t = a$$

$\therefore y_x = k \cdot a^x$ will be the general solution. To find the particular integral (PI), we write

$$y_x = \frac{1}{E - a} \cdot b = \frac{1}{1 + \Delta - a} \cdot b = \frac{1}{1 - a + \Delta} \cdot b$$

$$= \frac{1}{1 - a}\left(1 + \frac{\Delta}{1 - a}\right)^{-1} b = \frac{b}{1 - a}$$

The complete solution is: General solution + PI

$\therefore$ the complete solution is

$$y_x = k \cdot a^x + \frac{b}{1 - a} \quad (a \neq 1)$$

The constant k may be obtained from the given initial condition, say $y_0 = 2$.

Now

$$y_0 = k + \frac{b}{1 - a} = 2$$

$$\therefore k = 2 - \frac{b}{1 - a} = \frac{2 - 2a - b}{1 - a}$$

$$\therefore y_x = \frac{2 - 2a - b}{1 - a} a^x + \frac{b}{1 - a}$$

If b is not a constant but a function of x, we may proceed as under.

Example 12.4.1. Solve: $y_{x+1} - by_x = ca^x$.

Solution: For the complimentary function, we take $y_x = k \cdot t^x$ as the trial solution

$$\therefore kt^{x+1} - bkt^x = 0$$

$$\Rightarrow kt^x(t - b) = 0$$

$$\therefore t = b$$

Hence $y_x = k \cdot b^x$ is the complimentary function. PI will be obtained from:

$$y_x = \frac{1}{E-b} \cdot ca^x = a^x \cdot \frac{1}{aE-b} \cdot c = a^x \cdot \frac{1}{a(1+\Delta)-b} \cdot c$$

$$= a^x \cdot \frac{1}{(a-b)+a\Delta} \cdot c = \frac{a^x}{a-b}\left(1+\frac{a}{a-b}\Delta\right)^{-1} \cdot c$$

$$= \frac{a^x}{a-b} \cdot c \quad (a \neq b)$$

∴ the complete solution

$$y_x = k \cdot b^x + \frac{ca^x}{a-b}$$

Alternatively,

Case 1: When $b \neq a$.

Let $y_x = kca^x$ be a solution then

$$kca^{x+1} - bkca^x = ca^x,$$

or, $kca^x(a-b) = ca^x$

$\Rightarrow k = 1/(a-b)$

$$\therefore y_x = \frac{ca^x}{a-b}$$

is a particular solution of the equation.

Case 2: When $b = a$.

The equation is

$$y_{x+1} - ay_x = ca^x$$

Let $y_x = kcxa^x$ be a solution.

$\therefore kc(x+1)a^{x+1} - kcxa^{x+1} = ca^x$

$\Rightarrow kca^{x+1} = ca^x$

$\Rightarrow k = 1/a$

$\therefore y_x = cxa^{x-1}$

is a particular solution of the equation.

EXAMPLE 12.4.2. Solve: $y_{x+1} - y_x = (x^2 - 2x)2^x$.

Solution: Let $y_x = (a + bx + cx^2) \cdot 2^x$ be a solution

$$\therefore y_{x+1} - y_x = \left\{a + b(x+1) + c(x+1)^2\right\} 2^{x+1} - (a + bx + cx^2)2^x$$

$$= (a + bx + cx^2)2^x + (b + 2cx + c)2^{x+1}$$

$$= \left\{a + bx + cx^2 + 2b + 4cx + 2c\right\} \cdot 2^x$$

$$= \left\{(a + 2b + 2c) + (b + 4c)x + cx^2\right\} \cdot 2^x$$

Comparing coefficients, we have $c = 1$, $b = -6$, $a = 10$, whence

$$y_x = (10 - 6x + x^2)2^x + C(x)$$

Alternatively, the equation is $\Delta y_x = (x^2 - 2x)2^x$.

$$\begin{aligned}
\therefore\ y_x &= \frac{1}{\Delta}\left\{x^{(2)} - x^{(1)}\right\} \cdot 2^x + C(x) \\
&= \frac{1}{E-1}\left\{x^{(2)} - x^{(1)}\right\} \cdot 2^x + C(x) \\
&= 2^x \cdot \frac{1}{2E-1}\left\{x^{(2)} - x^{(1)}\right\} + C(x) \\
&= 2^x \cdot \frac{1}{1+2\Delta}\left\{x^{(2)} - x^{(1)}\right\} + C(x) \\
&= 2^x(1 - 2\Delta + 4\Delta^2)\left\{x^{(2)} - x^{(1)}\right\} + C(x) \\
&= 2^x\left[x^{(2)} - x^{(1)} - 2\left\{2x^{(1)} - 1\right\} + 4(2)\right] + C(x) \\
&= 2^x[x^2 - 2x - 4x + 2 + 8] + C(x) \\
&= 2^x(x^2 - 6x + 10) + C(x)
\end{aligned}$$

12.5 Linear difference equations

A difference equation is said to be linear if its degree is one. The most common form of a linear difference equation may be

$$y_{x+n} + p_1 y_{x+n-1} + p_2 y_{x+n-2} + \cdots + p_{n-1} y_{x+1} + p_n y_x = B(x) \tag{12.5.1}$$

where $p_1, p_2, \ldots, p_n$ may be absolute constants or periodic function of x. $B(x)$ is some function of the independent variable x.

If $B(x)$ is zero, the linear difference equation (12.5.1) will be called linear homogeneous difference equation of order n. If $B(x) \neq 0$, it will be called linear non-homogeneous or complete difference equation of order n :

$$y_{x+n} + p_1 y_{x+n-1} + \cdots + p_n y_x = 0 \tag{12.5.2}$$

where p_i's are functions of x.

Analogue to the theorems of ordinary differential equations, we may state the following results which may however be proved easily:

1. If $y_1(x)$ is a solution of (12.5.2) so is $C_1 y_1(x)$, where C_1 is an arbitrary constant.
2. If $y_1(x), y_2(x), \ldots, y_n(x)$ are n distinct solutions of (12.5.2), then

   $C_1 y_1(x) + C_2 y_2(x) + \cdots + C_n y_x(x)$

   is a solution.

Since linear equations with variable coefficients are solvable under some restrictions on the coefficients, we consider at the beginning, linear equations with constant coefficients which may be written in the form:

$$y_{x+n} + a_1 y_{x+n-1} + \cdots + a_n y_x = 0 \tag{12.5.3}$$

Using operator E, we have

$$(E^n + aE^{n-1} + \cdots + a_{n-1}E + a_n) y_x = 0 \tag{12.5.4}$$

Equation (12.5.4) may be written as $f(E)y_x = 0$, where

$f(E) \equiv E^n + a_1 E^{n-1} + \cdots + a_{n-1}E + a_n$

and the coefficients $a_1, \ldots, a_{n-1}, a_n$ are constants.

We shall call $f(E) = 0$ the auxiliary equation of the equation (12.5.4) and the function

$f(E) \equiv E^n + a_1 E^{n-1} + \cdots + a_n$

is known as the characteristic function of the given equation.

We shall now consider the following cases:

Case 1: The equation $f(E) = 0$ has *n distinct real roots* $\alpha_1, \alpha_2, \ldots, \alpha_n$.

$\therefore (E - \alpha_1)(E - \alpha_2)\cdots(E - \alpha_n)y_x = 0$

Hence

$(E - \alpha_1)y_x = 0, (E - \alpha_2)y_x = 0, \ldots (E - \alpha_n)y_x = 0$

Consider a typical equation

$(E - \alpha_r)y_x = 0,\ r = 1, 2, \ldots n \Rightarrow y_{x+1} - \alpha_r y_x = 0$

The solution is $y_x = C\alpha_r^x$, where C is a periodic function of period 1.

Hence the solution of the given equation is

$y_x = C_1\alpha_1^x + C_2\alpha_2^x + \cdots + C_n\alpha_n^x$

Case 2: $f(E) = 0$ has repeated real roots, say, α, α. Then $(E - \alpha)^2 y_x = 0$.

$$y_{x+2} - 2\alpha y_{x+1} + \alpha^2 y_x = 0 \tag{A}$$

To solve this equation, we make the substitution $y_x = \alpha^x \cdot u_x$, where u_x is a new dependent variable. Equation (A) may be written as

$\alpha^{x+2}u_{x+2} - 2\alpha^{x+2}u_{x+1} + \alpha^{x+2}u_x = 0$

$\Rightarrow u_{x+2} - 2u_{x+1} + u_x = 0 \quad (\alpha \neq 0)$

$\Rightarrow \Delta^2 u_x = 0.\ \therefore u_x = C_1 + C_2 x$

$$\therefore y_x = (C_1 + C_2 x)\cdot\alpha^x \tag{B}$$

where C_1, C_2 are periodic functions with period 1.

Similarly, we may show that if α is repeated r times, and the remaining $(n - r)$ roots are distinct real roots, then the solution of (B) is

$$y_x = (C_1 + C_2 x + \cdots + C_r x^{r-1})\cdot\alpha^x + b_1\alpha_{r+1}^x + b_2\alpha_{r+2}^x + \cdots + b_{n-r}\alpha_n^x \tag{12.5.5}$$

Case 3: $f(E) = 0$ has non-repeated complex roots.

Consider two roots $\alpha_1 = \alpha + i\beta$, $\alpha_2 = \alpha - i\beta$. Then the solution is

$$y_x = C_1(\alpha + i\beta)^x + C_2(\alpha - i\beta)^x = r^x(A_1\cos\theta x + A_2\sin\theta x) \tag{A}$$

where $r = +\sqrt{\alpha^2 + \beta^2}$, $\theta = \tan^{-1}(\beta/\alpha)$, $A_1 = C_1 + C_2$, $A_2 = i(C_1 - C_2)$.

Case 4: $f(E) = 0$ has repeated complex roots, say, $\alpha_1 = \alpha + i\beta$, $\alpha_2 = \alpha + i\beta$, $\alpha_3 = \alpha - i\beta$, $\alpha_4 = \alpha - i\beta$.

In such a case the solution is

$y_x = r^x[(C_1 + C_2 x)\cos\theta x + (C_3 + C_4 x)\sin\theta x]$

where $r = +\sqrt{\alpha^2 + \beta^2}$, $\theta = \tan^{-1}(\beta/\alpha)$.

EXAMPLE 12.5.1. Homogeneous linear equations with constant coefficients. Solve:

(a) $y_{x+3} - 3y_{x+1} - 2y_x = 0$

(b) $y_{x+2} - 6y_{x+1} + 9y_x = 0$

(c) $y_{x+2} + 4y_x = 0.$

Solution: (a) The equation is

$(E^3 - 3E - 2)y_x = 0$

The auxiliary function is

$$f(E) = E^3 - 3E - 2 = E^3 + E^2 - E^2 - E - 2E - 2$$
$$= E^2(E + 1) - E(E + 1) - 2(E + 1)$$
$$= (E + 1)(E^2 - E - 2)$$
$$= (E + 1)(E - 2)(E + 1)$$

∴ the auxiliary equation has two equal roots −1, −1 and one distinct root 2.

∴ the solution is

$$y_x = (C_1 + C_2x)(-1)^x + C_3 \cdot 2^x$$

where C_1, C_2, C_3 are arbitrary constants.

(b) The equation is

$$(E^2 - 6E + 9)y_x = 0 \Rightarrow (E - 3)^2 y_x = 0$$

The auxiliary equation $f(E) = 0$ has two equal real roots 3, 3.

∴ the solution is

$$y_x = (C_1 + C_2x) \cdot 3^x$$

where C_1, C_2 are arbitrary constants.

(c) The equation is

$$(E^2 + 4)y_x = 0$$

Here $f(E) = 0$ has a pair of complex roots $\pm 2i$.

∴ the solution is

$$y_x = 2^x[A_1 \cos \theta x + A_2 \sin \theta x]$$

When $\theta = +\pi/2$,

$$y_x = 2^x \left[A_1 \cos \frac{\pi}{2}x + A_2 \sin \frac{\pi}{2}x\right]$$

12.6 Non-homogeneous linear equations

Analogue to the case of linear differential equations, the general solution of the non-homogeneous difference equation is found by adding a particular solution of the non-homogeneous to the general solution of the corresponding homogeneous equation.

In other words,

General Solution = Complementary Function + Particular Integral.

12.7 Methods of evaluating the PI $1/f(E)$

1. when $B(x) = V_x$, (V_x is a polynomial in x of degree m, say);
2. when $B(x)$ is of the form a^x;
3. when $B(x) = a^x \cdot V_x$;

where V_x is a polynomial in x of degree m, say.

1. For simplicity, let $V_x = x^m$, where m is zero or a positive integer.
   In this case we write

   $$y_x = \frac{1}{f(E)} V_x = \frac{1}{f(1+\Delta)} V_x$$

   Also $1/f(1 + \Delta)$ is expanded in the form:

   $$(b_0 + b_1\Delta + b_2\Delta^2 + \cdots + b_m\Delta^m)$$

   Hence

   $$(b_0 + b_1\Delta + b_2\Delta^2 + \cdots + b_m\Delta^m)x^m$$

   is the solution of $f(E)y_x = V_x$.

• The above method works if the characteristic equation $f(E) = 0$ does not admit the root, unity.
  If, however, the characteristic equation admits the root unity of *multiplicity* r, then $f(E)y_x = x^m$ becomes

  $$f_1(E) \cdot (E-1)^r \cdot y_x = x^m$$

  where $f_1(\rho)$ is a polynomial of degree $(n - r)$ in ρ.
  We may put

  $$(E-1)^r y_x = \Delta^r y_x = v_x$$

  and write the equation as

  $$f_1(E) \cdot v_x = x^m$$

  Now

  $$f_1(1+\Delta) = C_0 + C_1\Delta + C_2\Delta^2 + \cdots + C_{n-r}\Delta^{n-r}, \text{ when } C_0 \neq 0$$

  Thus,

  $$V_x = (C_0 + C_1\Delta + \cdots + C_{n-r}\Delta^{n-r})x^m$$

  which is a polynomial of degree m in x.
  Writing this as

  $$V_x = d_0\,{}^xC_m + d_1\,{}^xC_{m-1} + \cdots + d_m\,{}^xC_0$$

  a particular solution of the given equation is obtained from

  $$\Delta^r y_x = d_0\,{}^xC_m + d_1\,{}^xC_{m-1} + \cdots + d_m\,{}^xC_0$$

  Hence the particular solution is

  $$y_x = d_0\,{}^xC_{m+r} + d_1\,{}^xC_{m+r-1} + \cdots + d_m\,{}^xC_r$$

2. If $B(x) = a^x$, we have $f(E)y_x = a^x$.

   $$\therefore\ y_x = \frac{1}{f(E)} a^x = \frac{a^x}{f(a)}$$

   provided $f(a) \neq 0$.

If $f(E) = f_1(E)(E-a)^r$, then $y_x = a^x v_x$.

$\therefore f_1(E)(E-a)^r \cdot a^x v_x = a^x$,

or, $a^x \cdot f_1(aE)(aE-a)^r v_x = a^x$,

or, $a^r \cdot \Delta^r v_x = \dfrac{1}{f_1(aE)} \cdot 1 = \dfrac{1}{f_1(a+a\Delta)} = \dfrac{1}{f_1(a)}$

$\therefore \Delta^r v_x = \dfrac{a^{-r} \cdot r!}{f^{(r)}(a)} = \dfrac{a^{-r} \cdot r!}{f^{(r)}(a)} {}^x C_0$

so that

$$v_x = {}^x C_r \cdot \frac{a^{-r} \cdot r!}{f^{(r)}(a)}$$

Hence

$$y_x = a^x v_x = {}^x C_r \cdot \frac{a^{-r} \cdot a^x \cdot r!}{f^{(r)}(a)}$$

is the particular solution.

3. If $B(x) = a^x V_x$, where V_x is a polynomial in x of degree m.
The equation is

$f(E)y_x = a^x V_x$

We put $y_x = a^x v_x$, so that

$$a^x f(aE)v_x = a^x V_x \Rightarrow f(aE)v_x = V_x. \quad \therefore v_x = \frac{1}{f(aE)} V_x$$

This is the case discussed in 1.

Example 12.7.1. Solve: (a) $y_{x+2} - 4y_x = 5 \cdot 3^x$

(b) $y_{x+2} - 4y_x = 2^x$

(c) $y_{x+1} - 2y_x = x \cdot 2^x$

(d) $y_{x+4} - 5y_{x+3} + 9y_{x+2} - 7y_{x+1} + 2y_x = x^3 + 1$.

Solution: (a) For the complementary function, we consider the homogeneous equation

$y_{x+2} - 4y_x = 0$

Let $y = t^x$ be the trial solution ($t^x \neq 0$).

$t^x(t^2 - 4) = 0 \Rightarrow t^2 - 4 = 0. \therefore t = \pm 2$

$\therefore$ the complementary function is

$y_x = C_1 \cdot 2^x + C_2(-2)^x$

To find the PI, we have

$$y_x = \frac{1}{E^2-4} 5.3^x = 5\left[\frac{1}{(E+2)(E-2)}\right] 3^x$$

$$= 5\left[\frac{1}{(3+2)(3-2)}\right] 3^x = 3^x$$

$\therefore$ the complete solution is

$y_x = C_1 \cdot 2^x + C_2(-2)^x + 3^x$

(b) Here the complementary function is

$y_x = C_1 \cdot 2^x + C_2(-2)^x$

as before.

For the particular Integral, we have

$$y_x = \frac{1}{E^2-4} \cdot 2^x = \frac{1}{(E-2)(E+2)} \cdot 2^x = \frac{1}{2+2} \cdot \frac{1}{E-2} \cdot 2^x$$

$$= \frac{1}{4} \cdot 2^x \frac{1}{2E-2} \cdot 1 = \frac{1}{4} 2^x \cdot \frac{1}{2\Delta} 1 = \frac{1}{8} x \cdot 2^x$$

$\therefore$ the complete solution is

$$y_x = C_1 \cdot 2^x + C_2(-2)^x + \frac{1}{8} x \cdot 2^x$$

(c) For the complementary function, we consider the homogeneous equation

$y_{x+1} - 2y_x = 0$

Let $y_x = t^x$ be the trial solution.

$\therefore t^x(t-2) = 0 \Rightarrow t - 2 = 0. \quad \therefore t = 2$

$\therefore$ complementary function is $y_x = C \cdot 2^x$, where C is an arbitrary constant.

For the particular Integral, we have

$$y_x = \frac{1}{E-2} \cdot x \cdot 2^x = 2^x \cdot \frac{1}{2E-2} \cdot x = 2^{x-1} \frac{1}{\Delta} x = \frac{1}{2} x^2 \cdot 2^{x-1} = x^2 \cdot 2^{x-2}$$

$\therefore$ the complete solution is

$$y_x = C \cdot 2^x + x^2 \cdot 2^{x-2} = 2^x \left(C + \frac{1}{4} x^2\right)$$

(d) For the complementary function, we consider the homogeneous equation

$y_{x+4} - 5y_{x+3} + 9y_{x+2} - 7y_{x+1} + 2y_x = 0$

We take $y_x = t^x$ as the trial solution.

$\therefore t^x(t^4 - 5t^3 + 9t^2 - 7t + 2) = 0 \Rightarrow t^4 - 5t^3 + 9t^2 - 7t + 2 = 0$

or, $t^4 - t^3 - 4t^3 + 4t^2 + 5t^2 - 5t - 2t + 2 = 0$

or, $t^3(t-1) - 4t^2(t-1) + 5t(t-1) - 2(t-1) = 0$

or, $(t-1)(t^3 - 4t^2 + 5t - 2) = 0$

or, $(t-1)^2(t-1)(t-2) = 0. \quad \therefore t = 1, 1, 1, 2$

The complementary function is

$y_x = (C_1 + C_2 x + C_3 x^2)1^x + C_4 \cdot 2^x = C_1 + C_2 x + C_3 x^2 + C_4 2^x$

For the particular Integral

$$y_x = \frac{1}{(E-1)^3(E-2)}(x^3+1) = \frac{1}{\Delta^3} \cdot \frac{1}{E-2}(x^3+1) \equiv \frac{1}{\Delta^3} v_x \text{ (say)}$$

where

$$v_x = \frac{1}{E-2}\left\{x^{(3)} + 3x^{(2)} + x^{(1)} + 1\right\} - (1-\Delta)^{-1}\left\{x^{(3)} + 3x^{(2)} + x^{(1)} + 1\right\}$$
$$= -(1 + \Delta + \Delta^2 + \Delta^3)\left\{x^{(3)} + 3x^{(2)} + x^{(1)} + 1\right\}$$
$$= -(x^3 + 1) - \left\{3x^{(2)} + 6x^{(1)} + 1\right\} - \left\{6x^{(1)} + 6\right\} - 6$$
$$= -x^{(3)} - 6x^{(2)} - 13x^{(1)} - 14$$

Thus,

$$\Delta^3 y_x = -3!\,{}^xC_3 - 2!6\,{}^xC_2 - 1!13\,{}^xC_1 - 14\,{}^xC_0$$
$$= -6\,{}^xC_3 - 12\,{}^xC_2 - 13\,{}^xC_1 - 14$$

The particular solution is

$$-6\,{}^xC_6 - 12\,{}^xC_5 - 13\,{}^xC_4 - 14\,{}^xC_3$$

Hence the general solution reduces to

$$y_x = C_1 + C_2x + C_3x^2 + C_4x^2 - \frac{17}{24}x^3 - \frac{1}{4}x^4 + \frac{1}{40}x^5 - \frac{1}{120}x^6$$

12.8 Non-homogeneous linear equation of the first order

Let us illustrate this method by the following examples.

EXAMPLE 12.8.1. Given equation: $y(x+1) - ay(x) = \cos mx$.

Solution: The given equation can be written as

$$Ey(x) - ay(x) = \cos mx$$

Let $y_2(x)$ be the particular function (PI)

$$\therefore\ y_2(x) = \frac{1}{E-a}\cdot\cos mx = \text{Real Part of}\left[\frac{1}{E-a}\cdot e^{imx}\right]$$
$$= \text{Real Part of}\left[e^{imx}\cdot\frac{1}{e^{im}-a}\cdot 1\right]$$
$$= \text{Real Part of}\left[\frac{e^{imx}(e^{-imx}-a)}{1 - a(e^{im} + e^{-im}) + a^2}\right]$$
$$= \frac{\cos(m-1)x - a\cos mx}{1 - 2a\cos mx + a^2}$$

Noting that $y_1(x)$ is the CF, we may write

$$y(x) = y_1(x) + y_2(x)$$

where $Ey_1(x) - ay_1(x) = 0 \Rightarrow y_1(x) = c(x)a^x$.

Hence the solution is

$$y(x) = c(x)\cdot a^x + \frac{\cos(m-1)x - a\cos mx}{1 - 2a\cos mx + a^2}$$

EXAMPLE 12.8.2. **Homogeneous Linear Equation with Rational Coefficients. Solution of the type $y(x+1) = f(x)y(x)$, where**

$$f(x) = d\frac{(x-a_1)(x-a_2)\cdots(x-a_k)}{(x-b_1)(x-b_2)\cdots(x-b_k)}$$

Solution: We may see that

$$y(x) = f(x-1)y(x-1) = f(x-1)\cdot f(x-2)\cdot y(x-2)$$

$$= \cdots\cdots\cdots\cdots\cdots\cdots\cdots$$

$$= f(x-1)f(x-2)\cdots f(2)\cdot f(1)\cdot y(1)$$

It can be easily shown that

$$y(x) = C(x)d^x \frac{\Gamma(x-a_1)\Gamma(x-a_2)\cdots\Gamma(x-a_k)}{\Gamma(x-b_1)\Gamma(x-b_2)\cdots\Gamma(x-b_k)}$$

EXAMPLE 12.8.3. Solve:

$$(lx+m)y(x+1) - (l'x+m')y(x) = 0$$

$$\Rightarrow y(x+1) = \frac{l'x+m'}{lx+m}y(x) = \frac{l'}{l}\cdot\frac{x+(m'/l')}{x+(m/l)}y(x)$$

Solution: The solution is therefore

$$y(x) = C(x)\left(+\frac{l'}{l}\right)^x \cdot \frac{\Gamma[x+(m'/l')]}{\Gamma[x+(m/l)]}$$

12.9 Equation of the type $y(x+1) - \phi(x)y(x) = f(x)$

Let us write

$y(x) = y_1(x)y_2(x)$

where $y_1(x)$ is the CF and $y_2(x)$ is the PI

The given equation becomes

$y_1(x+1)y_2(x+1) - \phi(x)y_1(x)y_2(x) = f(x)$

The CF of the given equation is given by

$$y(x+1) - \phi(x)y(x) = 0, \text{ or, } \frac{y(x+1)}{y(x)} = \phi(x)$$

$\therefore$ $\log y(x+1) - \log y(x) = \log\phi(x)$, or, $\Delta \log y(x) = \log\phi(x)$,

or, $\log y(x) = \Delta^{-1}\{\log\phi(x)\}$ + constant, or, $y(x) = C_1(x)\cdot e^{\Delta^{-1}\{\log\phi(x)\}}$

Thus, if $y_1(x)$ be the CF, then

$y_1(x) = C_1(x)e\Delta^{-1}\{\log\phi(x)\}$

Again if $y_1(x)$ be the CF, then

$y_1(x+1)y_2(x+1) - y_1(x+1)y_2(x) = f(x)$, or, $y_1(x+1)\Delta y_2(x) = f(x)$,

$$\text{or, } y_2(x) = \Delta^{-1}\left\{\frac{f(x)}{y_1(x+1)}\right\} \text{Constant} = \Delta^{-1}\left\{\frac{f(x)}{y_1(x+1)}\right\} C_2(x)$$

Hence the complete solution is

$$y(x) = C(x)\Delta^{-1}\left\{\frac{f(x)}{y_1(x+1)}\right\} e^{[\Delta^{-1}\{\log\phi(x)\}]}$$

EXAMPLE 12.9.1. Solve: $y(x+1) - e^{2x-1}y(x) = 5xe^{x^2}$.

Solution: The complementary function is given by

$$y(x+1)e^{2x-1}y(x) = 0, \text{ or, } \frac{y(x+1)}{y(x)} = e^{2x-1},$$

or, $\log y(x+1) - \log y(x) = (2x-1)$, or, $\Delta \log y(x) = (2x-1)$,

or, $\log y(x) = \frac{1}{\Delta}(2x-1)$, or, $y_1(x) = C_1(x) \cdot e^{[\Delta^{-1}(2x-1)]} = C_1(x)e^{(x-1)^2}$.

Putting $y(x) = e^{(x-1)^2} \cdot y_2(x)$ in the equation, we have

$$e^{x^2} y_2(x+1) - e^{2x-1} \cdot e^{x^2-2x+1} y_2(x) = 5x \cdot ex^2,$$

or, $y_2(x+1) - y_2(x) = 5x$, or, $\Delta y_2(x) = 5x$,

or, $y_2(x) = \Delta^{-1}(5x)$ constant $= C_2(x)\Delta^{-1}(5x) = C_2(x)\frac{5}{2}x(x-1)$

$\therefore$ the complete solution is

$$y(x) = \left[C(x) \cdot \frac{5}{2}x(x-1)\right] e^{(x-1)^2}$$

EXERCISES

1. Eliminate Δ from the following difference equations:

   (a) $\Delta y_x = 0$; [*Ans:* y_x = constant.]

   (b) $\Delta^2 y_x + 3\Delta y_x - 3y_x = x$; [*Ans:* $y_{x+2} + y_{x+1} - 5y_x = x$.]

   (c) $\Delta^3 y_x + \Delta^2 y_x + \Delta y_x + y_x = 0$; [*Ans:* $y_{x+3} - 2y_{x+2} + 2y_{x+1} = 0$.]

   (d) $(\Delta y_x)^2 + (y_x)^2 = -7$; [*Ans:* $(y_{x+1} - y_x)^2 + (y_x)^2 = -7$.]

   (e) $y_x \Delta^2 y_x = 2$. [*Ans:* $y_x y_{x+2} - 2y_x y_{x+1} + (y_x)^2 = 2$.]

2. (a) Solve: $y_{x+1} - y_x = 0$. [*Ans:* $y_x = k$ (= constant).]

   (b) Solve: $y_{x+1} - y_x = 2$. [*Ans:* $y_x = 2x + k$.]

   (c) Show that $y_x = (1/2)x(x-1) + C$ is the solution of $y_{x+1} - y_x = x$.

   (d) Solve: $y_{x+2} - y_x = 0$. [*Ans:* $y_x = c_1 + c_2(-1)^x$.]

   (e) Solve: $y_{x+2} - 2y_{x+1} - 3y_x = 0$. [*Ans:* $y_x = c_1 \cdot 3^x + c_2(-1)^x$.]

   (f) Solve: $y_{x+2} + 2y_{x+1} - 4y_x = 0$. [*Ans:* $y_x = c_1(\sqrt{5}-1)^x + c_2(-\sqrt{5}-1)^x$.]

   (g) Solve: $y_{x+2} - 7y_{x+1} + 12y_x = 0$. [*Ans:* $y_x = c_1 4^x + c_2 3^x$.]

3. Eliminate the constants from $y_x = A \cdot 2^x + B \cdot 3^x$ and derive the corresponding difference equation. [*Ans:* $y_{x+2} - 5y_{x+1} + 6y_x = 0$.]

4. (a) Show that $y_x = C_1 \cos\left(\frac{2\pi x}{3}\right) + C_2 \sin\left(\frac{2\pi x}{3}\right)$ is a solution of

   $y_{x+2} + y_{x+1} + y_x = 0$

   (b) Show that $y_x = 3^{x/2}\{C_1 \cos\theta x + C_2 \sin\theta x\}$ is a solution of

   $y_{x+2} + 2y_{x+1} - y_x = 0$

   where $\theta = \tan^{-1}(-\sqrt{2})$.

5. Solve the following equations by the method of difference:

(a) $y_{x+1} - y_x = x^2$; [*Ans:* $y_x = \frac{1}{6}(2x^3 - 3x^2 + x) + C$.]

(b) $y_{x+1} - 3y_x = x$; [*Ans:* $y_x = c_1 \cdot 3^x - \frac{1}{2}x - \frac{1}{4}$.]

(c) $3y_{x+1} - y_x = x$; [*Ans:* $y_x = c_1(13)^x + \frac{1}{2}x - \frac{3}{4}$.]

(d) $y_{x+1} - xy_x = 0$; [*Ans:* $y_x = c \cdot \{(x-1)!\}$.]

(e) $y_{x+1} - x^2 y_x = 0$; [*Ans:* $y_x = c\prod_{x=1}^{x-1} x^2 = c\{(x-1)!\}^2$.]

(f) $(x+1)y_{x+1} - xy_x = 0$; [*Ans:* $y_x = c \cdot \frac{1}{x}$.]

(g) $y_{x+2} + a^2 y_x = \cos ax$.

[*Ans:* $y_x = a^x(c_1 \cos \frac{\pi}{2}x + c_2 \sin \frac{\pi}{2}x) + \frac{\cos a(x-2) + a^2 \cos ax}{1 + a^4 + 2a^2 \cos 2a}$.]

6. Solve the following equations by the method of differences:

(a) $y_{x+2} - 5y_{x+1} + 6y_x = 5^x$; [*Ans:* $y_x = c_1 \cdot 2^x + c_2 \cdot 3^x + \frac{1}{6}5^x$.]

(b) $y_{x+2} - 3y_{x+1} - 4y_x = 3^x$; [*Ans:* $y_x = c_1 4^x + c_2(-1)^x - \frac{1}{4} \cdot 3^x$.]

(c) $y_{x+2} - 2y_{x+1} + y_x = x(x-1)(x-2) + x(-1)^x$;

[*Ans:* $y_x = (c_1 + c_2 x)(-1)^x + \frac{1}{20}x^{(5)} + \frac{1}{4}(x-1)(-1)^x$.]

(d) $y_{x+2} - 4y_x = 2^x$; [*Ans:* $y_x = c_1 \cdot 2^x + c_2(-2)^x + \frac{1}{8} \cdot x2^x$.]

(e) $y_{x+1} - 2y_x = x^2 \cdot 2^x$; [*Ans:* $y_x = c \cdot 2^x + \frac{1}{3}x(x-1)(2x-1)2^{x-2}$.]

(f) $y_{x+3} - 5y_{x+2} + 8y_{x+1} - 4y_x = x \cdot 2^x$;

[*Ans:* $y_x = c_1 \cdot 1^x + (c_2 + c_3 x)2^x + \frac{x}{24}(x-1)(x-2) \cdot 2^x$.]

(g) $y_{x+1} - ay_x = (2x+1)a^x$; [*Ans:* $y_x = c \cdot a^x + x^2 a^{x-1}$.]

(h) $y_{x+1} - e^{2x-1} y_x = 8x^3 e^{x^2}$; [*Ans:* $y_x = \{c(x) \cdot 2x^2(x-1)^2\}e^{(x-1)^2}$.]

(i) $y_x y_{x+1} - ay_x + b = 0$;

[*Hints:* Put $y_x = (u_{x+1})/u_x$.]

[*Ans:* $y_x = \frac{u_{x+1}}{u_x}$, where $u_x = c_1(a + \frac{\sqrt{a^2 - 4b}}{2})^x + c_2(a - \frac{\sqrt{a^2 - 4b}}{2})^x$, $a^2 > 4b$.]

(j) $y_{x+1}y_x + (x+2)y_{x+1} + xy_x = -2 - 2x - x^2$. [*Ans:* $y_x = \frac{c_1 + c_2 \cdot 2^{x+1}}{c_1 + c_2 \cdot 2^x} - (x+2)$.]

7. (a) Show that $(x-2)!$ is a PI of the equation

$$y_{x+2} - 3xy_{x+1} + 2x(x-1)y_x = 0$$

(b) Show that $(1/24)2^x \cdot x^{(3)}$ is the PI of the equation

$$y_{x+3} - 5y_{x+2} + 8y_{x+1} - 4y_x = x \cdot 2^x$$

8. Solve:

(a) $y_{x+1} - ay_x = \frac{a^x}{(x+1)^2}$; [*Ans:* $y_x = c_1 \cdot a^x - \frac{a^{x-1}}{x+1}$.]

(b) $y_{x+1} - e^{-2ax} y_x = e^{-ax^2}$, $a \neq 0$; [*Ans:* $y_x = \frac{e^{-a(x-1)^2}}{1-e^{-a}} + c_1(x) \cdot e^{-ax(x-1)}$, $a \neq 0$.]

(c) $y_{x+1} y_x - a^x (y_{x+1} - y_x) + 1 = 0$. [*Ans:* $y_x = \tan[c + \Delta^{-1}\{\tan^{-1}(a^x)\}]$.]

[*Hints:* The given equation is

$\frac{y_{x+1} - y_x}{1 + y_x y_{x+1}} = \frac{1}{a^x}$. $\therefore \tan \Delta u_x = a^x$

where $y_x = \tan u_x$.]

9. Solve:

(a) $(ax+b)y_{x+1} + (cx+d)y_x = 0$. [*Ans:* $y_x = c(x) \cdot (-\frac{c}{a})^x \cdot \frac{\Gamma(x+\frac{d}{c})}{\Gamma(x+\frac{b}{a})}$.]

(b) $y_x y_{x+2} = y_{x+1}^2$. [*Ans:* $y_x = c \cdot e^{kx}$.]

10. Find the general solution of the equation $y_{x+1} - 2y_x + y_{x-1} = f(x)$ in each of the following cases:

(a) $f(x) = a^x$, $a \neq 1$; [*Ans:* $y_x = c_1 + c_2 x + \frac{a^{x+1}}{(a-1)^2}$, $a \neq 1$.]

(b) $f(x) = e^{ax}$, $a \neq 0$; [*Ans:* $y_x = c_1 + c_2 x + \frac{e^{a(x+1)}}{(e^a-1)^2}$, $a \neq 0$.]

(c) $f(x) = \sin ax$; [*Ans:* $y_x = c_1 + c_2 x - \frac{\sin ax}{2(1-\cos a)}$, $a \neq 2n\pi$.]

(d) $f(x) = xe^{ax}$, $a \neq 0$. [*Ans:* $y_x = c_1 + c_2 x + \frac{e^{a(x+1)}}{(e^a-1)^3}\{(e^a-1)x - (e^a+1)\}$, $a \neq 0$.]

11. Solve: $y_{x+4} \cdot y_{x+2}^6 \cdot y_x = y_{x+1}^4 \cdot y_{x+3}^4$. [*Ans:* $y_x = A_0 \cdot e^{A_1x^3 + A_2x^2 + A_3x}$.]

[*Hints:* $\log y_{x+4} + 6\log y_{x+2} + \log y_x = 4\log y_{x+3} + 4\log y_{x+1}$.

Putting $u_x = \log y_x$, $u_{x+4} - 4u_{x+3} + 6u_{x+2} - 4u_{x+1} + u_x = 0$. Solve for u_x and note that $y_x = e^{u_x}$.]

Chapter 13

The Solution of Numerical Algebraic and Transcendental Equations

Summary

I. Equations in one unknown

13.1 Introduction

The method of solving literal equations of all degrees including the fourth has been discussed in most of the textbooks on algebra. But there is no discussion in these books about the solution of equations of the types:

$lx + m\log x = n, \quad le^{-x} + m\tan x = n$, etc.

These are transcendental equations and no general method exists for finding their solutions. It is, however, possible to find the numerical solution of these equations and no general method exists for finding their solutions. It is however possible to find the numerical solution of these equations up to any desired degree of accuracy when the coefficients of such equations are pure numbers.

The aim of the present chapter is to set fourth the most useful methods for finding the roots of any equation having numerical coefficients. Since Horner's method cannot be applied to transcendental equations, we shall not discuss this method in the present chapter. The method, however, is explained in most of the books of college algebra.

13.2 Finding approximate values of the roots

In order to find the real roots of a numerical equation by any method excepting that of Graeffe's, it is necessary to find an approximate value of the root from a *graph* or otherwise.

Let

$$f(x) = 0 \tag{13.2.1}$$

denote the equation whose roots are to be found.

If we take a pair of rectangular coordinate axes and plot the graph of

$$y = f(x) \tag{13.2.2}$$

it is evident that the abscissas of the points where the graph crosses the x-axis are the real roots of the given equation, for at these points $y = 0$ and therefore equation (13.2.1) is satisfied. Approximate real roots of the equation can therefore be found from the graph. It is, however, not necessary to draw the entire graph, only the portion of the curve near the neighbourhood of the abscissa-points where the function $f(x)$ changes sign from positive to negative or vice-versa, will suffice.

In most cases, the approximate values of the real roots of $f(x) = 0$ are obtained by rewriting the equation $f(x) = 0$ in the form:

$$f_1(x) = f_2(x) \tag{13.2.3}$$

and then drawing two graphs

$y_1 = f_1(x)$ and $y_2 = f_2(x)$

on the same axes.

The abscissa of the points of intersection of these two curves are the real roots of the equation (13.2.1), for at these points $y_1 = y_2$, or, $f_1(x) = f_2(x)$. Consequently $f(x) = 0$ is satisfied.

More useful and important than a graph is the following fundamental theorem in algebra:

If $f(x)$ and $f(b)$ have opposite signs, then there is at least one real root between a and b. The truth of the above theorem is evident from Fig. 13.1, for if $f(a)$ and $f(b)$ have opposite signs, the graph must cross the x-axis at least once between $x = a$ and $x = b$.

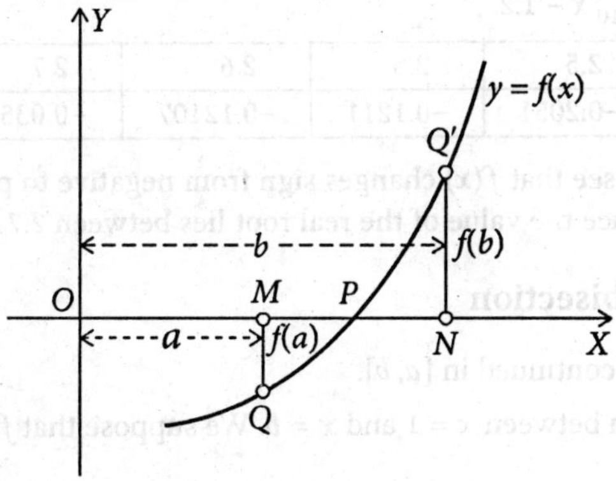

Fig. 13.1

We shall now apply the foregoing methods to the following example:

Example 13.2.1. Find approximate values of the real roots of $x \log_{10} x = 1.2$.

Solution: *Method* 1: First we draw the graphs of the following two functions.

1. $y_1 = 1.2/x \equiv f_1(x)$;

2. $y_2 = \log_{10} x \equiv f_2(x)$.

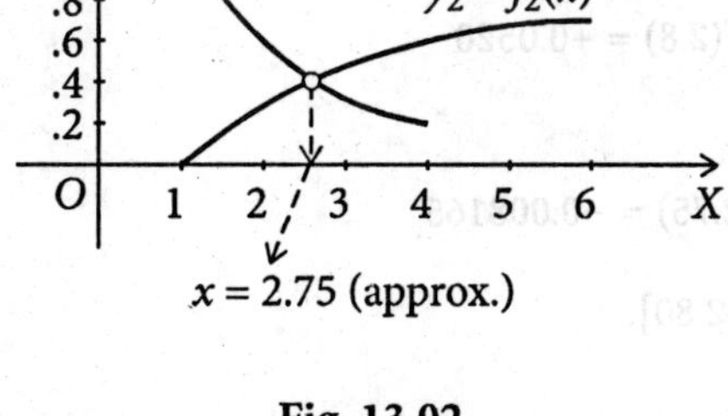

Fig. 13.02

The abscissa of the point of intersection of these two curves will be the real root of the given equation.

We assign positive integral values of x as under:

x	1	2	3	4	6
y_1	1.2	0.6	0.4	0.3	0.2
y_2	0	0.3010	0.4771	0.6021	0.7781

From the graph, we may find that the approximate real root is 2.75.

Method 2: $y = x \log_{10} x - 1.2$.

x	2.0	2.5	2.5	2.6	2.7	2.8
y	−0.5979	−0.2051	−0.1211	−0.12107	−0.0353	+0.0520

From the above we see that $f(x)$ changes sign from negative to positive in the interval $x = 2.7$ to $x = 2.8$. Hence the value of the real root lies between 2.7 and 2.8.

13.3 Method of bisection

We assume $f(x)$ to be continued in $[a, b]$.

Let $f(x)$ change sign between $x = 1$ and $x = b$. We suppose that $f(a) \cdot f(b) < 0$.

We define

$$\xi = \frac{a+b}{2}$$

If (i) $f(a) \cdot f(\xi) < 0$, the real root lies in $[a, \xi]$.

However, if (*ii*) $f(a) \cdot f(\xi) < 0$, the real root lies in $[\xi, b]$.

If it lies in $[a, \xi]$, we define

$$\xi' = \frac{a+\xi}{2}$$

and examine whether $f(a)f(\xi') < 0$, or, > 0.

If the root lies α in $[\xi, b]$, we define

$$\xi' = \frac{\xi+b}{2}$$

and examine whether $f(\xi')f(l) < 0$, or > 0. $[a, \xi]$, or $[\xi, b]$, then to $[a, \xi']$, or, $[\xi', b]$.

The process is repeated until the root is obtained to the desired accuracy.

To illustrate this point, we may consider the example in 13.2.1.

We have seen that

$$f(2.7) = -0.0353 \quad \text{and} \quad f(2.8) = +0.0520$$

We next take

$$\xi = \frac{2.7+2.8}{2} = 2.75, \quad f(2.75) = -0.008165$$

$\therefore$ the root lies in [2.75, 2.80].

Next we take

$$\xi = \frac{2.75+2.80}{2} = 2.775, \quad f(\xi') = 0.03005$$

$\therefore$ the root lies in [2.75, 2.775].

Next we take

$$\xi'' = \frac{2.75, 2.775}{2} = 2.7625, \quad f(\xi'') = 0.0191$$

Then root lies in [2.75, 2.7625].

We take next

$$\xi''' = \frac{2.75 + 2.7625}{2} = 2.75625, \quad f(\xi''') = 0.01363$$

∴ the root lies in [2.75, 2.75625].

Next take

$$\xi^{iv} = \frac{2.75 + 2.75625}{2} = 2.753125, \quad f(\xi^{iv}) = 0.0110$$

∴ the root lies in [2.75, 2.7531].

∴ the real root is approximately 2.75 (*correct to two decimal places*).

The method of bisection is also called the bracketing method, since the method reduces the two ends (*brackets*) of an interval surrounding the real root at each stage of operation.

13.4 Method of false position (Regula Falsi)

This is perhaps the oldest method for computing the real roots of a numerical equation $f(x) = 0$. In this method, we find two numbers x_0 and x_1 between which the root lies. x_0 and x_1 should be as close together as possible. Evidently $f(x_0)$ and $f(x_1)$ will be of opposite signs and therefore the graph of $y = f(x)$ must cross the x-axis at least once between $x = x_0$ and $x = x_1$.

If x_0 and x_1 be sufficiently close, we may assume that the portion of the curve $y = f(x)$ between $x = x_0$ and $x = x_1$ is practically a straight line. The method of false position is based on the validity of this assumption.

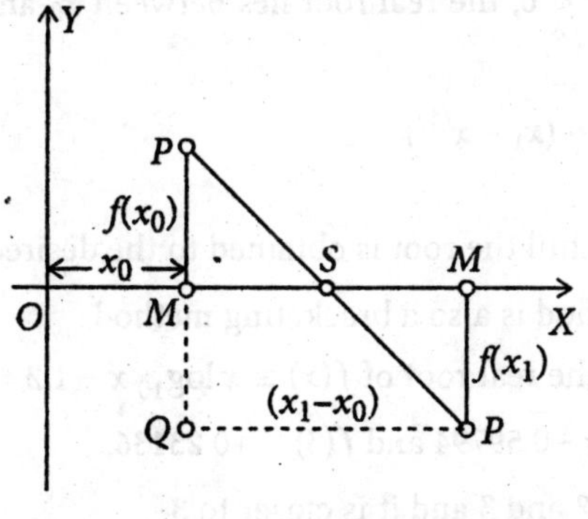

Fig. 13.3 The portion of the Curve between *M* and *M*′ is assumed to the linear

To derive a formula for computing the root, let Fig. 13.3 represent a magnified view of that portion of the graph between $\{x_0, f(x_0)\}$ and $\{x_1, f(x_1)\}$.

From Fig. 13.4, we have

$$\overline{OM} = x_0,$$
$$\overline{OM'} = x_1,$$
$$|\overline{PM}| = |f(x_0)|,$$
$$|\overline{P'M'}| = |f(x_1)|$$
$$\overline{QP'} = \overline{MM'} = x - x_0$$
$$\overline{MS} = \overline{OS} - \overline{OM} = x - x_0$$

$\Delta^S PMS$ and PQP' are similar. Hence

$$\frac{\overline{MS}}{\overline{MP}} = \frac{\overline{QP'}}{\overline{PQ}}, \text{ or, } \frac{x - x_0}{|f(x_0)|} = \frac{x_1 - x_0}{|f(x_0)| + |f(x_1)|}$$

$$\therefore\ x = x_0 + \frac{|f(x_0)|}{|f(x_0)| + |f(x_1)|} \cdot (x_1 - x_0) \tag{13.4.1}$$

This value of x is not, however, the true value of the root, because the graph $y = f(x)$ is not a perfectly straight line between the points P and P'. It is merely a closer approximation to the true root.

Denoting this solution of x, by $x^{(1)}$, the first approximation of the true root, we next examine the signs of $f(x_0)$, $f(x^{(1)})$ and $f(x_1)$.

Let $f(x_0) > 0$ and $f(x_1) < 0$.

Case 1: If $f(x^{(1)}) > 0$, then the real root lies between $x^{(1)}$ and x_1, and in that case, the second approximation of x will be

$$x^{(2)} = x^{(1)} + \frac{|f(x^{(1)})|}{|f(x^{(1)})| + |f(x_1)|} \cdot (x_1 - x^{(1)}) \tag{13.4.2}$$

Case 2: If, however, $f(x^{(1)}) < 0$, the real root lies between x_0 and $x^{(1)}$ and $x^{(2)}$ will be in that case

$$x^{(2)} = x_0 + \frac{|f(x_0)|}{|f(x_0)| + |f(x^{(1)})|} \cdot (x_1 - x^{(1)}) \tag{13.4.3}$$

The process is repeated until the root is obtained to the desired degree of accuracy.

Note: The Regula Falsi method is also a bracketing method.

Example 13.4.1. Compute the real root of $f(x) \equiv x \log_{10} x - 1.2 = 0$.

Solution: We see that $f(2) = -0.59794$ and $f(3) = +0.23136$.

$\therefore$ real root lies between 2 and 3 and it is closer to 3.

$$\therefore\ x^{(1)} = 2 + \frac{0.59794}{0.82930} \cdot (3 - 2)$$
$$= 2 + 0.72 = 2.72$$

Again $f(2.72) = -0.01797$.

$\therefore$ real root lies between 2.72 and 3

$$\therefore\ x^{(2)} = 2.72 + \frac{0.01797}{0.24933} \times (3 - 2.72)$$

$$= 2.72 + 0.02 = 2.74$$

Again $f(2.74) = -0.0005634$.

$\therefore$ the root lies between 2.74 and 3.

$$\therefore\ x^{(3)} = 2.74 + \frac{0.0005634}{0.2319234} \times 0.26$$

$$= 2.7406$$

Again $f(2.7406) = -0.0000264765$.

$\therefore$ real root lies between 2.7406 and 3.

$$\therefore\ x^{(4)} = 2.7406 + \frac{0.0000264765}{0.2313335}(3 - 2.7406)$$

$$= 2.7406 + 0.00002969$$

$$= 2.74063$$

Again $f(2.74063) = -0.000014038$.

$\therefore$ real root lies between 2.74063 and 3.

$$\therefore\ x^{(5)} = 2.74063 + \frac{0.000014038}{2.2313460} \times 0.25937$$

$$= 2.74063 + 0.0000157$$

$$= 2.740646$$

13.5 The Newton-Raphson method

When the derivative of $f(x)$ is a simple expression and easily found, the real root (*non-repeated*) of $f(x) = 0$ can be computed rapidly by a process known as the Newton-Raphson method.

Let x_0 be the approximate real root of $f(x) = 0$, found graphically or otherwise. Let h be the correction to be applied to x_0 to give the exact value of the root, so that $x = x_0 + h$.

Hence $f(x_0 + h) = 0$.

Using Taylor's theorem, we have

$$f(x_0 + h) = f(x_0) + hf'(x_0) + \frac{h^2}{2}f''(x_0 + \theta h), \quad 0 < 0 < 1$$

$$\therefore f(x_0) + hf'(x_0) + \frac{h^2}{2}f''(x_0 + \theta h) = 0$$

Now, if h is relatively small, as it should be, we may ignore the term containing h^2 and get the simple relation

$$f(x_0) + h_1 f'(x_0) \Rightarrow h_1 = -\frac{f(x_0)}{f'(x_0)} \tag{13.5.1}$$

The improved root will then be

$$x^{(1)} = x_0 - \frac{f(x_0)}{f'(x_0)} \tag{13.5.2}$$

The succeeding approximations are

$$x^{(2)} = x^{(1)} - \frac{f(x^{(1)})}{f'(x^{(1)})}$$

$$x^{(3)} = x^{(2)} - \frac{f(x^{(2)})}{f'(x^{(2)})}$$

$$x^{(n)} = x^{(n-1)} - \frac{f(x^{(n-1)})}{f'(x^{(n-1)})} \tag{13.5.3}$$

Note 1: It should be noted that the method will fail, if $f'(x) = 0$, or very nearly so (small), in the neighbourhood of the real root.

Note 2: The initial approximate root x_0, should be taken sufficiently close to the unknown root, otherwise the method may diverge.

Note 3: It is clear from the formula (13.5.3) that the larger the derivative $f'(x)$, the smaller is the correction necessary to be applied to the approximate root. This implies that the graph is nearly vertical where it crosses the x-axis and in that case, the correct value of the root can be founa with sufficient rapidity. If, on the other hand, $f'(x)$ is *numerically* small in the neighbourhood of the root, the amount of correction to be applied would be large and the computation of the root by this method will be a slow process and may even fail altogether.

Note 4: The Newton-Raphson method should never be used when the graph of $y = f(x)$ is nearly horizontal where it crosses the x-axis.

Note 5: If a root x_0 of $f(x) = 0$ is a multiple root, then the Newton-Raphson's method fails, since then both $f(x_0)$ and $f'(x_0)$ are zero. In such cases, a modification of the method is possible on the line suggested below.

Let x_0 be a multiple root of $f(x) = 0$ of multiplicity n. Then in the immediate neighbourhood of $x = x_0$, we may write $f(x) \simeq A(x - x_0)^n$, where $a = f^n(x_0)/n!$ ($\neq 0$) is effectively constant.

Then

$$f'(x) = A.n(x - x_0)^{n-1} \quad \text{and} \quad \frac{f(x)}{f'(x)} = \frac{x - x_0}{n}$$

where x_0 is a simple root of $f(x)/f'(x)$.

Thus,

$$x_0 \simeq x - n\frac{f(x)}{f'(x)}$$

in the neighbourhood of the multiple root x_0.

This is the modification of the Newton's method in such cases.

Rule: If x_1 is located in the neighbourhood of a real root α of multiplicity n, then

$$\hat{x}_2 = x_1 - n \cdot \frac{f(x_1)}{f'(x_1)}$$

is a better approximation of α than x_1.

Example 13.5.1. Find by Newton-Raphson's method the real root of

$$2x - 3\sin x - 5 = 0$$

Solution:

$$f(x) \equiv 2x - 3\sin x - 5 = 0$$
$$f'(x) = 2 - 3\cos x$$
$$f(2) = -3.72789228 < 0$$
$$f(3) = +0.57664$$

$\therefore$ real root lies between 2 and 3.

$$x^{(1)} = 2 - \frac{f(2)}{f'(2)} = 2 + \frac{3.72789228}{3.248440509} = 3.1476$$
$$x^{(2)} = 3.1476 - \frac{1.2952}{4.999946} = 3.1476 - 0.259043 = 2.8886$$
$$x^{(3)} = 2.8886 - \frac{0.0261677}{4.904503} = 2.8886 - 0.005335 = 2.883265$$
$$x^{(4)} = 2.883265 - \frac{0.000138}{4.9004557} = 2.883265 - 0.00002816 = 2.883237$$
$$x^{(5)} = 2.883237 - \frac{0.000000624}{4.9004342} = 2.883237 - 0.00000013 \simeq 2.883237$$

(correct to six decimal places).

13.6 Geometric significance of Newton-Raphson method

Let us take a magnified view of the graph $y = f(x)$ where it crosses the x-axis. Let PSP' be the portion of the curve where it crosses the x-axis at S. $|\overline{OS}|$ is the real root of $f(x) = 0$.

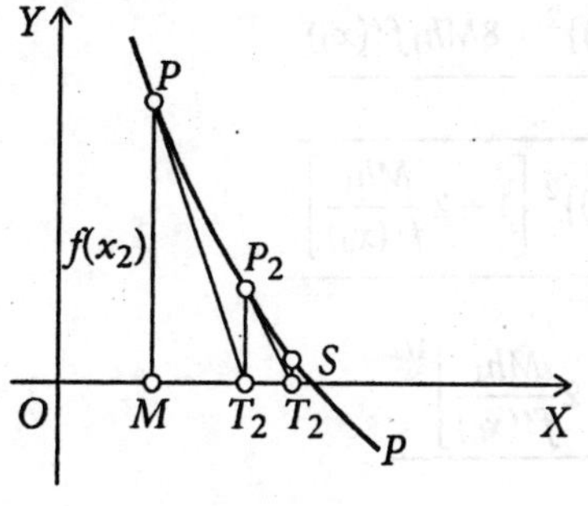

Fig. 13.4

Let $\overline{OM} = x_0$ be the approximate root. $\overline{PM} = fx_0$ is the ordinate at M. At P we draw a tangent PT_1 cutting the x-axis at T_1. Then $\overline{OT_1}$ is the first approximation of the root) $= x^{(1)}$). At T_1, let $\overline{P_2T_1}(= f(x^{(1)})$ be the ordinate and at P_2 we draw another tangent $P_2\,T_2$ cutting the x-axis at T_2. Then $\overline{OT_2}$ is the second approximation of the root.

It is to be noted that T_1 is a point lying between M and S, T_2 lying between T_1 and S, T_3 lying between T_2 and S and so on. It is evident intuitionally that if the curvature of the graph does not change sign between P and S, the points $T_1, T_2, \ldots$ will approach the point S as a limit. The lengths $\overline{OT_1}$, $\overline{OT_2}$, etc. are the successive approximations to the desired root. This is the geometric significance of the Newton-Raphson method.

[*Note:* $\overline{OT_1} = \overline{OM} + \overline{MT_1} = x_0 + \overline{PM}\cot(180^0 - \angle PT_1T_2) = x_0 - f(x_0)/f'(x_0)$.]

13.7 The inherent error in Newton-Raphson method

If x_0 be the approximate value of a root of $f(x) = 0$ and h is the correction to be applied, so that $f(x_0 + h) = 0$, then we have

$$f(x_0 + h) = f(x_0) + hf'(x_0) + \frac{h^2}{2}f''(x_0 + \theta h) = 0, \quad 0 < \theta < 1 \tag{13.7.1}$$

In this method, we neglected term containing h^2 and obtained

$$f(x_0) + h_1 f'(x_0) = 0 \tag{13.7.2}$$

Subtracting (2) from (1), we get

$$(h - h_1)f'(x_0) + \frac{h^2}{2}f''(x_0 + \theta h) = 0$$

$$\therefore\ h - h_1 = -\frac{h^2}{2}\cdot\frac{f''(x_0 + \theta h)}{f'(x_0)} \tag{13.7.3}$$

Since h is the true correction (actual) and h_1 is its approximate value, $h - h_1$ is the error in h_1. Let M denote the maximum value of $f''(x)$ in the neighbourhood of $x_0 + h$. Then

$$h - h_1 = -\frac{h^2 M}{2f'(x_0)} \tag{13.7.4}$$

We now propose to express the above in terms of the known quantity h_1.

From (4), we obtain

$$h^2 M + 2f'(x_0)\cdot h - 2f'(x_0)\cdot h_1 = 0$$

which is a quadratic equation in h, $(h > 0)$.

$$h = \frac{-2f'(x_0) + \sqrt{4\{f'(x_0)\}^2 + 8Mh_1 f'(x_0)}}{2M}$$

$$= \frac{-2f'(x_0) + \sqrt{4\{f'(x_0)\}^2\left[1 + 2\frac{Mh_1}{f'(x_0)}\right]}}{2M}$$

$$= \frac{-f'(x_0) + f'(x_0)\left[1 + 2\frac{Mh_1}{f'(x_0)}\right]^{1/2}}{M}$$

$$= \frac{-f'(x_0) + f'(x_0)\left[1 + \frac{Mh_1}{f'(x_0)} - \frac{1}{2}\frac{M^2h_1^2}{\{f'(x_0)\}^2} + \cdots\right]}{M}$$

(using Binomial expansion)

$$= \frac{1}{M}\left[-f'(x_0) + f'(x_0) + Mh_1 - \frac{1}{2}\frac{M^2h_1^2}{f'(x_0)}\right]$$

(neglecting higher powers of h_1)

$$= h_1 - \frac{1}{2}\frac{Mh_1^2}{f'(x_0)}$$

$$\therefore h - h_1 = -\frac{Mh_1^2}{2f'(x_0)} \tag{13.7.5}$$

($\because h_1$ is always small, the principal part of the error is contained in the first term only).

Thus, the formula for error reduces to

$$E_1 \le \left|\frac{Mh_1^2}{2f'(x_0)}\right| \tag{13.7.6}$$

This is the error in $x^{(1)}$. The error in $x^{(n)}$ is therefore

$$E_n \le \left|\frac{Mh_n^2}{2f'(x^{(n)})}\right| \tag{13.7.7}$$

Note: In most equations, which we propose to solve by the Newton-Raphson method, the quantity $M/[2f'(x_0)]$ is not greater than 1. We may therefore suppose that $|M/[2f'(x_0)]| \le 1$.

Hence (13.7.7) reduces to

$$|E_n| \le h_n^2 \tag{13.7.8}$$

This is an important result. If h_n, which is inevitably a small fraction, begins with k zeros, the error begins with $2k$ zeros.

13.8 The method of iteration

When a numerical equation $f(x) = 0$ can be expressed in the form:

$$x = \phi(x) \tag{13.8.1}$$

such that $|\phi'(x)| < 1$, in the immediate neighbourhood of the given approximate root, the real root of the equation $f(x) = 0$ can be found by the process of *iteration*. The approximate root x_0 (say) may be found graphically or otherwise. We next substitute this x_0 in the right-hand member of (13.7.8) and obtain a better approximation $x^{(1)}$ by the equation

$$x^{(1)} = \phi(x_0) \tag{13.8.2}$$

Then the succeeding approximations are

$$x^{(2)} = \phi(x^{(1)})$$

$$x^{(3)} = \phi(x^{(3)})$$

$$\dots\dots\dots\dots$$

$$x^{(n)} = \phi(x^{(n-1)})$$

EXAMPLE 13.8.1. Find by iteration method a real root of $f(x) \equiv 2x - \log_{10} x - 7 = 0$.

Solution: We draw the graphs of

$$y_1 = 2x - 7 \equiv f_1(x)$$

and $y_2 = \log_{10} x \equiv f_2(x)$

on the same graph with a common set of coordinate axes. It may be found that the abscissa of the point of intersection is a value between 3.5 and 4.

Also we see

$f(3.5) = -\log 3.5 < 0$

$f(4) = 1 - 2\log 2 = +0.39794 > 0$

We may take $x_0 = 3.5$ as the approximate root and rewrite the equation as

$$x = \frac{1}{2}(7 + \log_{10} x) \equiv \phi(x)$$

[Note that $\phi'(3.5) = \{1/2 \cdot 1/x \log_{10} e\}_{3.5} < 1.$]

$$x^{(1)} = \frac{1}{2}(7 + \log_{10} 3.5) = 3.77$$

$$x^{(2)} = \frac{1}{2}(7 + \log_{10} 3.77) = 3.788$$

$$x^{(3)} = \frac{1}{2}(7 + \log_{10} 3.788) = 3.7892$$

$$x^{(4)} = \frac{1}{2}(7 + \log_{10} 3.7892) = 3.78927 \simeq 3.7893$$

(correct to 4 decimal places).

EXAMPLE 13.8.2. Use of iterative process to find $\sqrt{N}$ for $N = 10$.

Step 1: We shall obtain the following iterative procedure

$$x^{(n)} = \frac{1}{2}\left[x^{(n-1)} + \frac{N}{x^{(n-1)}}\right]$$

for evaluating $\sqrt{N}$ as the solution of the equation $x^2 - N = 0$.

Solution: Here $f(x) \equiv x^2 - N, f'(x) = 2x$.

$\therefore$ Newton-Raphson's method gives

$$x^{(n)} = x^{(n-1)} - \frac{f(x^{(n-1)})}{f'(x^{(n-1)})} = x^{(n-1)} - \frac{\{x^{(n-1)^2}\}^2 - N}{2x^{(n-1)}}$$

$$= x^{(n-1)} - \frac{x^{(n-1)}}{2} + \frac{N}{2x^{(n-1)}}$$

$$= \frac{1}{2}\left[x^{(n-1)} + \frac{N}{x^{(n-1)}}\right], \quad N = 1, 2, \ldots$$

For $N = 10$, we may take $x_0 = 3$.

$$\therefore x^{(1)} = \frac{1}{2}\left(3 + \frac{10}{3}\right) = \frac{1}{2}(3 + 3.333) = 3.1667$$

$$x^{(2)} = \frac{1}{2}\left[3.1667 + \frac{10}{3.1667}\right] = 3.1623$$

$$x^{(3)} = \frac{1}{2}\left[3.1623 + \frac{10}{3.1623}\right] = 3.162278 = 3.1623$$

(correct to 4 decimal places).

EXAMPLE 13.8.3. Show that $N^{1/k}$ can be evaluated by using the iterative procedure

$$x^{(n)} = \frac{1}{k}\left[(k-1)x^{(n-1)} + \frac{N}{\{x^{(n-1)}\}^{k-1}}\right]$$

Solution: Here $\sqrt[k]{N}$ is a positive real root of $x^k - N = 0$.

Newton-Raphson method gives

$$x^{(n)} = x^{(n-1)} - \frac{f(x^{(n-1)})}{f'(x^{(n-1)})} = x^{(n-1)} - \frac{\{x^{(n-1)}\}^k - N}{k\{x^{(n-1)}\}^{k-1}}$$

$$= x^{(n-1)} - \frac{x^{(n-1)}}{k} + \frac{N}{k\{x^{(n-1)}\}^{k-1}}$$

$$= \frac{1}{k}\left\{(k-1)x^{(n-1)} + \frac{N}{\{x^{(n-1)}\}^{k-1}}\right\}$$

EXAMPLE 13.8.4. Compute $2^{1/5}$ by using the result in (Example 3).

Solution: Here $f(x) = x^5 - 2 = 0, f'(x) = 5x^4$.

We take $x_0 = 1.14$.

$$\therefore x^{(1)} = \frac{1}{5}\left[4x_0 + \frac{2}{x_0^4}\right] = \frac{1}{5}\left[4.56 + \frac{2}{1.925415}\right]$$

$$= 1.1197$$

$$x^{(2)} = \frac{1}{5}\left[4 \times 1.1197 + \frac{2}{1.759983}\right] = \frac{1}{5}[4.4788 + 1.136375]$$

$$= 1.1230$$

$$x^{(3)} = \frac{1}{5}\left[4 \times 1.123 + \frac{2}{1.78607}\right] = \frac{1}{5}[4.4920 + 1.119777]$$

$$= 1.1224$$

13.9 Convergence of the iteration process

We shall now obtain the condition under which the iteration process converges. The true value of the root satisfies the equation

$$x = \phi(x) \tag{13.9.1}$$

and the first approximation satisfies

$$x^{(1)} = \phi(x_0) \tag{13.9.2}$$

Subtracting (13.9.2) from (13.9.1)

$$(x - x^{(1)}) = \phi(x) - \phi(x_0)$$

$$= (x - x_0)\phi'(\xi_0), \quad x_0 \le \xi_0 \le x$$

(By the mean value theorem).

A similar equation holds for all succeeding approximations, so that we may get

$$x - x^{(2)} = (x - x^{(1)})\phi'(\xi_1)$$

$$x - x^{(3)} = (x - x^{(2)})\phi'(\xi_2)$$

....

$$x - x^{(n)} = (x - x^{(n-1)})\phi'(\xi_{n-1})$$

Multiplying these results and after some manipulations, we get

$$x - x^{(n)} = (x - x_0)\phi'(\xi_0)\phi'(\xi_1)\cdots\phi'(\xi_{n-1})$$

Now, *if the maximum value of $\phi'(x)$ is less than* 1 throughout the interval (x_0, x), so that each of the quantities

$$\phi'(\xi_0), \phi'(\xi_1), \ldots, \phi'(\xi_{n-1})$$

is less than a proper fraction m, then

$$|x - x^{(n)}| \leq |x - x_0| \cdot m^n \tag{13.9.3}$$

Since the right-hand member of the inequality in (13.9.3) approaches to zero as $n \to \alpha$, we can make the error $x - x^{(n)}$ as small as we please by repeating the iteration process a sufficient number of times.

13.10 Convergence of the Newton-Raphson method

The nth approximation $x^{(n)}$ is given by

$$x^{(n)} = x^{(n-1)} - \frac{f(x^{(n-1)})}{f'(x^{(n-1)})} \equiv \phi(x^{(n-1)})$$

which is a iterative relation.

The condition of convergence in the iteration process is $|\phi'(x)| < 1$, in the immediate neighbourhood of the root.

We now have

$$x^{(n+1)} \equiv \phi(x^{(n)}) = x^{(n)} - \frac{f(x^{(n)})}{f'(x^{(n)})}$$

$$\therefore \phi'(x^{(n)}) = \frac{d}{dx}\left[x - \frac{f(x)}{f'(x)}\right] \text{ at } x = x^{(n)}$$

$$= \left[1 - \frac{\{f'(x)\}^2 - f(x)f''(x)}{\{f'(x)\}^2}\right] \text{ at } x = x^{(n)}$$

$$= \left[\frac{f(x)f''(x)}{\{f'(x)\}^2}\right]_{x=x^{(n)}}$$

The sufficient condition for convergence is therefore

$$\left|\frac{f(x^{(n)})f''(x^{(n)})}{\{f'^{(n)}\}^2}\right| < 1, \quad \text{or,} \quad \left|f(x^{(n)})f''(x^{(n)})\right| < \left\{f'(x^{(n)})\right\}^2 \tag{13.10.1}$$

II. Simultaneous equations in several unknowns

13.11 Introduction

The real roots of simultaneous algebraic and transcendental equations in several unknowns can be found either by the Newton-Raphson method or by the method of iteration. We shall give an outline of both these methods for the cases of two unknowns and three unknowns. There is however no difficulty in extending both methods to the case of any number of unknowns.

13.12 The Newton-Raphson method for simultaneous equations

13.12.1 Case of two equations in two unknowns

Let the given equations be

$$f(x, y) = 0 \tag{13.12.1}$$

$$\phi(x, y) = 0 \tag{13.12.2}$$

We suppose that x_0, y_0 be approximate values of a pair of roots (*obtained graphically* or *otherwise*) and let h, k be the corrections, so that

$$x = x_0 + h$$

$$y = y_0 + k$$

Then (13.12.1) and (13.12.2) give

$$f(x_0 + h, y_0 + k) = 0 \tag{13.12.3}$$

$$\phi(x_0 + h, y_0 + k) = 0 \tag{13.12.4}$$

Expanding (13.12.3) and (13.12.4) by Taylor's theorem for a function of two variables, we have

$$f(x_0 + h, y_0 + k) = f(x_0, y_0) + h\left(\frac{\partial f}{\partial x}\right)_0 + k\left(\frac{\partial f}{\partial y}\right)_0 + \text{terms of higher powers of } h \text{ and } k = 0 \tag{13.12.5}$$

$$f(x_0 + h, y_0 + k) = \phi(x_0, y_0) + h\left(\frac{\partial \phi}{\partial x}\right)_0 + k\left(\frac{\partial \phi}{\partial y}\right)_0 + \text{terms of higher powers of } h \text{ and } k = 0, \tag{13.12.6}$$

where the notations

$\left(\frac{\partial \phi}{\partial x}\right)_0$ and $\left(\frac{\partial \phi}{\partial y}\right)_0$

mean the values of

$\left(\frac{\partial \phi}{\partial x}\right)$ and $\left(\frac{\partial \phi}{\partial y}\right)$

respectively at $x = x_0$, $y = y_0$. Similar are the meanings of

$\left(\frac{\partial f}{\partial x}\right)_0$ and $\left(\frac{\partial f}{\partial y}\right)_0$

Now since h and k are relatively small, their squares and higher powers and products may be neglected. Then (13.12.6) and (13.12.7) become simply

$$f(x_0, y_0) + h\left(\frac{\partial f}{\partial x}\right)_0 + k\left(\frac{\partial f}{\partial y}\right)_0 = 0 \tag{13.12.7}$$

$$\phi(x_0, y_0) + h\left(\frac{\partial \phi}{\partial x}\right)_0 + k\left(\frac{\partial \phi}{\partial y}\right)_0 = 0 \tag{13.12.8}$$

Solving by Cramer's Rule, we find the first corrections to be

$$h_1 = \frac{\begin{vmatrix} -f(x_0, y_0) & \left(\frac{\partial f}{\partial y}\right)_0 \\ -\phi(x_0, y_0) & \left(\frac{\partial \phi}{\partial y}\right)_0 \end{vmatrix}}{D}, \tag{13.12.9}$$

$$k_1 = \frac{\begin{vmatrix} \left(\frac{\partial f}{\partial x}\right)_0 & -f(x_0, y_0) \\ \left(\frac{\partial \phi}{\partial x}\right)_0 & -\phi(x_0, y_0) \end{vmatrix}}{D} \tag{13.12.10}$$

where

$$D = \begin{vmatrix} \left(\frac{\partial f}{\partial x}\right)_0 & \left(\frac{\partial f}{\partial y}\right)_0 \\ \left(\frac{\partial \phi}{\partial x}\right)_0 & \left(\frac{\partial \phi}{\partial y}\right)_0 \end{vmatrix} \tag{13.12.11}$$

Then

$$x^{(1)} = x_0 + h_1$$

and $y^{(1)} = y_0 + k_1$

The process may be iterated further by taking

$$x^{(2)} = x^{(1)} + h_2$$

$$y^{(2)} = y^{(1)} + k_2$$

and considering

$$f\left(x^{(1)} + h_2,\ y^{(1)} + k_2\right) = 0 \tag{13.12.12}$$

and $\phi\left(x^{(1)} + h_2,\ y^{(1)} + k_2\right) = 0$ (13.12.13)

The same operation as above will yield

$$h_2 = \frac{\begin{vmatrix} -f\left(x^{(1)},\ y^{(1)}\right) & \left(\frac{\partial f}{\partial y}\right)_1 \\ -\phi\left(x^{(1)},\ y^{(1)}\right) & \left(\frac{\partial \phi}{\partial y}\right)_1 \end{vmatrix}}{\begin{vmatrix} \left(\frac{\partial f}{\partial x}\right)_1 & \left(\frac{\partial f}{\partial y}\right)_1 \\ \left(\frac{\partial \phi}{\partial x}\right)_1 & \left(\frac{\partial \phi}{\partial y}\right)_1 \end{vmatrix}} \tag{13.12.14}$$

$$= \frac{\begin{vmatrix} \left(\frac{\partial f}{\partial x}\right)_1 & -f\left(x^{(1)}, y^{(1)}\right) \\ \left(\frac{\partial \phi}{\partial x}\right)_1 & -\phi\left(x^{(1)}, y^{(1)}\right) \end{vmatrix}}{\begin{vmatrix} \left(\frac{\partial f}{\partial x}\right)_1 & \left(\frac{\partial f}{\partial y}\right)_1 \\ \left(\frac{\partial \phi}{\partial x}\right)_1 & \left(\frac{\partial \phi}{\partial y}\right)_1 \end{vmatrix}} \tag{13.12.15}$$

Further iterations may be applied, if necessary, to obtain the values of the roots up to e desired degree of accuracy.

te 1: The two equations $f(x, y) = 0$ and $\phi(x, y) = 0$ can be written in the forms:

$= F_1(x, y)$,

$= F_2(x, y)$

several ways.

We shall show in the next section that the iteration process for two unknowns will nverge when the two conditions

$$\left|\frac{\partial F_1}{\partial x}\right| + \left|\frac{\partial F_2}{\partial x}\right| < 1$$

d $$\left|\frac{\partial F_1}{\partial y}\right| + \left|\frac{\partial F_2}{\partial y}\right| < 1$$

ld for all points in the neighbourhood of (x_0, y_0).

te 2: The convergence of the iteration process will be rapid enough to make it advisable apply, if each of the quantities

$$\left|\frac{\partial F_1}{\partial x}\right| + \left|\frac{\partial F_2}{\partial x}\right|$$

d $$\left|\frac{\partial F_1}{\partial y}\right| + \left|\frac{\partial F_2}{\partial y}\right|$$

much less than 1.

Therefore, we must choose only those forms of F_1 and F_2 for which the above two con-ions hold.

te 3: Iteration process should not be applied at all in those examples, where the conver-nce is too slow, because in that case, we may make the roots decidedly worse instead mproving them.

.13 Convergence of the iteration process in the case of several unknowns

consider the case of two unknowns. Let the given equations be rewritten in the forms:

$F_1(x, y)$,

$F_2(x, y)$

These equations are satisfied by exact values of the pair of roots x, y.

The first approximations satisfy the equations

$$x^{(1)} = F_1(x_0, y_0),$$

$$y^{(1)} = F_2(x_0, y_0)$$

Subtracting these from the two earlier equations, we have

$$x - x^{(1)} = F_1(x, y) - F_1(x_0, y_0) \qquad (13.13.1)$$

$$y - y^{(1)} = F_2(x, y) - F_2(x_0, y_0) \qquad (13.13.2)$$

The right-hand side of (13.13.1) may be written by mean value theorem for a function of two variables,

$$F_1(x, y) - F_1(x_0, y_0) = (x - x_0)\frac{\partial \bar{F}_1}{\partial x} + (y - y_0)\frac{\partial \bar{F}_1}{\partial y}$$

where

$$\frac{\partial \bar{F}_1}{\partial x} = \frac{\partial F_1[x_0 + \phi(x - x_0), y_0 + \phi(y - y_0)]}{\partial x}, \quad 0 < \theta < 1$$

and
$$\frac{\partial \bar{F}_1}{\partial y} = \frac{\partial F_1[x_0 + \phi(x - x_0), y_0 + \phi(y - y_0)]}{\partial y}$$

Similarly, from (13.13.2), we have

$$F_2(x, y) - F_2(x_0, y_0) = (x - x_0)\frac{\partial \bar{F}_2}{\partial x} + (y - y_0)\frac{\partial \bar{F}_2}{\partial y}$$

Hence

$$(x - x^{(1)}) = (x - x_0)\frac{\partial \bar{F}_1}{\partial x} + (y - y_0)\frac{\partial \bar{F}_2}{\partial y}$$

$$(y - y^{(1)}) = (x - x_0)\frac{\partial \bar{F}_2}{\partial x} + (y - y_0)\frac{\partial \bar{F}_2}{\partial y}$$

$$\Rightarrow |x - x^{(1)}| + |y - y^{(1)}| \le |x - x_0| \left\{\left|\frac{\partial \bar{F}_1}{\partial x}\right| + \left|\frac{\partial \bar{F}_2}{\partial x}\right|\right\} + |y - y_0| \left\{\left|\frac{\partial \bar{F}_1}{\partial y}\right| + \left|\frac{\partial \bar{F}_2}{\partial y}\right|\right\} \qquad (13.13.3)$$

Now we suppose that the maximum value of either

$$\left|\frac{\partial F_1}{\partial x}\right| + \left|\frac{\partial F_2}{\partial x}\right|, \quad \text{or,} \quad \left|\frac{\partial F_1}{\partial y}\right| + \left|\frac{\partial F_2}{\partial y}\right|$$

be a proper fraction m for all points in the region (x_0, x) and (y_0, y), then (13.13.3) becomes

$$\left|x - x^{(1)}\right| + \left|y - y^{(1)}\right| \le m\{|x - x_0| + |y - y_0|\}$$

Similar relations hold for other approximations

$$\left|x - x^{(2)}\right| + \left|y - y^{(2)}\right| \le m\left\{\left|x - x^{(1)}\right| + \left|y - y^{(1)}\right|\right\}$$

$$\left|x - x^{(3)}\right| + \left|y - y^{(3)}\right| \le m\left\{\left|x - x^{(2)}\right| + \left|y - y^{(2)}\right|\right\}$$

....

$$\left|x - x^{(n)}\right| + \left|y - y^{(n)}\right| \le m\left\{\left|x - x^{(n-1)}\right| + \left|y - y^{(n-1)}\right|\right\}$$

Now multiplying together all these inequalities, we have after some manipulation

$$\left|x - x^{(n)}\right| + \left|y - y^{(n)}\right| \leq m^n \{|x - x_0| + |y - y_0|\} \tag{13.13.4}$$

Since m is a proper fraction, we may make the right-hand member of (13.13.4) as small as we please by repeating the iteration process a large number of times. This implies that the errors $|x - x^{(n)}|$ and $|y - y^{(n)}|$ can be made as small as we like.

Note: The sufficient condition for convergence of iteration process are

$$\left|\frac{\partial F_1}{\partial x}\right| + \left|\frac{\partial F_2}{\partial x}\right| < 1 \quad \text{and} \quad \left|\frac{\partial F_1}{\partial y}\right| + \left|\frac{\partial F_2}{\partial y}\right| < 1$$

13.13.1 Case of three equations in three unknowns

Let the three equations in three unknowns be $f(x, y, z) = 0$

$$\phi(x, y, z) = 0$$

$$\psi(x, y, z) = 0$$

Let x_0, y_0, z_0 be the approximate values of x, y and z respectively and h, k, l be the corresponding corrections. Then proceeding exactly in the manner given for two unknowns, we get the three simple equations,

$$f(x_0, y_0, z_0) + h\left(\frac{\partial f}{\partial x}\right)_0 + k\left(\frac{\partial f}{\partial y}\right)_0 + l\left(\frac{\partial f}{\partial z}\right)_0 = 0$$

$$\phi(x_0, y_0, z_0) + h\left(\frac{\partial \phi}{\partial x}\right)_0 + k\left(\frac{\partial \phi}{\partial y}\right)_0 + l\left(\frac{\partial \phi}{\partial z}\right)_0 = 0$$

$$\Psi(x_0, y_0, z_0) + h\left(\frac{\partial \psi}{\partial x}\right)_0 + k\left(\frac{\partial \psi}{\partial y}\right)_0 + l\left(\frac{\partial \psi}{\partial z}\right)_0 = 0$$

for determining the first corrections h_1, k_1 and l_1. The process may be repeated as many times as desired.

EXAMPLE 13.13.1. Use Newton-Raphson method to find a solution of the following simultaneous equations

$$x^2 + y - 11 = 0 \tag{i}$$

$$x + y^2 - 7 = 0 \tag{ii}$$

given the approximate values of the roots: $x_0 = 3$, $y_0 = -2$.

Solution: Let

$$f(x, y) \equiv x^2 + y - 11 = 0$$

and $f(x, y) \equiv x + y^2 - 7 = 0$

Let h, k be the corrections necessary. Then

$$f(x_0 + h, y_0 + k) = f(3 + h, -2 + k) = 0$$

$$\Rightarrow f(3, -2) + h(2x)_{x=3} + k = 0$$

$$\Rightarrow -4 + 6h + k = 0 \tag{iii}$$

and $\phi(x_0 + h, y_0 + k) = \phi(3 + h, -2 + k) = 0$

$\Rightarrow \phi(3, -2) + h + k(2y)_{y=-2} = 0$

$\Rightarrow 0 + h - 4k = 0$

$\Rightarrow h = 4k$

For first corrections from (iii) $k_1 = 4/25 = 0.16$, $h_1 = 16/25 = 0.64$.

$\therefore x^{(1)} = 3 + h_1 = 3.64$

$y^{(1)} = -2 + 0.16 = -1.84$

Again writing

$x^{(2)} = x^{(1)} + h_2 = 3.64 + h_2$

$y^{(2)} = y^{(1)} + k_2 = 1.84 + k_2$

we have

$$f(3.64, -1.84) + h_2(2x)_{x=3.64} + k_2 = 0$$

$$\phi(3.64, -1.84) + h_2 + k_2(2y)_{y=-1.84} = 0$$

$$\Rightarrow 0.4096 + 7.28h_2 + k_2 = 0$$

$$0.0256 + h_2 - 3.68k_2 = 0$$

$$\therefore\ h_2 = \frac{\begin{vmatrix} -0.4096 & 1 \\ -0.0256 & -3.68 \end{vmatrix}}{\begin{vmatrix} 7.28 & 1 \\ 1 & -3.68 \end{vmatrix}} = \frac{1.532928}{-27.7904} = -0.0551$$

$$\text{and } k_2 = \frac{\begin{vmatrix} 7.28 & -0.4096 \\ 1 & -0.0256 \end{vmatrix}}{\begin{vmatrix} 7.28 & 1 \\ 1 & -3.68 \end{vmatrix}} = \frac{0.223232}{-27.7904} = -0.0080$$

$\therefore x^{(2)} = 3.64 - 0.055 = 3.685$

$y^{(2)} = -1.84 - 0.008 = -1.848$

13.14 The method of iteration for simultaneous equations

13.14.1 Case of two unknowns

Let the two equations be

$f(x, y) = 0, \quad \phi(x, y) = 0$

We first rewrite the given equations in the forms:

$x = F_1(x, y), \quad y = F_2(x, y)$

Then, if x_0, y_0 be approximate values of a pair of roots, improved values are found by the steps indicated below:

First approx. $\begin{cases} x^{(1)} = F_1(x_0, y_0) \\ y^{(1)} = F_2\left(x^{(1)}, y_0\right) \end{cases}$

Second approx. $\begin{cases} x^{(2)} = F_1\left(x^{(1)}, y^{(1)}\right) \\ y^{(2)} = F_2\left(x^{(2)}, y^{(1)}\right) \end{cases}$

. . . .

nth approx. $\begin{cases} x^{(n)} = F_1\left(x^{(n-1)}, y^{(n-1)}\right) \\ y^{(n)} = F_2\left(x^{(n)}, y^{(n-1)}\right) \end{cases}$

13.14.2 Case of three unknowns

Let the three equations in three unknowns be

$f(x, y, z) = 0$

$\phi(x, y, z) = 0$

$\Psi(x, y, z) = 0$

We may rewrite these in the forms:

$x = F_1(x, y, z)$

$y = F_1(x, y, z)$

$z = F_1(x, y, z)$

Let x_0, y_0, z_0 be the approximate values of the roots. The improved values are found by the steps indicated below:

First approx. $\begin{cases} x^{(1)} = F_1(x_0, y_0, z_0) \\ y^{(1)} = F_2(x_1, y_0, z_0) \\ z^{(1)} = F_3(x_1, y_1, z_0) \end{cases}$

Second approx. $\begin{cases} x^{(2)} = F_1(x^{(1)}, y^{(1)}, z^{(1)}) \\ y^{(2)} = F_2(x^{(2)}, y^{(1)}, z^{(1)}) \\ z^{(2)} = F_3(x^{(2)}, y^{(2)}, z^{(1)}) \end{cases}$, etc.

13.15 Construction of convergent iterative functions

We have seen in the earlier sections that the initial step in the method of iteration process is to rewrite the equations $f(x, y) = 0$ and $\phi(x, y) = 0$ as

$$x = F_1(x, y) \tag{13.15.1}$$

$$y = F_2(x, y) \tag{13.15.2}$$

We have also noted in this connection that there are a number of ways of doing this.

To ensure convergence of the iteration process, we give below a technique by which (13.15.1) and (13.15.2) may be restructured.

Let

$$f(x, y) = x + \alpha\Psi(x, y) + \beta g(x, y) \tag{13.15.3}$$

and $$\phi(x, y) = y + \gamma\Psi(x, y) + \delta g(x, y) \tag{13.15.4}$$

such that $\alpha, \beta, \gamma, \delta$ satisfy the following relations:

$$\left.\begin{aligned} 1 + \alpha\Psi_x(x_0, y_0) + \beta g_x(x_0, y_0) &= 0 \\ \alpha\Psi_x(x_0, y_0) + \beta g_x(x_0, y_0) &= 0 \\ \gamma\Psi_x(x_0, y_0) + \delta g_x(x_0, y_0) &= 0 \\ 1 + \gamma\Psi_y(x_0, y_0) + \beta g_y(x_0, y_0) &= 0 \end{aligned}\right\}$$

The above equations are obtained by taking partial derivations of (13.15.3) and (13.15.4) w.r.t. x and y respectively.

III. Graeffe's root-squaring process for solving algebraic equations (non-Linear)

13.16 Introduction

Graeffe's Root Squaring Process works on the following Principle: The given equation $f(x) = 0$ is transformed into another equation whose roots are the squares of the roots of the given equation. The first transformed equation is again similarly transformed into the second equation whose roots are the squares of the first transformed equation or the fourth power of the roots of the original equations. The second transformed equation is, in its turn, transformed into the another equation whose roots are the squares of the second transformed equation or eighth power of the roots of the original equation $f(x) = 0$. Let the root squaring process be repeated so many times that the roots of the last transformed equation is 2^mth power of the roots of the original equation. It m is sufficiently large, then we may say the roots are widely separated. The roots are said to be widely separated, when the ratio of any root to the next larger in the last transformed equation is negligible in comparison with unity.

For example, it two roots of an equation by 3 and 2 and the root squaring process is repeated to raise the power of the roots by $k = 2^6$, say.

The two roots of the transformed equation will be 3^{64} and 2^{64}.

We may easily see that the ratio $2^{64}/3^{64}$ is negligibly small

$$\left(\frac{2^{64}}{3^{64}} < 10^{-11}\right)$$

We shall now explain the mechanism of the Root Squaring process by considering a polynomial equation of nth degree.

Let

$$f(x) = a_0x^n + a_1x^{n-1} + a_2x^{n-2} + a_3x^{n-3} + a_4x^{n-4} + \cdots + a_{n-1}x + an = 0. \tag{13.16.1}$$

Then, if $x_1, x_2, \ldots, x_n$ be the roots of (13.16.1), we may write it in the form:

$$f(x) = a_0(x - x_1)(x - x_2)\cdots(x - x_n) = 0 \tag{13.16.2}$$

Multiplying (13.16.2) by the function

$$(-1)^n f(-x) = (-1)^n a_0(-x - x_1)(-x - x_2)\cdots(-x - x_n)$$
$$= a_0(x + x_1)(x + x_2)\cdots(x + x_n)$$

the result is

$$(-1)^n f(-x)f(x) = a_0^2(x^2 - x_1^2)(x^2 - x_2^2)\cdots(x^2 - x_n^2) \qquad (13.16.3)$$

Writing $y = x^2$, (13.16.3) becomes

$$\phi(y) = a_0^2(y - x_1^2)(y - x_2^2)\cdots(y - x_n^2) = 0 \qquad (13.16.4)$$

The roots of this equations are $x_1^2, x_2^2, \ldots, x_n^2$.

Hence to form an equation whose roots are the squares of those of $f(x) = 0$, we just multiply $f(x) = 0$ by $(-1)^n f(-x)$.

This multiplication can however be carried out in a simple routine manner, which we shall show by considering the fifth degree equation:

$$f(x) = a_0x^5 + a_1x^4 + a_2x^3 + a_3x^2 + a_4x + a_5 = 0$$

Then

$$(-1)^5 f(-x) = (-1)^5\left\{-a_0x^5 + a_1x^4 - a_2x^3 + a_3x^2 - a_4x + a_5\right\}$$
$$= a_0x^5 - a_1x^4 + a_2x^3 - a_3x^2 + a_4x - a_5$$

The multiplication $(-1)^5 f(-x)\cdot f(x)$ is carried out with detached coefficients as indicated below.

(It is to be noted that the law of formation of the coefficients in the squared equation is the same irrespective of the degree of the given equation.)

a_0	a_1	a_2	a_3	a_4	a_5
a_0	$-a_1$	a_2	$-a_3$	a_4	$-a_5$
a_0^2	$-a_1^2$	a_2^2	$-a_3^2$	a_4^2	$-a_5^2$
	$+2a_0a_2$	$-2a_1a_3$	$+2a_2a_4$	$-2a_3a_5$	
		$+2a_0a_4$	$-2a_1a_5$		
b_0	b_1	b_2	b_3	b_4	b_5

The coefficients in the new transformed equation are the sums $b_0, b_1, b_2, b_3, b_4, b_5$.

The coefficients can be written down according to the following rules:

1. The numbers in the top row are the squares of the coefficients directly above them with alternating signs—the *second, fourth, sixth,* etc. squared numbers being negative.
2. The quantities directly under these squared numbers are the doubled products of the coefficient equally removed from the one directly overhead.
3. The signs of the doubled products are changed alternately in going along the rows and also in going down the columns. The sign of the first doubled product in each row is not changed.

When the roots are widely separated, we stop this process.

Now we shall apply Graeffe's method to three cases of algebraic equations.

Case 1: **Roots all real and unequal:** Suppose in the last transformed equation, the roots are the *m*th power of the roots of the original equation.

Let

$|x_1| > |x_2| > \cdots > |x_n|$

be the roots of the original equation. Therefore, the roots of the last transformed equation

$b_0(x^m)^n + b_1(x^m)^{n-1} + \cdots + b_{n-1}(x^m) + b_n = 0$

will be

$|x_1| > |x_2| > \cdots > |x_n^m|$

Using the relations between the roots and the coefficients, we have

$$x_1^m + x_2^m + \cdots + x_n^m = -\frac{b_1}{b_0} \tag{13.16.5}$$

$$x_1^m x_2^m + x_1^m x_3^m + \cdots = +\frac{b_2}{b_0} \tag{13.16.6}$$

$$x_1^m x_2^m x_2^m + x_1^m x_2^m x_4^m + \cdots = -\frac{b_3}{b_0} \tag{13.16.7}$$

....

$$x_1^m x_2^m x_3^m \cdots x_n^m = (-1)^n \cdot \frac{b_n}{b_0} \tag{13.16.8}$$

From (13.16.5)

$$-\frac{b_1}{b_0} = x_1^m\left(1 + \frac{x_2^m}{x_1^m} + \frac{x_2^m}{x_1^m} + \cdots + \frac{x_n^m}{x_1^m}\right) = x_1^m \tag{13.16.9}$$

From (13.16.6)

$$\frac{b_2}{b_0} = x_1^m x_2^m\left(1 + \frac{x_2^m}{x_2^m} + \frac{x_4^m}{x_2^m} + \cdots\right) = x_1^m x_2^m, \text{etc.} \tag{13.16.10}$$

From (13.16.8)

$$b_0 x_1^m + b_1 = 0 \tag{13.16.11}$$

Dividing (13.16.9) by (13.16.10)

$$-\frac{b_2}{b_1} = x_2^m \Rightarrow \left.\begin{aligned} b_1 x_2^m + b_2 &= 0 \\ b_2 x_3^m + b_3 &= 0 \\ &\cdots\cdots\cdots\cdots \\ b_{n-1} x_n^m + b_n &= 0 \end{aligned}\right\} \tag{13.16.12}$$

Thus, the root squaring process breaks up the original equation into n simple equations from which the desired roots can be found easily.

Now the question is: *How long should the root squaring process be continued and when to stop it?* In practice we continue the root squaring process until the doubled products in the second row have no effect on the coefficients of the next transformed equation.

Note 1. The coefficients in the given equation are not in general, all positive. The signs of the doubled products will not therefore occur in regular order as in the literal equation which we used to illustrate the root-squaring process. There is a strong possibility of making a mistake in the signs of these products. Special precaution is to be taken in this regard.

Note 2. The root squaring process necessarily increases the coefficients in the transformed equations. We shall always write them as simple numbers multiplied by powers of 10.

We now compute all the real roots (*unequal case*) of an algebraic equation by Graeffe's method. The cases of real but equal roots will be taken up in the next section. The case of complex roots is left out from the scope of our present discussion.

Example 13.16.1. Find all the roots of the equation

$x^3 - 8x^2 + 17x - 10 = 0$

by Graeffe's method.

Solution: We apply root-squaring process (*with detached coefficients*).

	1	-8	$+17$	-10
	1	$+8$	$+17$	$+10$
	1	-64	$+289$	-100
		$+34$	-160	
2nd P.	1	-30	129	-100
	1	$+30$	129	$+100$
	1	-900	16641	-10^4
		$+258$	-6000	
4th P.	1	-642	10641	-10^4
	1	$+642$	10641	$+10^4$
	1	-412164	$1.13230881.10^8$	-10^8
		$+21282$	$-0.1284.10^8$	
8th P.	1	$-3.90882.10^5$	$1.00390881.10^8$	-10^8
	1	$+3.90882.10^5$	$1.00390881.10^8$	$+10^8$
	1	$-1.52788.10^{11}$	$1.0078329.10^{16}$	-10^{16}
		$+0.0020078.10^{11}$	$-0.00781784.20^{16}$	
16th P.	1	$-1.5258722.10^{11}$	$1.00001506.10^{16}$	-10^{16}
	1	$+1.5258722.10^{11}$	$1.0000150.10^{16}$	$+10^{16}$
	1	$-2.328286.10^{22}$	$1.00003012.10^{32}$	-10^{32}
	1	$+0.00003012.10^{22}$	$-0.000030517.10^{32}$	
32nd P.	1	$-2.3282558.10^{22}$	$0.999999603.10^{32}$	-10^{32}

Let

$|x_1| > |x_2| > |x_3|$

be the three roots. Then

$$x_3^{32} = (0.999999603)^{-1} = 1.000000397$$

$$x_2^{32} = 0.4295.10^{10}$$

$$x_1^{32} = 2.3282558.10^{22}$$

$$\therefore x_1 = 4.9999 \simeq 5$$

$$x_2 = 2.000000 \simeq 2$$

$$x_3 = 1$$

As regards sign, we may apply Descartes' rule of signs and see that the equation has all the three positive roots.

Case 2: **Roots Real and numerically Equal: If two roots of an equation are numerically equal, the root-squaring process will never be able to break up the equation into linear fragments. One of the doubled products will always remain in the first row. The product will be just half the squared term above it. To examine this, let us consider an equation of degree 3, in x.**

Let the roots of the equation

$$x_3 + a_1x^2 + a_2x + a_3 = 0 \tag{13.16.13}$$

be x_1, x_2, x_3.

Let

$|x_1| > |x_2| = |x_3|$ $(x_2 = x_3)$

The equation whose roots are the mth power of the roots of (13.16.13) is

$$(y - x_1^m)(y - x_2^m)(y - x_3^m) = 0, \text{ where } y = x^m$$

$$\text{or,} y^3 - (x_1^m x_2^m x_3^m)y^2 + (x_1^m x_2^m + x_1^m x_3^m + x_2^m x_3^m)y - x_1^m x_2^m x_3^m = 0$$

$$\text{or,} y^3 - x_1^m\left(1 + \frac{x_2^m}{x_1^m} + \frac{x_3^m}{x_1^m}\right)y^2 + x_1^m x_2^m\left(1 + \frac{x_3^2}{x_2^m} + \frac{x_3^2}{x_1^m}\right)y - x_1^m x_2^m x_3^m = 0$$

$$\Rightarrow y^3 - x_1^m y^2 + 2x_1^m x_2^m y - x_1^m x_1^{2m} = 0 \tag{13.16.14}$$

The roots of (13.16.13) are now separated as widely as possible. We now apply root-squaring process to (13.16.14) to see what happens.

mth P.	1	$-x^m$	$2x_1^m x_2^m$	$-x_1^m x_2^{2m}$
	1	$+x_1^m$	$2x_1^m x_2^m$	$+x_1^m x_2^{2m}$
	1	$-x_1^{2m}$	$4x_1^{2m} x_2^{2m}$	$-x_1^{2m} x_2^{4m}$
		$+4x_1^m x_2^m$	$-2x_1^{2m} x_2^{2m}$	
$2m$th P.	1	$-x_1^{2m}$	$+2x_1^{2m} x_2^{2m}$	$-x_1^{2m} x_2^{4m}$

We may note that the first doubled product is negligible in comparison with the squared term above it. The doubled product of the next column is of the same order of magnitude as the squared term above but just half as large. Further, we may see that the coefficients of the equation (except one) with $2m$th power of the roots are the squares of the corresponding coefficients of the equation with mth power of the roots. These peculiarities enable us to detect the presence of the equal roots.

We are, however, leaving out the case of complex roots from the scope of our present discussion.

EXERCISES

1. Find the real root of the equation $x \log_{10} x - 1.2 - 0$, correct to five places of decimals. [*Ans:* 2.74065.]
2. Use Regula Falsi method to find the real root of the equation (correct to 3 decimal places), $x^3 - x - 1 = 0$, which lies between 1 and 2. [*Ans:* 1.325.]

3. Apply:
   (a) Method of False Position
   (b) Newton-Raphson method to solve the equation $3x - \cos x - 1 = 0$.
   [*Ans:* 0.6071, 0.6071.]
4. Use Newton-Raphson method to solve the equation $\log x = \cos x$, correct to three decimal places. [*Ans:* 1.303.]
5. (a) Find a iterative formula to find the mth root of a positive number N.
   (b) Hence obtain the value of:
   i. $7^{1/5}$. [*Ans:* 1.476.]
   ii. $15^{1/7}$. [*Ans:* 1.472.]
6. Find the real root of the equation $x^2 + 4\sin x = 0$ correct to three decimal places by using Newton-Raphson method. [*Ans:* −1.934.]
7. Give the geometrical significance of the Newton-Raphson method and obtain the condition of convergence of the formula.
8. Solve by iteration method the equation $\sin x = 5x - 2$. [*Ans:* 0.5235988.]

   [*Hints:* Rewrite the equations as $x = \sin^{-1}(5x - 2)$.]
9. Find the real root of $2x - 3\sin x - 5 = 0$, correct to six decimal places.
   [*Ans:* 2.883238.]
10. Find by iteration the smallest root of
    $$1 - x + \frac{x^2}{(2!)^2} - \frac{x^3}{(3!)^2} + \frac{x^4}{(4!)^2} - \frac{x^5}{(5!)^2} + \cdots = 0$$ [*Ans:* 1.44575.]
11. Find the smallest positive root of $x^x + 2x = 6$. [*Ans:* 1.723.]
12. Find, to four places of decimals, a solution of
    $$\sin x = y - 1.32,$$
    $$\cos y = x - 0.85$$
    [*Ans:* $x = 0.567325$, $y = 1.857378$.]
13. Solve for a pair of real roots of
    $$x + 3\log_{10} x - y^2 = 0$$
    $$2x^2 - xy - 5x + 1 = 0$$
    by iteration method with $x_0 = 3.4$ and $y_0 = 2.2$. [*Ans:* $x = 3.4782$, $y = 2.26154$.

    [*Hints:* Write the equation as $x = y^2 - 3\log_{10} x$; $y = (1/x) + 2x - 5$.]
14. Give the geometric significance of the iteration process.
15. Obtain a set of sufficient conditions of convergence for the iteration process for solution of simultaneous equations with two unknowns.
16. Find all the real roots of the following equations by Graeffe's root-squaring process:
    (a) $x^3 - 5x^2 - 44x - 60 = 0$. [*Ans:* 10, −3, −2.]
    (b) $x^3 - 3.1x^2 - 13.1x - 30.6 = 0$. [*Ans:* 3.4, −4.5, −2.]
    (c) $1.23x^5 - 2.52x^4 - 16.1x^3 + 17.3x^2 + 29.4x - 1.34 = 0$.
    [*Ans:* 4.066, −2.991, 1.959, −1.0285, 0.0445.]
    (d) $5x^3 + 2x^2 - 15x - 6 = 0$. [*Ans:* −0.4, −1.732, +1.732.]
    (e) $x^4 - 2x^3 + 1.99x^2 - 2x + 0.99 = 0$. [*Ans:* 1.1, 0.9, $\pm i$ (imaginary roots).]
    (f) $x^4 - 3.3x^3 - 9.03x^2 + 26.025x - 10.455 = 0$. [*Ans:* 4.1, 1.7, −3, 0.5.]
    (g) $x^3 - 9.65x^2 - 8.65x - 21.3$. [*Ans:* 10.65, −0.5, $\pm 1.323i$ (imaginary roots).]
    (h) $x^4 - 1.1x^3 - 0.2x^2 - 1.2x + 0.9 = 0$. [*Ans:* 1.5, 0.6, $0.5 \pm 0.866i$.]

Chapter 14

Solution of System of Linear Equations

Summary

14.1 Introduction

Various methods have been advocated for the numerical solution of simultaneous linear equations. No single method can be regarded as the best in all the situations. In the present chapter we shall explain some of the widely used general methods and try to illustrate them with numerical examples.

14.2 Cramer's rule

A simple method of solving simultaneous linear equations by determinants was discovered by G. Cramer (1750).

Let us consider as a simple case a system of linear equations with three unknowns x_1, x_2, x_3.

$$a_1x_1 + a_2x_2 + a_3x_3 = k_1;$$
$$b_1x_1 + b_2x_2 + b_3x_3 = k_2;$$
$$c_1x_1 + c_2x_2 + c_3x_3 = k_3$$

We first write down the arbitrary determinant

$$\Delta \equiv \begin{vmatrix} (a_1x_1 + a_2x_2 + a_3x_3) & a_2 & a_3 \\ (b_1x_1 + b_2x_2 + b_3x_3) & b_2 & b_3 \\ (c_1x_1 + c_2x_2 + c_3x_3) & c_2 & c_3 \end{vmatrix} = \begin{vmatrix} k_1 & a_2 & a_3 \\ k_2 & b_2 & b_3 \\ k_3 & c_2 & c_3 \end{vmatrix} \quad \textbf{(14.2.1)}$$

The addition law states that the left member of (14.2.1) may be expressed as the sum of the following three determinants.

$$\begin{vmatrix} a_1x_1 & a_2 & a_3 \\ b_1x_1 & b_2 & b_3 \\ c_1x_1 & c_2 & c_3 \end{vmatrix} + \begin{vmatrix} a_2x_2 & a_2 & a_3 \\ b_2x_2 & b_2 & b_3 \\ c_2x_2 & c_2 & c_3 \end{vmatrix} + \begin{vmatrix} a_3x_3 & a_2 & a_3 \\ b_3x_3 & b_2 & b_3 \\ c_3x_3 & c_2 & c_3 \end{vmatrix}$$

$$= x_1\begin{vmatrix} a_1 & a_2 & a_3 \\ b_1 & b_2 & b_3 \\ c_1 & c_2 & c_3 \end{vmatrix} + x_2\begin{vmatrix} a_2 & a_2 & a_3 \\ b_2 & b_2 & b_3 \\ c_2 & c_2 & c_3 \end{vmatrix} + x_3\begin{vmatrix} a_3 & a_2 & a_3 \\ b_3 & b_2 & b_3 \\ c_3 & c_2 & c_3 \end{vmatrix} = x_1\begin{vmatrix} a_1 & a_2 & a_3 \\ b_1 & b_2 & b_3 \\ c_1 & c_2 & c_3 \end{vmatrix}$$

[∵ the two columns are identical in the other determinants, they are zero.—Rule of the determinant.]

$$\therefore x_1 = \frac{\begin{vmatrix} k_1 & a_2 & a_3 \\ k_2 & b_2 & b_3 \\ k_3 & b_2 & c_3 \end{vmatrix}}{D}$$

where

$$D = \begin{vmatrix} a_1 & a_2 & a_3 \\ b_1 & b_2 & b_3 \\ c_1 & c_2 & c_3 \end{vmatrix} \neq 0$$

Similarly,

$$x_2 = \frac{\begin{vmatrix} a_1 & k_1 & c_1 \\ b_1 & k_2 & c_2 \\ c_1 & k_3 & c_3 \end{vmatrix}}{D} \quad \text{and} \quad x_3 = \frac{\begin{vmatrix} a_1 & a_2 & k_1 \\ b_1 & b_2 & k_2 \\ c_1 & c_2 & k_3 \end{vmatrix}}{D}$$

Note 1: The condition of unique solution of the equations is that the determinant D is not zero, i.e. the matrix

$$\begin{bmatrix} a_1 & a_2 & a_3 \\ b_1 & b_2 & b_3 \\ c_1 & c_2 & c_3 \end{bmatrix}$$

is non-singular.

Note 2: The method is laborious, if the number of unknowns is more than 3 or 4.

EXAMPLE 14.2.1. Find x, y, z from the equations

$$\begin{aligned} 3x + 2y - z &= 1 \\ x - y - 2z &= 3 \\ 2x + 3y + z &= -2 \end{aligned}$$

Step 1: We first examine whether the matrix

$$\begin{bmatrix} 3 & 2 & -1 \\ 1 & -1 & -2 \\ 2 & 3 & 1 \end{bmatrix}$$

is non-singular.

$$D = \begin{vmatrix} 3 & 2 & -1 \\ 1 & -1 & -2 \\ 2 & 3 & 1 \end{vmatrix} = \begin{vmatrix} 3 & 2 & -1 \\ 0 & -5/3 & -5/3 \\ 0 & 5/3 & 5/3 \end{vmatrix} \begin{aligned} r_1' &= r_1 \\ r_2' &= r_2 - (1/3)r_1 \\ r_3' &= r_3 - (2/3)r_1 \end{aligned}$$

$$= \begin{vmatrix} 3 & 0 & -3 \\ 0 & -5/3 & -7/3 \\ 0 & 0 & 0 \end{vmatrix} \begin{aligned} r_1'' &= r_1' + (6/5)r_2' \\ r_2'' &= r_2' \\ r_3'' &= r_2' + r_2' \end{aligned}$$

$$= 0$$

Hence the matrix of the coefficients is singular and therefore the equations have *no unique solution.*

Example 14.2.2. Solve:

$$\begin{aligned} 3x + 2y - z + t &= 1 \\ x - y - 2z + 4t &= 3 \\ 2x + 3y + z - 2t &= -2 \\ 5x - 2y + 3z + 2t &= 0 \end{aligned}$$

Here

$$D = \begin{vmatrix} 3 & 2 & -1 & 1 \\ 1 & -1 & -2 & 4 \\ 2 & 3 & 1 & -2 \\ 5 & -2 & 3 & 2 \end{vmatrix} = \begin{vmatrix} 0 & 5 & 5 & -11 \\ 1 & -1 & -2 & 4 \\ 0 & 5 & 5 & -10 \\ 0 & 3 & 13 & -18 \end{vmatrix} = \begin{vmatrix} 0 & 0 & 0 & -1 \\ 1 & -1 & -2 & 4 \\ 0 & 5 & 5 & -10 \\ 0 & 3 & 13 & -18 \end{vmatrix}$$

$$= \begin{vmatrix} 1 & -1 & -2 \\ 0 & 5 & 5 \\ 0 & 3 & 13 \end{vmatrix} = \begin{vmatrix} 1 & 0 & 0 \\ 0 & 5 & 5 \\ 0 & 3 & 13 \end{vmatrix} = 50$$

$$\Delta_1 = \begin{vmatrix} 1 & 2 & -1 & 1 \\ 3 & -1 & -2 & 4 \\ -2 & 3 & 1 & -2 \\ 0 & -2 & 3 & 2 \end{vmatrix} = \begin{vmatrix} 1 & 2 & -1 & 1 \\ 0 & -7 & 1 & 1 \\ 0 & 7 & -1 & 0 \\ 0 & -2 & 3 & 2 \end{vmatrix} = \begin{vmatrix} -7 & 1 & 1 \\ 7 & -1 & 0 \\ -2 & 3 & 2 \end{vmatrix}$$

$$= \begin{vmatrix} -7 & 1 & 1 \\ 0 & 0 & 1 \\ -2 & 3 & 2 \end{vmatrix} = -7 \times (-3) - 2 \times 1 = 19$$

$$\therefore x = \frac{\Delta_1}{D} = \frac{19}{50}$$

$$\Delta_2 = \begin{vmatrix} 3 & 1 & -1 & 1 \\ 1 & 3 & -2 & 4 \\ 2 & -2 & 1 & -2 \\ 5 & 0 & 3 & 2 \end{vmatrix} = \begin{vmatrix} 3 & 1 & -1 & 1 \\ 0 & 8/3 & -5/3 & 11/3 \\ 0 & -8/3 & 5/3 & -8/3 \\ 0 & -5/3 & 14/3 & 1/3 \end{vmatrix} = \begin{vmatrix} 3 & 1 & -1 & 1 \\ 0 & 8/3 & -5/3 & 11/3 \\ 0 & 0 & 0 & 1 \\ 0 & 1 & 3 & 4 \end{vmatrix}$$

$$= 3 \cdot \frac{1}{3} \begin{vmatrix} 8 & -5 & 11 \\ 0 & 0 & 1 \\ 1 & 3 & 4 \end{vmatrix} = \begin{vmatrix} 0 & -29 & -21 \\ 0 & 0 & 1 \\ 1 & 3 & 4 \end{vmatrix} = \begin{vmatrix} -29 & -21 \\ 0 & 1 \end{vmatrix} = -29$$

$$\therefore y = \frac{\Delta_2}{D} = \frac{29}{50}$$

$z = \Delta_3/D$, when

$$\Delta_3 = \begin{vmatrix} 3 & 2 & 1 & 1 \\ 1 & -1 & 3 & 4 \\ 2 & 3 & -2 & -2 \\ 5 & -2 & 0 & 2 \end{vmatrix} = \begin{vmatrix} 0 & 5 & -8 & -11 \\ 1 & -1 & 3 & 4 \\ 0 & 5 & -8 & -10 \\ 0 & 3 & -15 & -18 \end{vmatrix}$$

$$= \begin{vmatrix} 0 & 0 & 0 & -1 \\ 1 & -1 & 3 & 4 \\ 0 & 5 & -8 & -10 \\ 0 & 3 & -15 & -18 \end{vmatrix} = \begin{vmatrix} 1 & -1 & 3 \\ 0 & 5 & -8 \\ 0 & 3 & -15 \end{vmatrix} = -51$$

$$\therefore z = -\frac{51}{50}$$

$t = \Delta_4/D$, where

$$\Delta_4 = \begin{vmatrix} 3 & 2 & -1 & 1 \\ 1 & -1 & -2 & 3 \\ 2 & 3 & 1 & -2 \\ 5 & -2 & -3 & 0 \end{vmatrix} \Rightarrow \Delta_4 = \begin{vmatrix} 0 & 5 & 5 & -8 \\ 1 & -1 & -2 & 3 \\ 0 & 5 & 5 & -8 \\ 0 & 3 & 13 & -15 \end{vmatrix} = 0 \quad \text{[}\because \text{ 1st row and 3rd row are indentical.]}$$

$$\therefore t = \frac{0}{D} = 0$$

The solution is $x = 19/50$, $y = -29/50$, $z = -51/50$, $t = 0$.

14.3 The elimination method by Gauss

Let there be n unknowns in n linear equations

$$a_{11}x_1 + a_{12}x_2 + \cdots + a_{1n}x_n = k_1$$
$$a_{21}x_1 + a_{22}x_2 + \cdots + a_{2n}x_n = k_2$$
$$\cdots\cdots\cdots\cdots\cdots\cdots\cdots\cdots$$
$$a_{n1}x_1 + a_{n2}x_2 + \cdots + a_{nn}x_n = k_n$$

We assume that the set of equations is solvable. In Gauss's method, the unknowns x_1, $x_2, \ldots, x_n$ are eliminated successively, by solving some unknown (say x_1) in terms of the others, and then substituting this value for the same unknown in all the remaining equations. This eliminates x_1 from the new set of linear equations with $(n-1)$ unknowns.

The process is repeated on the new set of equations, thus eliminating another unknown, and so on, until the system is reduced to a single equation with one unknown.

We shall call the equation which expresses one unknown explicitly in terms of all the others as the pivotal equation. We shall select that linear equation with the highest coefficient of the unknown we desire to eliminate, as the first pivotal equation.

Finally, $x_n, x_{n-1}, x_{n-2}, \ldots, x_2$ are obtained by back substitution.

Example 14.3.1. Solve the equations by Gauss Method

$$2.63x + 5.21y - 1.694z + 0.938t - 4.23 = 0 \tag{i}$$

$$3.16x - 2.95y + 0.813z - 4.21t + 0.716 = 0 \tag{ii}$$

$$5.36x + 1.88y - 2.15z - 4.95t - 1.28 = 0 \tag{iii}$$

$$1.34x + 2.98y - 0.432z - 1.768t - 0.419 = 0 \tag{iv}$$

Solution: To eliminate x, we choose equation (iii) with highest coefficient of $x(= 5.36)$ as the first pivotal equation. Solving it for x, we obtain

$$x = -\frac{1.88}{5.36}y + \frac{2.15}{5.36}z + \frac{4.95}{5.36}t + \frac{1.28}{5.36}$$

$$= -0.3507y + 0.4011z + 0.9235t + 0.2388 \tag{v}$$

Substituting this value of x in (i), (ii) and (iv) successively, we get

$$4.2877y - 0.63911z + 3.3668t - 3.6020 = 0 \tag{vi}$$

$$-4.0582y + 2.0805z - 1.2917t + 1.4706 = 0 \tag{vii}$$

$$2.5101y + 0.1055z - 0.5305t - 0.0990 = 0 \tag{viii}$$

To eliminate y, we select (v), being the highest coefficient of y and write

$$y = \frac{0.63911}{4.2877}z - \frac{3.3668}{4.2877}t + \frac{3.6020}{4.2877}$$

$$= 0.1491z - 0.7853t + 0.8401 \tag{ix}$$

Substituting this in (vi) and (vii) successively, we get

$$1.4757z + 1.8951t - 1.9388 = 0 \tag{x}$$

$$0.4796z - 2.5014t + 2.0097 = 0 \tag{xi}$$

To eliminate z, we choose (ix) as the pivotal equation and write

$$z = -\frac{1.8951}{1.4757}t + \frac{1.9388}{1.4757} = -1.2842t + 1.3138 \tag{xii}$$

Substituting this in (x)

$$-3.1173t + 2.6398 = 0 \Rightarrow t = \frac{2.6398}{3.1173} = 0.847 \tag{xiii}$$

Substituting t in (xi)

$z = 0.226$

Substituting t and z in (viii)

$y = 0.209$

Finally substituting t, z and y in (A)

$x = 1.038$

Hence the required solution

$x = 1.038$

$y = 0.209$

$z = 0.226$

$t = 0.847$

Note: We may note that the systematic elimination of the unknowns by Gauss elimination method will reduce the augmented matrix into an equivalent upper triangular matrix from which the solution is obtained.

14.4 Jordan's method for solution

Jordan's method is a modification of Gauss elimination method and is often named as Gauss-Jordan's Method. This modification suggests that the coefficient matrix be transformed into a diagonal matrix from which the unknowns are obtained directly without much computational labour.

The difference between Jordan's method and Gauss' elimination method can better be explained by taking up a simple example.

EXAMPLE 14.4.1. Solve:

$$x_1 + 2x_2 + 3x_3 = 14$$
$$2x_1 + 5x_2 + 2x_3 = 18$$
$$3x_1 + x_2 + 5x_3 = 20$$

Method I: Gauss-Elimination Method

Step 1: We take the first equation as the first pivotal equation and multiply this successively by −2 and −3 and add to equations 2 and 3 respectively. The augmented matrix

$$\begin{bmatrix} 1 & 2 & 3 & 14 \\ 2 & 5 & 2 & 18 \\ 3 & 1 & 5 & 20 \end{bmatrix} \sim \begin{matrix} \\ -2 \\ -3 \end{matrix} \begin{bmatrix} 1 & 2 & 3 & 14 \\ 0 & 1 & -4 & -10 \\ 0 & -5 & -4 & -22 \end{bmatrix}$$

The corresponding system of linear equations

$$x + 2y + 3z = 14$$
$$y - 4z = -10$$
$$-5y - 4z = -22$$

Step 2: Taking the second equation in the above as the next pivotal equation, we multiply this equation by 5 and add it to the third equation in the above.

The augmented matrix will reduce to

$$\sim 5 \begin{bmatrix} 1 & 2 & 3 & 14 \\ 0 & 1 & -4 & -10 \\ 0 & 0 & -24 & -72 \end{bmatrix}$$

The corresponding system of equations

$$\begin{aligned} x + 2y + 3z &= 14 \\ y - 4z &= -10 \\ -24z &= -72 \end{aligned}$$

whence $z = 3$ and by back substitution; $y = -10 + 12 = 2$; $x = 1$.

Ans: $x = 1, y = 2, z = 3$.

14.4.1 Method II: Jordan's method

Step 1: This is the same as in Gauss Method. The augmented matrix reduces to

$$\sim \begin{bmatrix} 1 & 2 & 3 & 14 \\ 0 & 1 & -4 & -10 \\ 0 & -5 & -4 & -22 \end{bmatrix}$$

Step 2: We next take equation

$y - 4z = -10$

as the next pivotal equation. Multiply this by 5 and add it to the third equation as in the earlier method (*Gauss*). Now Jordan's modification is introduced. We also multiply $y - 4z = -10$ by -2 and add it to the first equation to eliminate y. The augmented matrix is

$$\sim \begin{matrix} -2 \\ \\ 5 \end{matrix} \begin{bmatrix} 1 & 0 & 11 & 34 \\ 0 & 1 & -4 & -10 \\ 0 & 0 & -24 & -72 \end{bmatrix}$$

The corresponding system of linear equations:

$$\begin{aligned} x + 11z &= 34 \\ y - 4z &= -10 \\ -24z &= -72 \end{aligned}$$

By back substitution $z = 3$, $y = 2$, $x = 1$.

14.5 Solution of equations by matrix methods

Let us consider a set of linear equations with n unknowns $x_1, x_2, \ldots, x_n$:

$$a_{11}x_1 + a_{12}x_2 + \cdots + a_{1n}x_n = k_1$$
$$a_{21}x_1 + a_{22}x_2 + \cdots + a_{2n}x_n = k_2$$
$$\cdots\cdots\cdots\cdots\cdots\cdots\cdots\cdots$$
$$a_{n1}x_1 + a_{n2}x_2 + \cdots + a_{nn}x_n = k_n$$

Written in the matrix form, the above equations can be represented by the simple equation

$$Ax = B \tag{14.5.1}$$

where

$$A = \begin{bmatrix} a_{11} & a_{12} & \cdots & a_{1n} \\ a_{21} & a_{22} & \cdots & a_{2n} \\ \cdots & \cdots & \cdots & \cdots \\ a_{n1} & a_{n2} & \cdots & a_{nn} \end{bmatrix}_{n\times n} \equiv \left[a_{ij}\right]_{n\times n}; \quad x = \begin{bmatrix} x_1 \\ x_2 \\ \vdots \\ x_n \end{bmatrix}_{n\times 1}; \quad B = \begin{bmatrix} k_1 \\ k_2 \\ \vdots \\ k_n \end{bmatrix}_{n\times 1}$$

Premultiplying (14.5.1) by A^{-1} (inverse of the matrix A), we have

$$x = A^{-1}B \tag{14.5.2}$$

provided det $|A| \neq 0$.

Equation (14.5.2) gives the solution of the given system. Hence the main problem in this case is to find the inverse of the given coefficient matrix A.

We shall now discuss some of the standard methods to find the inverse of A.

In all standard textbooks on Algebra we have the solution of the system $Ax = B$ in the form $x = A^{-1}B$, where A^{-1} is the inverse of the matrix A, defined by

$$A^{-1} = \frac{\text{Adj } A}{\det |A|}$$

where

$$\text{Adj } A = \begin{bmatrix} A_{11} & A_{21} & \cdots & A_{n1} \\ A_{12} & A_{22} & \cdots & A_{n2} \\ \vdots & \vdots & \vdots & \vdots \\ A_{1n} & A_{2n} & \cdots & A_{nn} \end{bmatrix}$$

and A_{ij} being the cofactor of a_{ij} in det $|A|$.

It is therefore evident that the solution exists provided det $|A| \neq 0$. In other words, the matrix A is *non-singular.*

The above method is not always suitable to find the inverse of A because it requires the evaluation of a number of determinants, depending on the number of unknowns, and no efficient method exists for evaluation of determinants. In simple terms, the computational labour will become prohibitive, if number of unknowns be more than 4 with large numerical coefficients. We shall advocate the use of the following two methods to find A^{-1}; one is the direct method and the other is the iterative method.

- Direct method is based on Gauss (*modified*) elimination method (*Gauss-Jordan Method for Matrix inversion*);
- Gauss-Seidel's iterative method.

14.6 Gauss-Jordan's method for matrix inversion

In this method we start by writing side by side the given matrix A and the unit matrix I of the same order. The augmented matrix is transformed into another matrix in consecutive steps (the number of steps being equal to the order of the matrix). A single column of the unit matrix is used in each step. The aim in each step is to reduce to zero all the elements in the first column, except one, and that element is reduced to unity by dividing its row throughout by such a number as will make it unity. The matrix at the beginning in each step is augmented by the appropriate column of the unit matrix, and all elements in each row of the augmented matrix are subjected to the same operations.

The different operations in each step may be explained by taking a non-singular matrix of order 3.

Let

$$A = \begin{bmatrix} a_{11} & a_{12} & a_{13} \\ a_{21} & a_{22} & a_{23} \\ a_{31} & a_{32} & a_{33} \end{bmatrix}$$

We augment this by a unit matrix I of order 3.

$$[A|I] = \left[\begin{array}{ccc|ccc} a_{11} & a_{12} & a_{13} & 1 & 0 & 0 \\ a_{21} & a_{22} & a_{23} & 0 & 1 & 0 \\ a_{31} & a_{32} & a_{33} & 0 & 0 & 1 \end{array}\right]$$

$$\sim \left[\begin{array}{ccc|ccc} 1 & \dfrac{a_{12}}{a_{11}} & \dfrac{a_{13}}{a_{11}} & \dfrac{1}{a_{11}} & 0 & 0 \\ 0 & a_{22} - a_{21}\cdot\dfrac{a_{12}}{a_{11}} & a_{23} - a_{21}\cdot\dfrac{a_{13}}{a_{11}} & 0 - a_{21}\cdot\dfrac{1}{a_{11}} & 1 & 0 \\ 0 & a_{32} - a_{31}\cdot\dfrac{a_{12}}{a_{11}} & a_{33} - a_{31}\cdot\dfrac{a_{13}}{a_{11}} & 0 - a_{31}\cdot\dfrac{1}{a_{11}} & 0 & 1 \end{array}\right]$$

Explanation

***Step* 1:** Pivotal row is the first row and a_{11} is the pivotal element.

1. In the first transformed matrix, the first row starts with element a_{12}/a_{11}, the second element is a_{13}/a_{11} and the third is $1/a_{11}$.
2. The first element of the second row is $a_{22} - a_{21} \cdot a_{12}/a_{11}$; the second element is $a_{23} - a_{21} \cdot a_{13}/a_{11}$ and the third is $0 - a_{21} \cdot 1/a_{11}$.
3. Elements of the third row are given: $a_{32} - a_{31} \cdot a_{12}/a_{11}$, $a_{33} - a_{31} \cdot a_{13}/a_{11}$, $0 - a_{31} \cdot 1/a_{11}$.

Let all these elements be denoted by dashes, and one column of the unit matrix is taken to the left.

$$\left[\begin{array}{c|ccc|cc} 1 & a'_{11} & a'_{12} & a'_{13} & 0 & 0 \\ 0 & a'_{21} & a'_{22} & a'_{23} & 1 & 0 \\ 0 & a'_{31} & a'_{32} & a'_{33} & 0 & 1 \end{array}\right]$$

a'_{21} is taken as the new pivotal element of the pivotal row 2. The same procedure is repeated.

$$\sim \left[\begin{array}{cc|cccc} 1 & 0 & a'_{12} - \dfrac{a'_{22}}{a'_{21}}a'_{11} & a'_{13} - a'_{11}\cdot\dfrac{a'_{23}}{a'_{21}} & 0 - a'_{11}\dfrac{1}{a'_{21}} & 0 \\ 0 & 1 & \dfrac{a'_{22}}{a'_{21}} & \dfrac{a'_{23}}{a'_{21}} & \dfrac{1}{a'_{21}} & 0 \\ 0 & 0 & a'_{32} - a'_{31}\cdot\dfrac{a'_{22}}{a'_{21}} & a'_{33} - a'_{31}\cdot\dfrac{a'_{23}}{a'_{21}} & 0 - a'_{31}\dfrac{1}{a'_{21}} & 1 \end{array}\right]$$

Let all these element be denoted by (″) dashes, another column of unit matrix is again taken to the left. We next write

$$\left[\begin{array}{ccc|ccc} 1 & 0 & a''_{11} & a''_{12} & a''_{13} & 0 \\ 0 & 1 & a''_{21} & a''_{22} & a''_{23} & 0 \\ 0 & 0 & a''_{31} & a''_{32} & a''_{33} & 1 \end{array}\right]$$

a''_{31} is now the pivotal element. The process is repeated.

$$\left[\begin{array}{ccc|ccc} 1 & 0 & 0 & a''_{12} - a''_{11} \cdot \dfrac{a''_{32}}{a''_{31}} & a''_{13} - a''_{11} \cdot \dfrac{a''_{33}}{a''_{31}} & 0 - \dfrac{a''_{11}}{a''_{31}} \\ 0 & 1 & 0 & a''_{22} - a''_{21} \cdot \dfrac{a''_{32}}{a''_{31}} & a''_{23} - a''_{21} \cdot \dfrac{a''_{33}}{a''_{31}} & 0 - \dfrac{a''_{21}}{a''_{31}} \\ 0 & 0 & 1 & \dfrac{a''_{32}}{a''_{31}} & \dfrac{a''_{33}}{a''_{31}} & \dfrac{1}{a''_{31}} \end{array}\right]$$

$$\therefore A^{-1} = \begin{bmatrix} a''_{12} - a''_{11} \cdot \dfrac{a''_{32}}{a''_{31}} & a''_{13} - a''_{11} \cdot \dfrac{a''_{33}}{a''_{31}} & 0 - \dfrac{a''_{11}}{a''_{31}} \\ a''_{22} - a''_{21} \cdot \dfrac{a''_{32}}{a''_{31}} & a''_{23} - a''_{21} \cdot \dfrac{a''_{33}}{a''_{31}} & 0 - \dfrac{a''_{21}}{a''_{31}} \\ \dfrac{a''_{32}}{a''_{31}} & \dfrac{a''_{33}}{a''_{31}} & \dfrac{1}{a''_{31}} \end{bmatrix}$$

EXAMPLE 14.6.1. Find the inverse of the matrix

$$A = \begin{bmatrix} 1 & -5 & 1 \\ 2 & 4 & 1 \\ 1 & 1 & 1 \end{bmatrix}$$

***Solution*: We consider the augmented matrix**

$$[A \mid I] = \left[\begin{array}{ccc|ccc} 1 & -5 & 1 & 1 & 0 & 0 \\ 2 & 4 & 1 & 0 & 1 & 0 \\ 1 & 1 & 1 & 0 & 0 & 1 \end{array}\right] \begin{array}{l} \\ \rightarrow \text{pivotal row} \\ \\ \end{array}$$

$$\sim \begin{bmatrix} 1 & -5 & 1 & 1 & 0 & 0 \\ 0 & 14 & -1 & -2 & 1 & 0 \\ 0 & 6 & 0 & -1 & 0 & 1 \end{bmatrix} \begin{array}{c} \rightarrow \\ -2 \\ -1 \end{array} \sim \begin{bmatrix} 1 & 0 & 9/14 & 4/14 & 5/14 & 0 \\ 0 & 1 & -1/14 & -2/14 & 1/14 & 0 \\ 0 & 0 & 6/14 & -2/14 & -6/14 & 1 \end{bmatrix} \begin{array}{c} 5 \\ \rightarrow \\ -6 \end{array}$$

$$\sim \begin{bmatrix} 1 & 0 & 0 & 7/14 & 1 & -3/2 \\ 0 & 1 & 0 & -7/42 & 0 & 1/6 \\ 0 & 0 & 1 & -1/3 & -1 & 14/6 \end{bmatrix} \begin{array}{c} -9/14 \\ 1/14 \\ \rightarrow \end{array}$$

$$\therefore A^{-1} = \begin{bmatrix} 7/14 & 1 & -3/2 \\ -7/42 & 0 & 1/6 \\ -1/3 & -1 & 14/6 \end{bmatrix} = 1/42 \begin{bmatrix} 21 & 42 & -63 \\ -7 & 0 & 7 \\ -14 & -42 & 98 \end{bmatrix}$$

Check:

$$A^{-1}A = 1/42\begin{bmatrix} 21 & 42 & -63 \\ -7 & 0 & 7 \\ -14 & -42 & 98 \end{bmatrix}\begin{bmatrix} 1 & -5 & 1 \\ 2 & 4 & 1 \\ 1 & 1 & 1 \end{bmatrix}$$

$$= 1/42\begin{bmatrix} 21+84-63 & -105+168-63 & 21+42-63 \\ -7+0+7 & 35+0+7 & -7+0+7 \\ -14-84+98 & 70-168+98 & -14-42+98 \end{bmatrix}$$

$$= 1/42\begin{bmatrix} 42 & 0 & 0 \\ 0 & 42 & 0 \\ 0 & 0 & 42 \end{bmatrix} = \begin{bmatrix} 1 & 0 & 0 \\ 0 & 1 & 0 \\ 0 & 0 & 1 \end{bmatrix} = I$$

Hence the result.

We shall consider an example, where we shall use all the three principal methods, namely:

- Gauss Elimination Method,
- Gauss-Jordan Method and
- Gauss's Matrix Inversion Method

and examine the extent of the computational labour involved.

Example 14.6.2. Solve:

$$6.1x_1 + 2.2x_2 + 1.2x_3 = 16.55 \quad \text{(i)}$$

$$2.2x_1 + 5.5x_2 - 1.5x_3 = 10.55 \quad \text{(ii)}$$

$$1.2x_1 - 1.5x_2 + 7.2x_3 = 16.80 \quad \text{(iii)}$$

by (i) Gauss Elimination Method, (ii) Gauss-Jordan's Method and (iii) Gauss's Matrix Inversion Method.

Gauss Elimination Method

Step 1: We take equation (i) as the first pivotal equation and write

$$x_1 = -\frac{2.2}{6.1}x_2 - \frac{1.2}{6.1}x_3 + \frac{16.55}{6.1} \quad \text{(iv)}$$

$$= 0.360656x_2 - 0.196721x_3 + 2.713115 \quad \text{(v)}$$

We next eliminate x_1 from (ii) and (iii) by substituting the value of x_1 in (iv) and obtain

$$4.706557x_2 - 1.932786x_3 - 4.581147 = 0 \quad \text{(vi)}$$

$$-1.932787x_2 + 6.963935x_3 - 13.544262 = 0 \quad \text{(vii)}$$

Step 2: We now take equation (v) as the next pivotal equation and write

$$x_2 = \frac{1.932786}{4.706557}x_3 + \frac{4.581147}{4.706557} = 0.410658x_3 + 0.973354 \quad \text{(viii)}$$

Substituting (vii) in (vi), we eliminate x_2 from (vi) and obtain

$$6.170221x_3 - 15.425548 = 0 \Rightarrow x_3 = \frac{15.425548}{6.170221} = 2.499999 \simeq 2.50 \quad \text{(ix)}$$

By successive back substitution

$x_2 = (0.410658)(2.50) + 0.973354 = 1.999999 \simeq 2.0$

$x_1 = (-0.360656)(2.0) - (0.196721)(2.5) + 2.713115$

$= -0.721312 - 0.4918025 + 2.713115 = 1.50000 = 1.5$

Ans: $x_1 = 1.5$, $x_2 = 2.0$, $x_3 = 2.5$.

Note: The above operations of the coefficients may be shown as under:

$$\begin{bmatrix} \underline{1} & 0.360656 & 0.196721 & -2.713115 \\ & 4.706557 & -1.932786 & -4.581147 \\ & -1.932787 & 6.963935 & -13.544262 \end{bmatrix} : \text{end of Step 1}$$

$$\Rightarrow \begin{bmatrix} 1 & 0.360656 & 0.196721 & -2.713115 \\ & \underline{1} & -0.410658 & -0.973354 \\ & & 6.170221 & -15.425548 \end{bmatrix} : \text{end of Step 2}$$

Gauss-Jordan Method

Step 1: Gauss's Method is followed totally in this step.

Step 2: We write, taking (v) as the next pivotal equation, $x_2 = 0.410658x_3 + 0.973354$.

We eliminate x_2 from (vi) as in Gauss method and also eliminate x_2 from (iv) and write

$$x_1 = -0.344827x_3 + 2.362069 \tag{14.6.1}$$

Final coefficient matrix:

$$\begin{bmatrix} 1 & 0 & 0.344827 & -2.362069 \\ & 1 & -0.410658 & -0.973358 \\ & & 6.170221 & -15.425548 \end{bmatrix}$$

Gauss's Inversion of Matrix

The augmented matrix

$$\left[\begin{array}{ccc|ccc} 6.1 & 2.2 & 1.2 & 1 & 0 & 0 \\ 2.2 & 5.5 & -1.5 & 0 & 1 & 0 \\ 1.2 & -1.5 & 7.2 & 0 & 0 & 1 \end{array}\right]$$

$$\sim \begin{bmatrix} 1 & 0.36066 & 0.19672 & 0.16393 & 0 & 0 \\ 0 & 4.70656 & -1.93279 & -0.36065 & 1 & 0 \\ 0 & -1.93279 & 6.96393 & -0.19672 & 0 & 1 \end{bmatrix} \begin{matrix} \\ -2.2 \\ -1.2 \end{matrix}$$

$$\sim \begin{bmatrix} 1 & 0 & 0.34483 & 0.19157 & -0.07663 & 0 \\ 0 & 1 & -0.41066 & -0.07663 & 0.21247 & 0 \\ 0 & 0 & 6.17021 & -0.34483 & 0.41066 & 1 \end{bmatrix} \begin{matrix} -0.34483 \\ \\ 1.93279 \end{matrix}$$

$$\sim \begin{bmatrix} 1 & 0 & 0 & 0.21084 & -0.09958 & -0.05589 \\ 0 & 1 & 0 & -0.09958 & 0.23980 & 0.06656 \\ 0 & 0 & 1 & -0.05589 & 0.06656 & 0.16207 \end{bmatrix}$$

$$\therefore A^{-1} = \begin{bmatrix} 0.21084 & -0.09958 & -0.05589 \\ -0.09958 & 0.23980 & 0.06656 \\ -0.05589 & 0.06656 & 0.16207 \end{bmatrix}$$

$$\therefore \begin{bmatrix} x_1 \\ x_2 \\ x_3 \end{bmatrix} = \begin{bmatrix} 0.021084 & -0.09958 & -0.55890 \\ -0.099580 & 0.23980 & 0.06656 \\ -0.055890 & 0.06656 & 0.16207 \end{bmatrix} \begin{bmatrix} 16.55 \\ 10.55 \\ 16.80 \end{bmatrix} = \begin{bmatrix} 1.4999 \\ 2.0000 \\ 2.5000 \end{bmatrix}$$

$\therefore x_1 = 1.4999,\ x_2 = 2.0000,\ x_3 = 2.5000$

14.7 Solution of linear equations by iteration

- Jacobi's Method and
- Gauss-Seidel's Method.

The iteration method is not applicable to all systems of linear equations. This iteration method will succeed, if each of the system contains one large coefficient (*larger than the others in that equation*) and the large coefficient must be attached to a different unknown in each equation. The above requirements are met when the large coefficients are present along the leading diagonal of the coefficient matrix, as we may find sometimes.

In solving a system of equations by iteration, each equation is first solved for the unknown having the large coefficient, expressing it explicitly in terms of the other unknowns. We may explain the others steps in this process by examples.

14.7.1 Jacobi's method of iteration

Let us consider a system of n linear equations:

$$\left.\begin{aligned} a_{11}x_1 + a_{12}x_2 + \cdots + a_{1n}x_n &= b_1 \\ a_{21}x_1 + a_{22}x_2 + \cdots + a_{2n}x_n &= b_2 \\ \cdots\cdots\cdots\cdots\cdots\cdots\cdots\cdots \\ a_{n1}x_1 + a_{n2}x_2 + \cdots + a_{nn}x_n &= b_n \end{aligned}\right\} \tag{14.7.1}$$

where the diagonal elements a_{ii}, $(i = 1, 2, \ldots, n)$ are not zero, and they are dominant (*large in comparison to other coefficients*).

Rewriting equations in (14.7.1), we get

$$\left.\begin{aligned} x_1 &= \frac{b_1}{a_{11}} - \frac{a_{12}}{a_{11}}x_2 - \frac{a_{13}}{a_{11}}x_3 \cdots - \frac{a_{1n}}{a_{11}}x_n \\ x_2 &= \frac{b_2}{a_{22}} - \frac{a_{21}}{a_{22}}x_1 - \frac{a_{23}}{a_{22}}x_3 \cdots - \frac{a_{2n}}{a_{22}}x_n \\ x_3 &= \frac{b_3}{a_{33}} - \frac{a_{31}}{a_{33}}x_1 - \frac{a_{32}}{a_{33}}x_2 \cdots - \frac{a_{3n}}{a_{33}}x_n \\ &\cdots\cdots\cdots\cdots\cdots\cdots\cdots\cdots \\ x_n &= \frac{b_n}{a_{nn}} - \frac{a_{n1}}{a_{nn}}x_1 - \frac{a_{n2}}{a_{nn}}x_2 \cdots - \frac{a_{nn-1}}{a_{nn}}x_{n-1} \end{aligned}\right\} \tag{14.7.2}$$

To solve these equations, we may make an initial guess $x_1^{(0)}, x_2^{(0)}, \ldots, x_n^{(0)}$ for the unknowns $x_1, x_2, \ldots, x_n$ (*initial guess may be taken as zero in absence of any other clue*). We next substitute these initial values in the right-hand side of (14.7.2) and obtain the first approximation as noted below:

$$\left.\begin{aligned}
x_1^{(1)} &= \frac{b_1}{a_{11}} - \frac{a_{12}}{a_{11}}x_2^{(0)} - \frac{a_{13}}{a_{11}}x_3^{(0)} \cdots - \frac{a_{1n}}{a_{11}}x_n^{(0)}\\
x_2^{(1)} &= \frac{b_2}{a_{22}} - \frac{a_{21}}{a_{22}}x_1^{(0)} - \frac{a_{23}}{a_{22}}x_3^{(0)} \cdots - \frac{a_{2n}}{a_{22}}x_n^{(0)}\\
x_3^{(1)} &= \frac{b_3}{a_{33}} - \frac{a_{31}}{a_{33}}x_1^{(0)} - \frac{a_{32}}{a_{33}}x_2^{(0)} \cdots - \frac{a_{3n}}{a_{33}}x_n^{(0)}\\
&\cdots\cdots\cdots\cdots\cdots\cdots\cdots\cdots\cdots\cdots\\
x_n^{(1)} &= \frac{b_n}{a_{nn}} - \frac{a_{n1}}{a_{nn}}x_1^{(0)} - \frac{a_{n2}}{a_{nn}}x_2^{(0)} \cdots - \frac{a_{nn-1}}{a_{nn}}x_{n-1}^{(0)}
\end{aligned}\right\} \tag{14.7.3}$$

Substituting $x_1^{(1)}, x_2^{(1)}, \ldots, x_n^{(1)}$ in the right side of (14.7.3), we get second approximation and the process is continued until the iterations have given desired number of significant figures in the approximate roots. This iterative process is due to Jacobi. The method converges if the coefficient matrix is diagonally dominant and further

$$\left.\begin{aligned}
&\sum_{\substack{j=1\\ j\neq i}}^{n} |a_{ij}| < |a_{ii}|, \quad (i = 1, 2, \ldots, n),\\
\text{or,}\quad &\sum_{\substack{i=1\\ i\neq j}}^{n} |a_{ij}| < |a_{jj}|, \quad (i = 1, 2, \ldots, n)
\end{aligned}\right\} \tag{14.7.4}$$

14.7.2 Gauss-Seidel's method

A simple modification of Jacobi's iteration process has been given in Gauss-Seidel's iteration method which gives faster convergence.

In the first equation of (14.7.2)

$$x_1 = -\frac{b_1}{a_{11}} - \frac{a_{12}}{a_{11}}x_2 - \frac{a_{13}}{a_{11}}x_3 - \cdots - \frac{a_{1n}}{a_{11}}x_n \equiv F_1(x_2, x_3, \ldots, x_n)$$

We substitute the initial values $x_2^{(0)}, x_3^{(0)}, \ldots, x_n^{(0)}$ and obtain the first approximation $x_1^{(1)}$ given by

$$x_1^{(1)} = F_1\left(x_2^{(0)}, x_3^{(0)}, \ldots, x_n^{(0)}\right)$$

We next substitute the improved values $x_1^{(1)}$, and $x_3^{(0)}, x_4^{(0)}, \ldots, x_n^{(0)}$ in the second equation of (14.7.2)

$$x_2 = -\frac{b_2}{a_{22}} - \frac{a_{21}}{a_{22}}x_1 - \frac{a_{31}}{a_{22}}x_3 - \cdots - \frac{a_{n1}}{a_{22}}x_n \equiv F_2(x_1, x_3, \ldots, x_n)$$

and obtain the first approximation of x_2, given by

$$x_2^{(1)} = F_2\left(x_1^{(1)}, x_3^{(0)}, \ldots, x_n^{(0)}\right)$$

In the third equation of (14.7.2)

$$x_3 = -\frac{b_3}{a_{33}} - \frac{a_{31}}{a_{33}}x_1 - \frac{a_{32}}{a_{33}}x_2 \cdots - \frac{a_{3n}}{a_{33}}x_n \equiv F_3(x_1, x_2, x_4, \ldots, x_n)$$

We substitute the two improved values $x_1^{(1)}$, $x_2^{(1)}$ and the other initial values $x_4^{(0)}$, $x_5^{(0)}$, $\ldots$, $x_n^{(0)}$ to obtain the first approximation of x_3, given by

$$x_3^{(1)} = F_3\left(x_1^{(1)}, x_2^{(1)}, x_4^{(0)}, \ldots, x_n^{(0)}\right)$$

Finally

$$\begin{aligned} x_n^{(1)} &= -\frac{b_n}{a_{nn}} - \frac{a_{n1}}{a_{nn}} x_1^{(1)} - \frac{a_{n2}}{a_{nn}} x_2^{(1)} - \cdots - \frac{a_{nn-1}}{a_{nn}} x_{n-1}^{(1)} \\ &\equiv F_n\left(x_1^{(1)}, x_2^{(1)}, \ldots, x_{n-1}^{(1)}\right) \end{aligned}$$

The process is continued until we get the values of $x_1, x_2, \ldots, x_n$ with sufficient degree of accuracy.

14.8 Conditions for the convergence of the iteration process

We may rewrite the equations discussed in (14.7.3) as

$$\begin{aligned} x_1 &= -\frac{b_1}{a_{11}} + 0 \cdot x_1 - \frac{a_{12}}{a_{11}} x_2 - \frac{a_{13}}{a_{11}} x_3 - \cdots - \frac{a_{1n}}{a_{11}} x_n \\ &\equiv F_1(x_1, x_2, \ldots, x_n) \end{aligned}$$

Similarly,

$$x_2 \equiv F_2(x_1, x_2, \ldots, x_n)$$

$$\cdots\cdots\cdots\cdots\cdots\cdots$$

$$x_n = F_n(x_1, x_2, \ldots, x_n)$$

The conditions for convergence are

$$\left|\frac{\partial F_1}{\partial x_1}\right| + \left|\frac{\partial F_2}{\partial x_1}\right| + \cdots + \left|\frac{\partial F_n}{\partial x_1}\right| < 1$$

and $(n-1)$ similar inequalities involving partial derivatives w.r.t. $x_2, x_3, \ldots, x_n$ respectively.

Adding the n inequalities and grouping the terms suitably, we may readily see that the following n conditions are satisfied.

$$\left.\begin{aligned} &\left|\frac{\partial F_1}{\partial x_1}\right| + \left|\frac{\partial F_1}{\partial x_2}\right| + \cdots + \left|\frac{\partial F_1}{\partial x_n}\right| < 1 \\ &\left|\frac{\partial F_2}{\partial x_1}\right| + \left|\frac{\partial F_2}{\partial x_2}\right| + \cdots + \left|\frac{\partial F_2}{\partial x_n}\right| < 1 \\ &\cdots\cdots\cdots\cdots\cdots\cdots\cdots\cdots \\ &\left|\frac{\partial F_n}{\partial x_1}\right| + \left|\frac{\partial F_n}{\partial x_2}\right| + \cdots + \left|\frac{\partial F_n}{\partial x_n}\right| < 1 \end{aligned}\right\} \qquad (14.8.1)$$

We may easily see that the above conditions are in essence equivalent to the sufficiency conditions:

$$\sum_{\substack{j=1 \\ j \neq 1}}^{n} |a_{ij}| < |a_{ii}|, \quad i = 1, 2, \ldots, n \quad \text{[given in (14.7.4)]}$$

EXAMPLE 14.8.1. Solve the following system by Gauss-Seidel's iteration process:

$0.89x + 4.32y - 0.47z + 0.95t = 3.36$ (i)

$1.13x - 0.89y + 0.61z + 5.63t = 4.27$ (ii)

$6.32x - 0.73y - 0.65z + 1.06t = 2.95$ (iii)

$0.74x + 1.01y + 5.28z - 0.88t = 1.97$ (iv)

Solution: We rewrite the equations (i), (ii), (iii) and (iv) respectively as

$$x = \frac{2.95}{6.32} + \frac{0.73}{6.32}y + \frac{0.65}{6.32}z - \frac{1.06}{6.32}t$$

$$= 0.4668 + 0.1155y + 0.1028z - 0.1677t \quad \text{(v)}$$

$$y = \frac{3.36}{4.32} - \frac{0.89}{4.32}x + \frac{0.47}{4.32}z - \frac{0.95}{4.32}t$$

$$= 0.7778 - 0.2060x + 0.1088z - 0.2199t \quad \text{(vi)}$$

$$z = \frac{1.97}{5.28} - \frac{0.74}{5.28}x - \frac{1.01}{5.28}y + \frac{0.88}{5.28}t$$

$$= 0.3731 - 0.1402x - 0.1913y + 0.1667t \quad \text{(vii)}$$

$$t = \frac{4.27}{5.63} - \frac{1.13}{5.63}x + \frac{0.89}{5.63}y - \frac{0.61}{5.63}z$$

$$= 0.7584 - 0.2007x + 0.1581y - 0.1083z \quad \text{(viii)}$$

Taking initial values $y = 0, z = 0, t = 0$ in (v)

$x^{(1)} = 0.4668$

Taking $x = 0.4668, z = 0, t = 0$ in (vi)

$y^{(1)} = 0.7778 - 0.0962 = 0.6816$

Taking $x = 0.4668, y = 0.6816$ and $t = 0$ in (vii)

$z^{(1)} = 0.3731 - 0.0654 - 0.1304 = 0.1773$

Taking $x = 0.4668, y = .6816$ and $z = 0.1773$ in (viii)

$t^{(1)} = 0.7584 - 0.0937 + 0.1078 - 0.0192 = 0.7533$

Similarly,

$x^{(2)} = 0.4668 + 0.1155y^{(1)} + 0.1028z^{(1)} - 0.1677t^{(1)}$

$= 0.4668 + 0.0787 + 0.0182 - 0.1263 = 0.4374$

$g^{(2)} = 0.7778 - 0.2060x^{(2)} + 0.1088z^{(1)} - 0.2199t^{(1)}$

$= 0.7778 - 0.0901 + 0.0193 - 0.1657 = 0.5413$

$z^{(2)} = 0.3731 - 0.1402x^{(2)} - 0.1913y^{(2)} + 0.1667t^{(1)}$

$= 0.3731 - 0.613 - 0.1036 + 0.1256 = 0.3338$

$t^{(2)} = 0.7584 - 0.2007x^{(2)} + 0.1581y^{(2)} - 0.1083z^{(2)}$

$= 0.7584 - 0.0878 + 0.0856 - 0.0362 = 0.7200$

$$x^{(3)} = 0.4668 + 0.1155y^{(2)} + 0.1028z^{(2)} - 0.1677t^{(2)}$$
$$= 0.4668 + 0.0625 + 0.0343 - 0.1207 = 0.443$$
$$y^{(3)} = 0.7778 - 0.2060x^{(3)} + 0.1088z^{(2)} - 0.2199t^{(2)}$$
$$= 0.7778 - 0.0912 + 0.0363 - 0.1577 = 0.565$$
$$z^{(3)} = 0.3731 - 0.1402x^{(3)} - 0.1913y^{(3)} + 0.1667t^{(2)}$$
$$= 0.3731 - 0.0621 - 0.1081 + 0.1200 = 0.323$$
$$t^{(3)} = 0.7584 - 2.007x^{(3)} + 0.1581y^{(3)} - 0.1083z^{(3)}$$
$$= 0.7584 - 0.0889 + 0.0893 - 0.0350 = 0.723$$

Ans: $x = 0.443$, $y = 0.565$, $z = 0.323$, $t = 0.723$.

14.9 Triangularisation

A matrix which has all the elements above the principal diagonal zero is called a *lower triangular matrix.*

$$\begin{bmatrix} a_{11} & 0 & 0 \\ a_{21} & a_{22} & 0 \\ a_{31} & a_{32} & a_{33} \end{bmatrix} \rightarrow \text{lower triangular matrix}$$

A matrix which has all the elements below the principal diagonal zero is called an *upper triangular matrix*, where the diagonal elements are unity.

$$\begin{bmatrix} 1 & b_{12} & b_{13} \\ 0 & 1 & b_{23} \\ 0 & 0 & 1 \end{bmatrix} \rightarrow \text{upper triangular matrix}$$

We have seen that Gauss elimination in sec. 14.3 leads to an upper triangular matrix' where all diagonal elements are 1. We may interpret the elimination process as a multiplication of the original coefficient matrix A (*augmented*) by a suitable triangular matrix.

For this purpose, we set (*taking the case of* 3 *unknowns*)

$$\begin{bmatrix} l_{11} & 0 & 0 \\ l_{21} & l_{22} & 0 \\ l_{31} & l_{32} & l_{33} \end{bmatrix} \begin{bmatrix} a_{11} & a_{12} & a_{13} \\ a_{21} & a_{22} & a_{23} \\ a_{31} & a_{32} & a_{33} \end{bmatrix} = \begin{bmatrix} 1 & r_{12} & r_{13} \\ 0 & 1 & r_{23} \\ 0 & 0 & 1 \end{bmatrix}$$

Finally, an upper triangular matrix to be inverted giving a similar matrix is the result.

14.10 Factorisation result

A square matrix

$$A = \left[a_{ij}\right]_{n \times n} \quad (i, j = 1, 2, \ldots, n)$$

can be factorised into $A = LU$, where L is the lower triangular matrix and U is an upper triangular matrix provided that all the principal minors of A are non-singular, i.e. if

$$a_{11} \neq 0, \begin{vmatrix} a_{11} & a_{12} \\ a_{21} & a_{22} \end{vmatrix} \neq 0, \text{ etc. and finally } |A| \neq 0$$

It is also known from the textbooks of linear algebra that such factorization is unique.

Let the system of n linear equations be written in the matrix form as

$$Ax = B \tag{14.10.1}$$

where

$$A = \begin{bmatrix} a_{11} & a_{12} & \cdots & a_{1n} \\ a_{21} & a_{22} & \cdots & a_{2n} \\ \cdots & \cdots & \cdots & \cdots \\ a_{n1} & a_{n2} & \cdots & a_{nn} \end{bmatrix}, \quad x = \begin{bmatrix} x_1 \\ x_2 \\ \vdots \\ x_n \end{bmatrix}_{n\times 1} \quad \text{and } B = \begin{bmatrix} b_1 \\ b_2 \\ \vdots \\ b_n \end{bmatrix} \tag{14.10.2}$$

We may assume that A can be uniquely factored in the form $A = LU$, where

$$L = \begin{bmatrix} l_{11} & 0 & 0 & \cdots & \cdots & 0 \\ l_{21} & l_{22} & 0 & \cdots & \cdots & 0 \\ \cdots & \cdots & l_{33} & \cdots & \cdots & 0 \\ \cdots & \cdots & \cdots & \cdots & \cdots & \cdots \\ l_{n1} & l_{n2} & l_{n3} & \cdots & \cdots & l_{nn} \end{bmatrix} \tag{14.10.3}$$

and

$$U = \begin{bmatrix} 1 & r_{12} & r_{13} & \cdots & r_{1n} \\ 0 & 1 & r_{23} & \cdots & r_{2n} \\ \cdots & \cdots & 1 & \cdots & r_{3n} \\ \cdots & \cdots & \cdots & \cdots & \cdots \\ 0 & 0 & \cdots & \cdots & 1 \end{bmatrix} \tag{14.10.4}$$

Then equation (14.10.1) can be written as

$$LUX = B \tag{14.10.5}$$

If we write

$$UX = z_1 \tag{14.10.6}$$

then

$$Lz = B \tag{14.10.7}$$

$z_1, z_n, \ldots, z_n$ can be solved from (14.7.2) by forward substitution. Having obtained z_i's, we may obtain $x_1, x_2, \ldots, x_n$ from (14.7.2) by backward substitution.

Note: The procedure will not work, if $l_{ii} = 0$ for some i in (14.93).

We may illustrate this method of triangularisation and factorisation, and also of solution of equations by taking equations with 3 unknowns.

Example 14.10.1. Solve:

$$10x + 2y + z = 69$$

$$x + 8y + 2z = -43$$

$$2x - y + 20z = 76$$

Method 1: Triangularisation of matrix.

Let us write

$$\begin{bmatrix} l_{11} & 0 & 0 \\ l_{21} & l_{22} & 0 \\ l_{31} & l_{32} & l_{33} \end{bmatrix} \left[\begin{array}{ccc|c} 10 & 2 & 1 & 69 \\ 1 & +8 & 2 & -43 \\ 2 & -1 & 20 & 76 \end{array}\right] = \begin{bmatrix} 1 & r_{12} & r_{13} \\ 0 & 1 & r_{23} \\ 0 & 0 & 1 \end{bmatrix}$$

The relevant equations are

$$10l_{11} = 1 \Rightarrow l_{11} = 1/10$$

$$10l_{21} + l_{22} = 0 \quad \text{(i)}$$

$$10l_{31} + l_{32} + 2l_{33} = 0 \quad \text{(ii)}$$

$$2l_{21} + l_{22} = 1 \quad \text{(iii)}$$

$$2l_{11} = r_{12} = 2/10 = 1/5$$

From (i) and (iii),

$$l_{22} = \frac{5}{39}, \quad l_{21} = -\frac{1}{78}$$

Also

$$\left.\begin{array}{l} 2l_{31} + 8l_{32} - l_{33} = 0 \\ l_{31} + 2l_{32} + 20l_{33} = 1 \end{array}\right\} \Rightarrow l_{31} = -\frac{17}{1571}, \quad l_{32} = \frac{14}{1571}, \quad l_{33} = \frac{78}{1571}$$

$$l_{11} = r_{13} = \frac{1}{10}, \quad l_{21} + 2l_{22} = r_{23} = -\frac{1}{78} + \frac{20}{78} = \frac{19}{78}$$

$$\therefore \begin{bmatrix} 1/10 & 0 & 0 \\ -1/78 & 10/78 & 0 \\ -17/1571 & 14/1571 & 78/1571 \end{bmatrix} \left[\begin{array}{ccc|c} 10 & 2 & 1 & 69 \\ 1 & 8 & 2 & -3 \\ 2 & -1 & 20 & 76 \end{array}\right]$$

$$= \left[\begin{array}{ccc|c} 1 & 2/10 & 1/10 & 69/10 \\ 0 & 1 & 19/78 & -99/78 \\ 0 & 0 & 1 & 4713/1571 \end{array}\right]$$

The last column is obtained by the product of the first matrix with $\begin{bmatrix} 69 \\ -3 \\ 76 \end{bmatrix}$.

This means that the system is solved to obtain $z = 4713/1571 = 3$.

$$y + \frac{19}{78}z = -\frac{99}{78} \Rightarrow y = -\frac{99}{78} - \frac{57}{78} = -\frac{156}{78} = -2$$

$$x + \frac{2}{10}y + \frac{1}{10}z = \frac{69}{10} \Rightarrow x = \frac{69}{10} + \frac{4}{10} - \frac{3}{10} = \frac{70}{10} = 7$$

Ans: $x = 7, y = -2, z = 3$.

Method 2: Method of Matrix Factorisation: In matrix notation we have $Ax = B$, where

$$A = \begin{bmatrix} 10 & 2 & 1 \\ 1 & 8 & 2 \\ 2 & -1. & 20 \end{bmatrix}, \quad x = \begin{bmatrix} x \\ y \\ z \end{bmatrix}, \quad B = \begin{bmatrix} 69 \\ -3 \\ 76 \end{bmatrix}$$

Let

$$\begin{bmatrix} l_{11} & 0 & 0 \\ l_{21} & l_{22} & 0 \\ l_{31} & l_{32} & l_{33} \end{bmatrix} \begin{bmatrix} 1 & r_{12} & r_{13} \\ 0 & 1 & r_{23} \\ 0 & 0 & 1 \end{bmatrix} = \begin{bmatrix} 10 & 2 & 1 \\ 1 & 8 & 2 \\ 2 & -1 & 20 \end{bmatrix} \quad [\because LU = A]$$

Multiplying and equating

$$\left.\begin{aligned} l_{11} &= 10 \\ l_{11}\cdot r_{12} &= 2 \Rightarrow r_{12} = \tfrac{2}{10} = \tfrac{1}{5} \\ l_{11}\cdot r_{13} &= 1 \Rightarrow r_{13} = \tfrac{1}{10} \end{aligned}\right\}$$

$$\left.\begin{aligned} l_{21} &= 1 \\ l_{21}\cdot r_{12} + l_{22} &= 8 \end{aligned}\right\} \Rightarrow l_{22} = 8 - \frac{1}{5} = \frac{39}{5}$$

$$l_{31} = 2$$

$$l_{31}\cdot r_{12} + l_{32} = -1 \Rightarrow l_{32} = -1 - \frac{2}{5} = -\frac{7}{5}$$

$$l_{31}\cdot r_{13} + l_{32}\cdot r_{23} + l_{33} = 20$$

$$l_{21}\cdot r_{13} + l_{22}\cdot r_{23} = 2 \Rightarrow r_{23} = \frac{2-(1/10)}{(39/5)} = \frac{19}{10}\times\frac{5}{39} = \frac{19}{78}\Bigg\}$$

$$\therefore l_{33} = 20 - \frac{2}{10} + \frac{7}{5}\cdot\frac{19}{78} = \frac{1571}{78}$$

$$\therefore \begin{bmatrix} 10 & 0 & 0 \\ 1 & 39/5 & 0 \\ 2 & -7/5 & 1571/78 \end{bmatrix}\begin{bmatrix} 1 & 1/5 & 1/10 \\ 0 & 1 & 19/78 \\ 0 & 0 & 1 \end{bmatrix} = \begin{bmatrix} 10 & 2 & 1 \\ 1 & 8 & 2 \\ 2 & -1 & 20 \end{bmatrix}$$

Therefore, we write

$$LZ = B \Rightarrow \begin{bmatrix} 10 & 0 & 0 \\ 1 & 39/5 & 0 \\ 2 & -7/5 & 1571/78 \end{bmatrix}\begin{bmatrix} x_1 \\ y_1 \\ z_1 \end{bmatrix} = \begin{bmatrix} 69 \\ -3 \\ 76 \end{bmatrix}, \quad \text{where } z = \begin{bmatrix} x_1 \\ y_1 \\ z_1 \end{bmatrix}$$

$$\Rightarrow x_1 = \frac{69}{10} \quad x_1 + \frac{39}{5}y_1 = -3$$

$$\Rightarrow \frac{39}{5}y_1 = -3 - \frac{69}{10} = -\frac{99}{10}$$

$$\Rightarrow y_1 = -\frac{99}{78} \quad 2x_1 - \frac{7}{5}y_1 + \frac{1571}{78}z_1 = 76$$

$$\Rightarrow \frac{1571}{78}z_1 = 76 - \frac{69}{5} + \frac{7}{5}\left(-\frac{99}{78}\right) = \frac{23565}{390} = \frac{4713}{78}$$

$\Rightarrow z_1 = 3$ [by forward substitution]

Hence the solution of the original system is

$$\begin{bmatrix} 1 & 1/5 & 1/10 \\ 0 & 1 & 19/78 \\ 0 & 0 & 1 \end{bmatrix}\begin{bmatrix} x \\ y \\ z \end{bmatrix} = \begin{bmatrix} 69/10 \\ -99/78 \\ 3 \end{bmatrix} \quad \therefore z = 3$$

$$y + \frac{19}{78}z = -\frac{99}{78} \Rightarrow y = -\frac{156}{78} = -2$$

$x + \frac{1}{5}y + \frac{1}{10}z = \frac{69}{10} \Rightarrow x = \frac{69}{10} + \frac{2}{5} - \frac{3}{10} = 7$ [by backward substitution]

Ans: $x = 7, y = -2, z = 3.$

EXERCISES

1. Solve the system:

$$x + 3y + 2z = 17$$
$$x + 2y + 3z = 16$$
$$2x - y + 4z = 13$$

by (a) Gauss-elimination and (b) Gauss-Jordan method. [*Ans:* $x = 4, y = 3, z = 2.$]

2. Solve the system:

$$15x + 3y - 2z = 85$$
$$2x + 10y + z = 51$$
$$x - 2y + 8z = 5$$

by (a) Jacobi iteration and (b) Gauss-Seidel method. [*Ans:* $x = 5, y = 4, z = 1.$]

3. Solve the system:

$$x - 7y + 2z - t = 10$$
$$3x + 4y - z + t = 4$$
$$2x - y + 4z - 2t = 7$$
$$5x + 2y - 3z + 2t = 9$$

by (a) Gauss Matrix inversion method and (b) Factorisation of Matrix method.
[*Ans:* $x = 2, y = -1, z = 3, t = 5.$]

4. Solve the following system:

$$2.38x_1 + 1.95x_2 - 3.27x_3 + 1.58x_4 = 2.16$$
$$3.21x_1 - 0.86x_2 + 2.42x_3 - 3.20x_4 = 3.28$$
$$1.44x_1 + 2.95x_2 - 2.14x_3 + 1.86x_4 = 1.42$$
$$4.17x_1 + 3.62x_2 - 1.68x_3 - 2.26x_4 = 5.21$$

by (a) Gauss elimination and (b) Gauss-Jordan method.
[*Ans:* $x_1 = 0.8072, x_2 = 0.2372, x_3 = -0.1046, x_4 = -0.3581.$]

5. Solve the following system:

$$3x + 4y + 2z = 15$$
$$5x + 2y + z = 18$$
$$2x + 3y + 2z = 10$$

by (a) method of factorisation and (b) triangularisation method.
[*Ans:* $x = 3, y = 2, z = -1.$]

6. Apply Jordan's method to find the inverse of the matrix:

$$A = \begin{bmatrix} -2 & 4 & 8 \\ -4 & 18 & -16 \\ -6 & 2 & -20 \end{bmatrix}$$ [*Ans:* $\begin{bmatrix} -41/190 & 6/95 & -13/95 \\ 1/95 & 11/190 & -4/95 \\ 5/76 & -1/76 & -1/76 \end{bmatrix} = A^{-1}.$]

7. Factorise the above matrix into two matrices of the same order, one an upper triangular matrix U, and the other a lower triangular matrix. Hence find the inverse of A.

8. Apply Gauss-Seidel method to solve the system:

$$3x + y + z = 3$$
$$2x + y + 5z = 5$$
$$x + 4y + z = 2$$

Give the condition of convergence of Gauss-Seidel's iteration method and show that it is satisfied in this case.

[*Ans:* $x = 34/47$, $y = 7/47$, $z = 32/47$.]

Chapter 15

The Numerical Solution of Ordinary Differential Equations

Summary

15.1 Introduction

An ordinary differential equation is an equation involving one independent variable, the dependent variable and its derivative(s) with respect to the independent variable. It is, however, not necessary that the two variables should enter into the equation explicitly.

All the standard textbooks on Calculus and Differential equations deal with certain types of differential equations and methods have been developed in these texts for solving them. We may find, by the techniques developed in these books, the solution of many such equations in explicit forms. But equations of this type are only limited.

Just as we have found roots of an algebraic or transcendental equation with numerical coefficients, correct only to a number of decimal places—but not exact, we may likewise find to any desired degree of accuracy the numerical solution of any ordinary differential equation having numerical coefficients and given initial conditions. Starting from the initial values, the solutions are thence constructed step by step through a series of equal intervals in the independent variable so that as soon as the solution has been carried to $x = x_i$; the next step will be to evaluate the change in the solution through the interval $\Delta x(= h)$ of x from x_i to x_{i+1}. The most important of the several methods for solving differential equations numerically will be explained in this chapter.

We may note in this connection that a general equation of the first order and first degree is

$$\frac{dy}{dx} = f(x, y)$$

15.2 Euler's method

It is perhaps the oldest and simplest method, though it may appear to be a crude one. This was devised by Euler.

Let the differential equation of the first order and degree one be

$$\frac{dy}{dx} = f(x, y) \tag{15.2.1}$$

Let $y = F(x)$ be its solution and its graph be shown in the xy-plane in the adjoining graph. Let it be required to find the value of y at $x = l$.

For this we divide the interval $l - x_0$ in n equal parts of width h. Let the intermediate points be $x_1, x_2, \ldots, x_{n-1}$.

In Fig. 15.2(a), let

$$\begin{aligned} \overline{OM_0} &= x_1, \\ \overline{OM_1} &= x_2, \text{ etc.} \\ \text{and } \overline{P_0M_0} &= y_0, \\ \overline{P_1M_1} &= y_1, \\ \overline{P_2M_2} &= y_2, \text{ etc.} \end{aligned}$$

Let us take a magnified view of the section $P_0M_0M_1P_1$ [Fig. 15.2(b)].

From the figure,

$$\tan\theta = \frac{\overline{NQ}}{\overline{P_0Q}} = \left(\frac{dy}{dx}\right)_{x=x_0}$$

If $x_1 - x_0 = h$ be small, we may see that

$$\overline{NQ} \simeq \overline{P_1Q} = \overline{P_1M_1} - \overline{QM_1} = \overline{P_1M_1} - \overline{P_0M_0} = y_1 - y_0$$

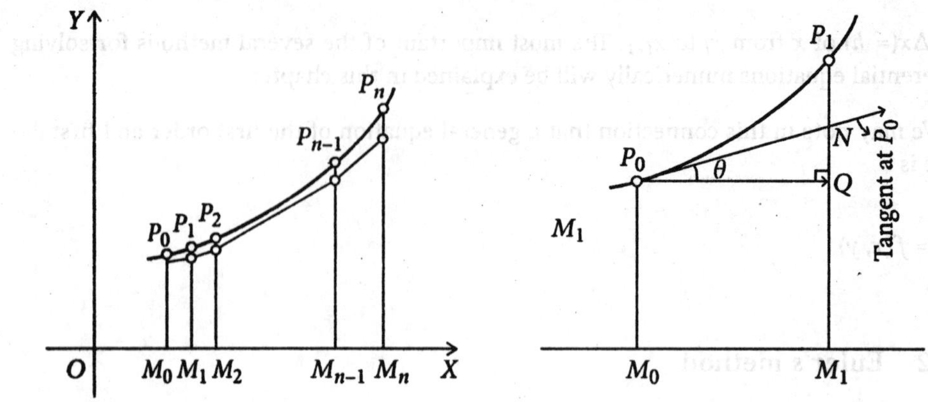

Fig. 15.1(a) Fig. 15.1(b)

$$\therefore \tan\theta = \frac{\overline{NQ}}{\overline{P_0Q}} \simeq \frac{\overline{P_1Q}}{\overline{P_0Q}} = \frac{y_1 - y_0}{x_1 - x_0}$$

$$\therefore \left(\frac{dy}{dx}\right)_{x=0} = \frac{y_1 - y_0}{h} \Rightarrow y_1 = y_0 + h\left(\frac{dy}{dx}\right)_{x=0}$$

$$= y_0 + hf(x_0, y_0)$$

This y is the approximate value of y at $x = x_1$. Proceeding in the same way, we may find the approximate value of y at $x = x_2$; ($\overline{OM_2} = x_2$) as

$y_2 = y_1 + hf(x_1, y_1)$

and finally the value of y at $x = x_n(= l)$ as

$y_n = y_{n-1} + hf(x_{n-1}, y_{n-1})$

NOTE 15.2.1. A major defect of this method lies in the fact that if dy/dx changes rapidly over an interval, then its value at the beginning of the interval may give a poor approximation as compared to the average value of dy/dx over the interval. Hence the value of y calculated by Euler's method may involve appreciable error. These errors may accumulate in the succeeding intervals and the value of y becomes quite erroneous ultimately at $x = x_n(= l)$.

NOTE 15.2.2. It may be seen that in Euler's method the actual solution curve $y = F(x)$ is approximated by the sequence of short lines and quite often the sequence of short lines deviates materially frɔm the curve $y = F(x)$. To reduce this error, we may prefer Euler's improved method. Here in this case, we consider the curvature of the actual curve in place of approximating the curve by the sequence of short lines.

15.3 Euler's modified method

Starting with the initial value y_0, an approximate value for y_1 is computed from the relation

$$y_1^{(1)} \simeq y_0 + h\left(\frac{dy}{dx}\right)_0 \tag{15.3.1}$$

Then this approximate value is substituted in (15.2.1) to get an approximate value of dy/dx at the end of the first interval or

$$\left(\frac{dy}{dx}\right)_1^{(1)} = f\left(x_1, y_1^{(1)}\right)$$

An improved value of Δy is obtained by multiplying h by the average value of dy/dx at the ends of the interval x_0 to x_1 or

$$\Delta y \simeq h \cdot \frac{\left(\frac{dy}{dx}\right)_0 + \left(\frac{dy}{dx}\right)_1^{(1)}}{2}$$

The second approximation for y_1 is now given as

$$y_1^{(2)} = y_0 + h \cdot \frac{\left(\frac{dy}{dx}\right)_0 + \left(\frac{dy}{dx}\right)_1^{(1)}}{2} \qquad (15.3.2)$$

This improved value of $y_1^{(2)}$ is now substituted into the given equation (15.2.1) to obtain a second approximation for

$$\left(\frac{dy}{dx}\right)_1, \quad \text{or,} \quad \left(\frac{dy}{dx}\right)_1^{(2)} = f\left(x_1, y_1^{(2)}\right) \qquad (15.3.3)$$

The third approximation for y_1 is then

$$y_1^{(3)} = y_0 + h \cdot \frac{\left(\frac{dy}{dx}\right)_0 + \left(\frac{dy}{dx}\right)_1^{(2)}}{2} \qquad (15.3.4)$$

This process is repeated until there is no change in the value of y_1 to the number of digits retained.

The computation for the next interval x_1 to $x_2(= x_1 + h)$ is then carried out in similar fashion.

Although the modified Euler method is slow and of moderate accuracy, its simplicity makes it a method of great value and also we may use this method, when no other method will work.

Example 15.3.1. Use modified Euler's method to compute $y(0.01)$ where $dy/dx = x^2 + y$ with initial condition $y_0 = 1$, when $x = 0$.

Solution: Here

$$y_1^{(0)} = y_0 + hf(x_0, y_0) = y_0 + 0.01(0 + 1) = 1.01$$

$$f\left(x_1, y_1^{(0)}\right) = 0.0001 + 1.01 = 1.0101$$

$$\therefore\ y_1^{(1)} = y_0 + \frac{h}{2}\cdot\left\{f(x_0, y_0) + f\left(x_1, y_1^{(0)}\right)\right\}$$

$$= 1 + \frac{0.01}{2}\{1 + 1.0101\}$$

$$\Rightarrow Y_1^{(1)} = 1 + 0.005 \times 2.0101 = 1 + 0.01005 = 1.01005$$

$$f\left(x_1, y_1^{(1)}\right) = 0.0001 + 0.001005 = 1.01005$$

$$\therefore\ y_1^{(2)} = y_0 + \frac{h}{2}\left\{f(x_0, y_0) + f\left(x_1, y_1^{(1)}\right)\right\}$$

$$= 1 + \frac{0.01}{2}\{1 + 1.01015\} = 1.01005$$

$$\therefore\ y(0.01) \simeq 1.01005$$

Example 15.3.2. Use Euler's modified method to compute a few values of y for the differential equation $dy/dx = x + y$ with the initial conditions $x_0 = 0$, $y_0 = 1$ with $h = 0.05$.

Solution:

$$y_1^{(1)} = y_0 + \left(\frac{dy}{dx}\right)_0 h = 1 + (0 + 1) \times 0.05 = 1.05$$

Then

$$\left(\frac{dy}{dx}\right)_1^{(1)} = x_1 + y_1^{(1)} = 0.05 + 1.05 = 1.10 \quad [x_1 = x_0 + h]$$

$$\therefore\ y_1^{(2)} = y_0 + h\cdot\frac{\left(\frac{dy}{dx}\right)_0 + \left(\frac{dy}{dx}\right)_1^{(1)}}{2} = 1 + 0.05\frac{1 + 1.10}{2} = 1.0525$$

Then the second approximation is

$$\left(\frac{dy}{dx}\right)_1^{(2)} = 0.05 + 1.0525 = 1.1025$$

The third approximation to y_1 is

$$y_1^{(3)} = 1 + \frac{1 + 1.1025}{2} \times 0.05 = 1 + 1.05125 \times 0.05 = 1.05256$$

Continuing the process, we obtain

$$\left(\frac{dy}{dx}\right)_1^{(3)} = 0.05 + 1.0 - 5256 = 1.10256$$

$$\therefore\ y_1^{(4)} = 1 + \frac{1 + 1.05256}{2} \times 0.05 = 1.05256$$

Since this $y_1^{(4)}$ is the same as $y_1^{(3)}$, we do not get any further change in y. Hence there is no use in continuing the process of approximations. We therefore take $y_1 = 1.05526$ and

$$d\left(\frac{dy}{dx}\right)_1 = 1.1026$$

Now first approximations to succeeding y's can be found more accurately from the formula

$$y_{n+1} = y_{n-1} + 2hy'_n \qquad \left[y'_n \text{ means } \left(\frac{dy}{dx}\right)_{x=x_n}\right] \tag{15.3.5}$$

To derive formula (15.3.5), we may note that the function y can be represented in the neighbourhood of x_n by Taylor's series. Then we write

$$f(x_n + h) = f(x_n) + hf'(x_n) + \frac{h^2}{2!}f''(x_n) + \cdots \tag{15.3.6}$$

$$f(x_n - h) = f(x_n) + hf'(x_n) - \frac{h^2}{2!}f''(x_n) - \cdots \tag{15.3.7}$$

Hence subtracting, we obtain

$$f(x_n + h) - f(x_n - h) = 2hf'(x_n) + \frac{h^3}{3}f'''(x_n) + \cdots \tag{15.3.8}$$

In terms of y, (15.3.8) can be written as

$$y_{n+1} - y_{n-1} = 2hy'_n$$

[$\because$ h is small, the term containing h^3 and above are neglected].

$$\Rightarrow y_{n+1} = y_{n-1} + 2hy'_n = y_{n-1} + 2h\left(\frac{dy}{dx}\right)_{x=x+n} \tag{15.3.9}$$

Using formula (15.3.9), the first approximation to y_2 is given by

$$y_2^{(1)} = 1 + 2.05(1.1026) = 1 + 0.1103 = 1.1103$$

Hence

$$\left(\frac{dy}{dx}\right)_2^{(1)} = 0.1 + 1.1103 = 1.2103$$

Then

$$y_2^{(2)} = 1.0526 + 0.05\left\{\frac{1.1026 + 1.2103}{2}\right\} = 1.1104$$

and $\left(\dfrac{dy}{dx}\right)_2^{(2)} = 0.1 + 1.1104 = 1.2104$

Hence

$$y_2^{(3)} = 1.0526 + 0.05\left\{\frac{1.1026 + 1.2104}{2}\right\} = 1.1104$$

which is same as $y_2^{(2)}$.

We therefore take $y_2 = 1.1104$ and

$$\left(\frac{dy}{dx}\right)_2 = 1.2104$$

Thus, we obtain the results:

x	y	dy/dx	
0.00	1.0000	1.0000	(initial point)
0.05	1.0526	1.1026	
0.10	1.1104	1.2104	

Similarly, we may derive $y_3, y_4, \ldots$, etc. and

$$\left(\frac{dy}{dx}\right)_3, \left(\frac{dy}{dx}\right)_4, \ldots, \text{etc.}$$

15.4 Picard's method of successive approximations

Let us consider the differential equation of first order given by

$$\frac{dy}{dx} = f(x, y)$$

Hence

$$dy = f(x, y)dx = \left(\frac{dy}{dx}\right) dx$$

Integrating the above between corresponding limits for x and y, we have

$$\int_{y_0}^{y} dy = \int_{x_0}^{x} f(x, y)dx = \int_{x_0}^{x} \left(\frac{dy}{dx}\right) dx$$

$$\Rightarrow y = y_0 + \int_{x_0}^{x} f(x, y)dx = y_0 + \int_{x_0}^{x} \left(\frac{dy}{dx}\right) dx \qquad (15.4.1)$$

Let us confine our attention to the first form in (15.4.1) given by

$$y = y_0 + \int_{x_0}^{x} f(x, y)dx \qquad (15.4.2)$$

We may notice that the equation (15.4.2) is complicated by the presence of y under the integral sign as well as outside it. An equation of this type can be solved by a process of successive approximations or iterations.

Therefore, the first approximation of y may be written as

$$y^{(1)} = y_0 + \int_{x_0}^{x} f(x, y_0)dx$$

The integral is now a function of x along and can therefore be integrated at least theoretically. The successive approximations of y will be:

Second approximation: $y^{(2)} = y_0 + \int_{x_0}^{x} f\left(x, y^{(1)}\right) dx$

....

....

....

nth approximation: $y^{(n)} = y_0 + \int_{x_0}^{x} f\left(x, y^{(n-1)}\right) dx$

The process of iteration is stopped when the values of $y^{(n-1)}$ and $y^{(n)}$ are nearly the same.

Let us now apply Picard's successive approximations to the same problem in Example 15.3.2.

Example 15.4.1. Solve $dy/dx = x + y$ with the initial conditions $x_0 = 0$, $y_0 = 1$.

Solution: The first approximation of y will be obtained from the formula

$$\int_{y_0}^{y} dy = \int_{x_0}^{x} \left(\frac{dy}{dx}\right) dx \quad \text{(note: } x_0 = 0, y_0 = 1\text{)}$$

$$\Rightarrow y - 1 = \int_{0}^{x} (x + y)dx$$

The first approximation is

$$y^{(1)} = 1 + \int_0^x (x+1)dx = 1 + \frac{d^2}{2} + x$$

For the second approximation

$$y^{(2)} = 1 + \int_0^x \left(x + 1 + \frac{x^2}{2} + x\right) dx = 1 + x + x^2 + \frac{d^3}{6}$$

For the third approximation

$$y^{(3)} = 1 + \int_0^x \left(x + 1 + x + x^2 + \frac{x^3}{6}\right) dx = 1 + x + x^2 + \frac{x^3}{3} + \frac{x^4}{24}$$

For $x = 0.1$

$$y^{(3)} = 1 + (0.1) + (0.10)^2 + \frac{(0.10)^3}{3} + \frac{(0.10)^4}{24}$$

$$= 1 + 0.1 + 0.01 + 0.0003 + \frac{0.0001}{24}$$

$$= 1.1103 + 0.000004 = 1.110304$$

$$\therefore y = 1.1103$$

For $x = 0.2$, the corresponding value of $y^{(3)}$ will be

$$y^{(3)} = \frac{(0.2)^4}{24} + \frac{(0.2)^3}{P}3 + (0.2)^2 + 1$$

$$= \frac{0.0016}{24} + \frac{0.008}{3} + \frac{0.04}{1} + 0.2 + 1$$

$$= 0.00007 + 0.0027 + 0.04 + 0.2 + 1 \simeq 1.2427$$

We may expect a better value by containing the approximations to $y^{(4)}$ and $y^{(5)}$, etc. We shall, however, advocate to move up to the point $x = 0.1$ and start afresh with the revised initial value.

Taking $x = 0.1$, we obtained $y = 1.1103$.

$$y^{(1)} = 1.1103 + \int_{0.1}^x (x + 1.1103)dx$$

$$= 1.1103 + \left[\frac{x^2}{2} + 1.1103x\right]_{0.1}^x$$

$$= \frac{x^2}{2} + 1.1103x + 0.9943$$

Then the second approximation is

$$y^{(2)} = 1.1103 + \int_{0.1}^x \left(x + \frac{x^2}{2} + 1.1103x + 0.9943\right) dx$$

$$= \frac{x^3}{6} + 1.0552x^2 + 0.09943x + 1.0001$$

Similarly,

$$y^{(3)} = 1.1103 + \int_{0.1}^{x}\left(x + \frac{x^3}{6} + 1.0552x^2 + 0.9943x + 1.0001\right)dx$$

$$= \frac{x^4}{24} + 0.3517x^3 + 0.9972x^2 + 1.0001x + 1.0000$$

$$= 0.000067 + 0.002814 + 0.03989 + 0.20002 + 1.0000$$

For $x = 0.2$

$$y^{(3)} = 1.24279 \simeq 1.2428$$

Example 15.4.2. Use Picard's method to approximate y, then $x = 0.1$, $x = 0.2$ for $dy/dx = x + y^2$, where $y = 0$, when $x = 0$.

Solution: Let $y^{(1)}$ be the first approximation, then

$$y^{(1)} = 0 + \int_0^x (x + 0)dx = \frac{x^2}{2}$$

The second approximation $y^{(2)}$ is then

$$y^{(2)} = 0 + \int_0^x \left(x + \frac{x^4}{4}\right)dx = \frac{x^2}{2} + \frac{1}{20}x^5$$

The third approximation $y^{(3)}$ is then

$$y^{(3)} = 0 + \int_0^x \left\{x + \left(\frac{x^2}{2} + \frac{1}{20}x^5\right)^2\right\}dx$$

$$= \int_0^x \left(x + \frac{x^4}{4} + \frac{1}{400}x^{10} + \frac{1}{20}x^7\right)dx$$

$$= \frac{1}{2}x^2 + \frac{1}{20}x^5 + \frac{1}{4400}x^{11} + \frac{1}{160}x^8$$

$$= \frac{1}{2}x^2 + \frac{1}{20}x^5 + \frac{1}{160}x^8 + \frac{1}{4400}x^{11}$$

For $x = 0.1$,

$$y^{(2)} = \frac{0.01}{2} + \frac{0.00001}{20} = 0.005 + 0.0000005 = 0.00500$$

$$y^{(3)} = \frac{0.01}{2} + \frac{0.00001}{20} + \frac{0.00000001}{160} + \cdots = 0.00500$$

There is no difference between $y^{(2)}$ and $y^{(3)}$ (up to 5 decimal places).

$\therefore\ y = 0.00500$ for $x = 0.1$

For $x = 0.2$, we may take $x = 0.1$, $y^{(1)} = 0.005$ as the initial values, we may write the first approximation $y^{(1)}$ as

$$y^{(1)} = 0.005 + \int_{0.1}^{x} (x + 0.000025)dx$$

$$= 0.005\left[\frac{x^2}{2} + 0.000025x\right]_{0.1}^{x}$$

$$= 0.005 - 0.005 - 0.0000025 + \frac{1}{2}x^2 + \frac{25}{10^6}x$$

$\therefore$ for $x = 0.2$

$$y^{(1)} = \frac{0.04}{2} + \frac{25}{10^6} \times 0.2 = 0.02 + 0.000005 \simeq 0.0200$$

$$y^{(2)} = 0.005 + \int_{0.1}^{x} (x + 0.0004)dx = 0.005 + \left[\frac{x^2}{2} + 0.0004x\right]_{0.1}^{x}$$

$$= \frac{x^2}{2} + \frac{4}{10^4}x - \frac{4}{10^5}$$

$\therefore$ for $x = 0.2$

$$y^{(2)} = 0.02 + 0.00008 - 0.00004 = 0.02004$$

There is therefore no difference between $y^{(1)}$ and $y^{(2)}$ up to the 4 decimal places. Thus, we obtain

x	y	dy/dx
0.0	0	0
0.1	0.0050	0.100025
0.2	0.0200	0.200400

NOTE 15.4.1. The practical difficulties inherent in Picard's method as discussed in this section lie in the fact that we may very often face difficult and sometimes impossible integrations which we could have to perform repeatedly.

Fortunately, such difficulties and, in fact, all direct integrations, can be avoided by the methods which we propose to discuss in the next two sections.

15.5 Use of approximating polynomials

The difficulties of the Picard's method mentioned above may be avoided, if we replace dy/dx by a polynomial and then integrate this polynomial over any desired interval. The appropriate polynomial in this case is the Newton's backward interpolation formula because in the numerical integration of differential equations, we start with the given initial conditions and construct the solution from that point onward. Therefore, the values of the function immediately behind us are known but the values ahead are unknown. The problem is to find the next value ahead.

Writing

$$\frac{dy}{dx} = y'$$

we get

$$y' = y'_n + u\Delta_1 y'_n + \frac{u(u+1)}{2!}\Delta_2 y'_n + \frac{u(u+1)(u+2)}{3!}\Delta_3 y'_n + \frac{u(u+1)(u+2)(u+3)}{4!}\Delta_4 y'_n$$

$$= y'_n + \Delta_1 y'_n \cdot u + \Delta_2 y'_n \cdot \frac{u^2+u}{2} + \Delta_3 y'_n \cdot \frac{u^3+3u^2+2u}{6}$$

$$+ \Delta_4 y'_n \cdot \frac{u^4+6u^3+11u^2+6u}{24} \qquad (15.5.1)$$

where

$$u = \frac{x - x_n}{h}, \quad \text{or,} \quad x = x_n + uh$$

Since the change in y in any interval is given by the formula:

$$\Delta y = \int_{x_r}^{x_{r+1}} \left(\frac{dy}{dx}\right) dx = \int_{x_r}^{x_{r+1}} y' dx$$

By means of (15.5.1) the change in y over any interval, where dy/dx is continuous may be found. For any interval $x_{r+1} - x_r$

$$\Delta y = \int_{x_r}^{x_{r+1}} h\left[y'_n + \Delta_1 y'_n \cdot u + \Delta_2 y'_n \cdot \frac{u^2+}{2} + \Delta_3 y'_n \cdot \frac{u^3+3u^2+2u}{6} + \Delta_4 y'_n \cdot \frac{u^4+6u^3+11u^2+6u}{24}\right] du$$

$$\text{or, } \Delta y = h\left[uy'_n + \frac{u^2}{2}\Delta_1 y'_n + \frac{\frac{u^3}{3}+\frac{u^2}{2}}{2}\Delta_2 y'_n + \frac{\frac{u^4}{4}+u^3+u^2}{6}\Delta_3 y'_n + \frac{\frac{u^5}{5}+\frac{3}{2}u^4+\frac{11}{3}u^3+3u^2}{24}\Delta_4 y'_n\right]_{x_r}^{x_{r+1}} \quad (15.5.2)$$

Let us now compute the value of Δy for the intervals $x_{n1} - x_n, x_n - x_{n-1}, x_{n-1} - x_{n-2}$, etc.

Then, by substituting in (15.5.2) proper limits for u are

$$u_{r+1} = \frac{(x_{n+1} - x_n)}{h} \equiv \frac{h}{h} = 1$$

$$u_r = \frac{x_n - x_n}{h} = 0$$

Substituting these in (15.5.2) and on simplification, we get

$$\Delta y = I_n^{n+1} = h\left[y'_n + \frac{1}{2}\Delta_1 y'_n + \frac{5}{12}\Delta_2 y'_n + \frac{3}{8}\Delta_3 y'_n + \frac{251}{720}\Delta_4 y'_n\right]$$

For the interval $x_n - x_{n-1}$, the limits for u are

$$u_{r+1} = \frac{x_n - x_n}{h} = 0,$$

$$u_r = \frac{x_{n-1} - x_n}{h} = \frac{-h}{h} = -1$$

$$\therefore \Delta y = I_{n-1}^{n} = h\left[y'_n - \frac{1}{2}\Delta_1 y'_n - \frac{1}{12}\Delta_2 y'_n - \frac{1}{24}\Delta_3 y'_n - \frac{19}{720}\Delta_4 y'_n\right]$$

Proceeding in the same way for other intervals, we get the formulae for the changes in y in those intervals. The results for the serveral intervals are given

$$I_n^{n+1} = h\left[y'n + \frac{1}{2}\Delta_1 y'_n + \frac{5}{12}\Delta_2 y'_n + \frac{3}{8}\Delta_3 y'_n + \frac{251}{720}\Delta_4 y'_n\right] \quad (15.5.3)$$

$$I_{n-1}^{n} = h\left[y'_n - \frac{1}{2}\Delta_1 y'_n - \frac{1}{12}\Delta_2 y'_n - \frac{1}{24}\Delta_3 y'_n - \frac{19}{720}\Delta_4 y'_n\right] \quad (15.5.4)$$

$$I_{n-2}^{n-1} = h\left[y'_n - \frac{3}{2}\Delta_1 y'_n + \frac{5}{12}\Delta_2 y'_n + \frac{1}{24}\Delta_3 y'_n + \frac{11}{720}\Delta_4 y'_n\right] \tag{15.5.5}$$

$$I_{n-3}^{n-2} = h\left[y'_n - \frac{7}{2}\Delta_1 y'_n - \frac{53}{12}\Delta_2 y'_n - \frac{55}{24}\Delta_3 y'_n + \frac{251}{720}\Delta_4 y'_n\right] \tag{15.5.6}$$

In the practical application of formulae (15.5.3) and (15.5.4), the coefficients 251/720 and 19/720 may be replaced by 1/3 and 1/38 respectively.

By adding (15.5.3) and (15.5.4), we get

$$I_{n-1}^{n+1} = h\left[2y'_n + \frac{1}{3}\Delta_2 y'_n + \frac{1}{3}\Delta_3 y'_n + \frac{232}{720}\Delta_4 y'_n\right]$$

$$= 2h\left[y'n + \frac{1}{6}(\Delta_2 y'_n + \Delta_2 y'_n + \Delta_4 y'_n) - \frac{1}{180}\Delta_4 y'n\right] \tag{15.5.7}$$

By adding (15.5.4) and (15.5.5), we get

$$I_{n-2}^{n} = 2h\left[y'_n - \Delta_1 y'_n + \frac{1}{6}\Delta_2 y'_n - \frac{1}{180}\Delta_4 y'_n\right] \tag{15.5.8}$$

We may replace the first and the second differences in (15.5.8) by their values in terms of the y''s,

i.e. $\Delta_1 y'_n = y'_n - y'_{n-1}$

$\Delta_2 y'_n = y'_n - 2y'_{n-1} + y'_{n-2}$

and simplifying, we get the equivalent form of (15.5.8)

$$I_{n-2}^{n} = \frac{h}{3}(y'_n + 4y'_{n-1} + y'_{n-2}) - \frac{h}{90}\Delta_4 y'_n \tag{15.5.9}$$

which one may recognise as Simpson's one-third rule with its remainder term.

Also noting that $I_n^{n+1} = y_{n+1} - y_n$, we may write (15.5.3) as

$$\bar{y}_{n+1} = y_n + h\left[y'_n + \frac{1}{2}\Delta_1 y'_n + \frac{5}{12}\Delta_2 y'_n + \frac{3}{8}\Delta_3 y'_n + \frac{1}{3}\Delta_4 y'_n\right] \tag{15.5.10}$$

(Adams-Bashforth Method)

A bar is placed above y_{n+1} to indicate that it is a first approximation of y_{n+1} obtained by extrapolation. When (15.5.4) is applied to the new interval, the n and $n-1$ in that formula become $n+1$ and n respectively. Hence the corrected value of y at the forward end of the new interval is given by

$$y_{n+1} = y_n + h\left[\bar{y}'_{n+1} - \frac{1}{2}\Delta_1 \bar{y}'_{n+1} - \frac{1}{12}\Delta_2 \bar{y}'_{n+1} - \frac{1}{24}\Delta_3 \bar{y}'_{n+1} - \frac{1}{38}\Delta_4 \bar{y}'_{n+1}\right] \tag{15.5.11}$$

where $\bar{y}_{n+1}$ is the value of y' obtained from the given equation when the y therein is replaced by $\bar{y}_{n+1}$ found from (15.5.10).

NOTE 15.5.1. The above technique of replacing dy/dx by a polynomial and integrating that polynomial over an interval was given by J.C. Adams.

15.6 Methods of starting the solution

Formulae (15.5.3) and (15.5.4) involve the first four differences of y'. They can be constructed only when five consecutive values of y' are known. The first value of y' is obtained from the initial conditions and the given equation, the remaining four can be found by one or more of the following methods.

15.7 Using Taylor's series

We have

$$f(x) = f(x_0) + (x - x_0)f'(x_0) + \frac{(x - x_0)^2}{2!} f''(x_0) + \cdots \tag{15.7.1}$$

which can also be written in the form:

$$y = y_0 + (x - x_0)y_0' + \frac{(x - x_0)^2}{2!} y_0'' + \cdots \tag{15.7.2}$$

where x_0, y_0 denote tne initial values of x and y.

To find y by (15.7.2), it is desirable to keep $x - x_0$ numerically as small as possible for rapid convergence of the series and consequent high accuracy in the y's.

Hence in general, we shall work on both sides of x_0, i.e. we shall compute y's both to the right and to the left of the point $x = x_0$.

Noting that $y_1 = f(x_0 + h)$,

$y_{-1} = f(x_0 - h)$,

$y_2 = f(x_0 + 2h)$,

$y_{-2} = f(x_0 - 2h)$, etc.

We have from (15.7.2)

$$y_1 = y_0 + hy_0' + \frac{h^2}{2!} y_0'' + \frac{h^3}{3!} y_0''' + \frac{h^4}{4!} y_0'''' + \cdots \tag{15.7.3}$$

$$y_{-1} = y_0 - hy_0' + \frac{h^2}{2!} y_0'' - \frac{h^3}{2!} y_0'' - \frac{h^3}{3!} y_0''' + \frac{h^4}{4!} y_0'''' + \cdots \tag{15.7.4}$$

$$y_2 = y_0 + y_0'(2h) + \frac{(2h)^2}{2!} y_0'' + \frac{(2h)^3}{3!} y_0''' + \cdots \tag{15.7.5}$$

$$y_{-2} = y_0 - y_0' + (2h)\frac{(2h)^2}{2!} y_0'' + \frac{(2h)^3}{3!} y_0''' + \cdots \tag{15.7.6}$$

where h = interval between the equidistant values of x.

NOTE 15.7.1. If y be non-existent for $x < x_0$, then we compute y's only to the right of x_0 by substituting in (15 7.2) the proper values of h.

EXAMPLE 15.7.1. Let us consider the same differential equation used in 15.4.1 $dy/dx = x + y$ with the initial conditions $x_0 = 0$, $y_0 = 1$.

Solution: We have

$y' = x + y$,

$y'' = 1 + y$,

$y''' = y''$,

$y^{iv} = y'''$,

$y^{v} = y^{iv}$

Hence

$y_0' = x_0 + y_0 = 1$,

$y_0'' = 1 + y_0 = 2$,

$y_0''' = y_0'' = 2$,

$y^{iv} = y''' = 2$,

$y^{v} = y^{iv} = 2$

Taking $h = 0.1$ and substituting in (15.7.3), (15.7.4), (15.7.5), (15.7.6), we obtain

$$y_1 = y_0 + 0.1 \times 1 + \frac{0.01}{2} \cdot 2 + \frac{0.001}{6} \cdot 2 + \frac{0.0001}{24} \cdot 2$$

$$= 1 + 0.1 + 0.01 + 0\ 00033 + 0.000008 = 1.1103$$

$$y_{-1} = y_0 - 0.1 \times 1 + \frac{0.01}{2} \cdot 2 - \frac{0.001}{6} \cdot 2 + \frac{0.0001}{24} \cdot 2$$

$$= 1 - 0.1 + 0.01 - 0.00033 + 0.000008$$

$$= 1.01 - 0.10033 = 0.0909678 = 0.9097$$

$$y_2 = y_0 + 0.2 \times 1 + \frac{4 \times 0.01}{2} \cdot 2 + \frac{8 \times 0.001}{6} \cdot 2 + \frac{16 \times 0.0001}{24} \cdot 2$$

$$= 1 + 0.2 + 0.04 + 0.00267 + \frac{0.0004}{3}$$

$$= 1.24267 + 0.00013 = 1.2428$$

$$y_{-2} = y_0 - 0.2 + 0.04 - 0.00267 + 0.00013 = 0.08375$$

These values are all correct to 4 decimal place. Five desired consecutive values of y' may be found now

$$y'_{-2} = x_{-2} + y_{-2} = -0.2 + 0.8375 = 0.6375$$

$$y'_{-1} = x_{-1} + y_{-1} = -0.1 + 0.9097 = 0.8097$$

$$y'_0 = x_0 + y_0 = 1$$

$$y'_1 = x_1 + y_1 = 0.1 + 1.1103 = 1.2103$$

$$y'_2 = x_2 + y_2 = 0.2 + 1.2428 = 1.4428$$

15.8 Using Milne's formulae

Frequently we may have cases where the higher derivatives of $y' = f(x, y)$ cannot be obtained without excessive labour or cannot be obtained at all. In such situations, case of Taylor's series as given in (15.3.6) cannot be made to solve numerical differential equations.

A reasonably accurate method of solving numerical differential equations has been developed by W. E. Milne. It uses two quadrature formulae, one for integrating ahead by extrapolation and the other for checking the extrapolated value. We may derive these formulae by Newton's forward formula. In terms of y' and u, we may write

$$y' = y'_0 + u\Delta y'_0 + \frac{u(u-1)}{2!}\Delta^2 y'_0 + \frac{u(u-1)(u-2)}{3!}\Delta^3 y'_0 + \frac{u(u-1)(u-2)(u-3)}{4!}\Delta^4 y'_0 + \cdots \qquad (15.8.1)$$

where

$$u = \frac{x - x_0}{h}, \quad \text{or,} \quad x = x_0 + hu$$

Integrating this formula over the range x_0 to $x_0 + 4h$ or $u = 0$ to $u = 4$, we have

$$\Delta y = \int_{x_0}^{x_0+4h} y'\,dx = h\int_0^4 \left\{ y'_0 + u\Delta y'_0 + \frac{u^2 - u}{2}\Delta^2 y'_0 + \frac{u^3 - 3u^2 + 2u}{6}\Delta^3 y'_0 + \frac{u^4 - 6u^3 + 11u^2 - 6u}{24} \cdot \Delta^4 y'_0 \right\} du$$

$$= h\left\{ \Delta y'_0 + 8\Delta y'_0 + \frac{20}{3}\Delta^2 y'_0 + \frac{8}{3}\Delta^3 y'_0 + \frac{28}{90}\Delta^4 y'_0 \right\} \text{ (on simplification).}$$

Replacing the differences (up to Δ^3) by their values and simplifying, we get

$$\Delta y = \frac{4h}{3}(2y_1' - y_2' + 2y_3') + \frac{28}{90}\Delta^4 y_0'$$

$\because \Delta y = y_4 - y_0$ (in this case)

$$\therefore y_4 = y_0 + \frac{4h}{3}(2y_1' - y_2' + 2y_3') + \frac{28}{90}\Delta^4 y_0' \qquad (15.8.2)$$

This is Milne's extrapolation formula or Predictor formula.

To get the other formula, which we called the *checking formula,* we integrate (15.8.1) over the interval x_0 to $x_0 + 2h$ or from $u = 0$ to $u = 2$, then

$$\Delta y = h\left(2y_0' + 2\Delta y_0' + \frac{1}{3}\Delta^2 y_0' - \frac{1}{90}\Delta^4 y_0'\right)$$

Replacing $\Delta y_0'$, $\Delta^2 y_0'$ by their values

$$\Delta y = \frac{h}{3}(y_0' + 4y_1' + y_2') - \frac{h}{90}\Delta^4 y_0' \qquad (15.8.3)$$

This the the Corrector formula of Milne.

For small h, ignoring the error, the final forms of the two formulae are

$$\Delta y \simeq \frac{\Delta h}{3}(2y_1' - y_2' + 2y_3') \quad \text{[Predictor formula]}$$

$$\Delta y \simeq \frac{h}{3}(y_0' + 4y_1' + y_2') \quad \text{[Corrector formula]}$$

True value of Δy in each case being given by (15.8.2) and (15.8.3) respectively.

15.9 Using Runge's method

Let us consider the equation

$$\frac{dy}{dx} = f(x, y)$$

with initial condition $y = y_0$, when $x = x_0$.

Let h be the interval of the equispaced values of x, then the first increment in y is obtained by the set of formulae given below:

$$k = f(x_0, y_0) = \left(\frac{dy}{dx}\right) = \text{slope at the beginning of interval } (x_0, y_0) \qquad (15.9.1)$$

$$k' = f(x_0 + h, y_0 + kh) \qquad (15.9.2)$$

$$k'' = f(x_0 + h, y_0 + k'h) = \text{slope at the end of the interval } (x_0, y_0) \qquad (15.9.3)$$

The slope at the middle point of the interval (x_0, y_0)

$$k_1 = f\left(x_0 + \frac{h}{2}, y_0 + k \cdot \frac{h}{2}\right) \qquad (15.9.4)$$

Now obtain the increment in the value of y in the interval by the formula given by

$$\Delta y = \frac{h}{6}(k + 4k_1 + k'') \qquad (15.9.5)$$

Result (15.8.8) is obtained by Simpson's rule

$$\Delta y = \int_{x_0}^{x_0+\delta x} \frac{dy}{dx} \cdot dx \text{ by the following table:}$$

x	x_0	$x_0 + \frac{h}{2}$	$x_0 + h$
$\frac{dy}{dx}$	k	k'	k''

The increment in y in the second interval is obtained by the following formulae

$k = f(x_1, y_1) = \left(\frac{dy}{dx}\right)$ = slope at the beginning of the second interval

$k' = f(x_1 + h, y_1 + kh) = f(x_1 + h, y_1 + \Delta y_1)$, (say)

$k'' = f(x_1 + h, y_1 + k'h) = f(x_1 + h, y_1 + \Delta y_2)$, (say)

= slope at the end of the second interval

$k_1 = f\left(x_1 + \frac{h}{2}, y_1 + k\frac{h}{2}\right)$ = slope at the middle of the second interval.

Then

$$\Delta y = \frac{h}{6}(k + 4k_1 + k'')$$

In the same way the increments in y can be obtained in the succeeding intervals.

15.10 Using Runge-Kutta method

The method devised by Runge was extended by Kutta a few years later. Here the increments of the function are calculated once for all by means of a definite set of formulae:

Let us consider again the equation

$\frac{dy}{dx} = f(x, y)$ (with initial values x_0, y_0).

Let h be the length of the interval between equidistant values of x.

The first increment in y is computed by the following set of formulae:

$$k_1 = f(x_0, y_0)h$$

$$k_2 = f\left(x_0 + \frac{h}{2}, y_0 + \frac{k_1}{2}\right)h$$

$$k_3 = f\left(x_0 + \frac{h}{2}, y_0 + \frac{k_2}{2}\right)h$$

$$k_4 = f(x_0 + h, y_0 + k_3)h$$

Then

$$\Delta y = \frac{1}{6}(k_1 + 2k_2 + 2k_3 + k_4)$$

and so on for the succeeding intervals.

Example 15.10.1. Use Runge-Kutta method to approximate y, when $x = 0.1$, given that $y = 1$, when $x = 0$ and

$$\frac{dy}{dx} = x + y$$

Solution: Here $x_0 = 0$, $y_0 = 1$, $f(x, y) = x + y$, $h = 0.1$.

Let

$k_1 = f(x_0, y_0)h = 1 \times 0.1 = 0.1$

$k_2 = f(x_0 + 0.05, y_0 + 0.05)h = 0.11$

$k_3 = f(x_0 + 0.05, y_0 + 0.055)h = 0.1(0 + 0.05 + 1.055)$

$= 0.1 \times 1.105 = 0.1105$

$k_4 = f(x_0 + 0.1 y_0 + 0.1105)h = 0.1\{0 + 0.1 + 1 + 0.1105\}$

$= 0.1 \times 1.2105 = 0.12105$

$\therefore \Delta y = \frac{1}{6}(k_1 + 2k_2 + 2k_3 + k_4) = \frac{1}{6}(0.1 + 0.22 + 0.2210 + 0.12105)$

$= \frac{1}{6} \times 0.66205 = 0.11034$

$x_1 = x_0 + h = 0.1$

$y_1 = y_0 + k = 1.1103$

15.11 Conditions for convergence

The conditions for convergence of the numerical solution of the differential equation by approximating polynomials can be easily obtained by means of Picard's process.

For the equation

$$\frac{dy}{dx} = y' = f(x, y)$$

the Picard's process converges to the true solution, if

$$h < \frac{1}{\left|\frac{\partial f}{\partial y}\right|_{max}}$$

in the region of integration.

Since a polynomial can be made to approximate a continuous function to some required degree of accuracy, the approximating polynomial can be made to approach the Picard's solution by taking h sufficiently small. Polynomial solution can therefore be made to coincide with the Picard's solution. Hence the condition of convergence holds for approximating polynomial method.

EXERCISES

1. Use Euler's method and an interval of $h = 0.1$ to find a solution of the differential equation $\frac{dy}{dx} = x^2 + y^2$ with initial condition $y = 0$ for $x = 0$ in the range $0 \le x \le 1$.
2. Use Euler's modified method to solve $\frac{dy}{dx} = x + y$ with initial condition $x_0 = 0$, $y_0 = 1$ in steps of 0.2.
3. Solve by Picard's method $\frac{dy}{dx} = 2x - y$ with $x_0 = 1$, $y_0 = 3$.
4. Use Taylor's series method to solve

   (a) $\frac{dy}{dx} = x^2 + y^2$ with $x_0 = 1$, $y_0 = 0$

   (b) $y' = y \sin x + \cos x$ with $x = 0$, $y = 0$

5. Using Runge-Kutta's method, solve $y' = \dfrac{1}{x+y}$ with $x = 0$, $y = 1$ for $x = 0.5(0.5)2$.
6. Use Runge-Kutta's method to approximate y, when $x = 0.1$, given that $y = 1$ at $x = 0$ and $\dfrac{dy}{dx} = 3x + y^2$.
7. Use Runge-Kutta's method to solve
   (a) $\dfrac{dy}{dx} = \dfrac{1}{x+y}$ for $x = 0.5$ with $x_0 = 0$, $y_0 = 1$.
   (b) $\dfrac{dy}{dx} = xy$ for $x = 1.4$ with $x_0 = 1$, $y_0 = 2$ (take $h = 0.2$).
8. Find y_2, if y_x is the solution of $\dfrac{dy}{dx} = \dfrac{1}{2}(x + y)$ assuming $y_0 = 2$, $y_{0.5} = 2.636$, $y_{1.0} = 3.595$, $y_{1.5} = 4.968$ (*use Milne's Predictor formula and Corrector formula*).
9. Tabulate by Predictor-Corrector formula the numerical solution of $\dfrac{dy}{dx} = x + y$ with initial conditions $x_0 = 0$, $y_0 = 1$ from $x = 0.20$ to $x = 0.30$.
10. Obtain Milne's *Predictor* and *Corrector* formulae. Why are they so called? At how many points x, the value of y must be initially known to apply the method? Can they be applied in any order?
11. Solve by Taylor Series method the following differential equation $\dfrac{dy}{dx} + \dfrac{y}{x} = \dfrac{1}{x^2}$ with $y = 0$, when $x = 1$, to obtain $y_{1.1}, y_{1.2}, y_{1.3}$ correct to three places of decimals.
12. Solve $\dfrac{dy}{dx} = 3x + \dfrac{1}{2}y$ with $y = 1$, when $x = 0$ by Runge-Kutta method.
13. Obtain $y_{0.1}$ by Euler's modified method, given $\dfrac{dy}{dx} = x + y$ with $y = 1$, when $x = 0$, taking $h = 0.05$.
14. Find four successive approximations in solving the following differential equations by Picard's method:
    (a) $\dfrac{dy}{dx} = x^2 - y^2$ with $y = 0$ at $x = 0$.
    (b) $\dfrac{dy}{dx} = xy + \sqrt{x}$ with $y = 0$ at $x = 0$.

Chapter 16

The Normal Law of Error

Summary

16.1 Introduction

All physical measurements are subject to error. No matter how much care one may take, whatever measuring instruments are used, there is bound to be variations in the observed measurements. Careful examinations may reveal that there is always some difference between the true measurement and recorded measurement of any physical quantity and this difference is called the error of measurements.

There may be three kinds of errors:

1. *Systematic errors,* which affect all measurements alike. They may be due to imperfections in the construction or adjustment of instruments, the "personal equation of the observer." This type of error can be rectified by applying appropriate corrective measures.
2. *Mistakes.* These are large errors due to careless reading of measuring instruments or faulty recording of the observations. Mistakes do not follow any law of chance and can be avoided only by exercising constant vigil and careful checking.
3. *Accidental errors.* These are errors, whose causes are unknown. They are usually small and they follow the laws of chance. The mathematical theory of errors deals with accidental errors only.

16.2 Law of error

Let X denote the physical measurement of a physical quantity, whose hypothetical true value is m. We shall assume the following linear model in our error theory $X = m + E$, where E denotes the random error (*accidental error*). Evidently X is a random variable and if we assume for a theoretical model that a measurement may yield any real number, X can then be regarded as a random variable defined in $-\infty < X < +\infty$.

The probability density function (pdf) of X should naturally contain the true value m as a parameter.

The random variable (r.v.) $E = X - m$ will be called the error of measurement. E has a probability distribution (pdf), say, $g(x - m)$. We shall assume that the probability distribution of error will be independent of the unknown true value of the quantity (m).

Hence we may write

$$f(x, m) = g(x - m) \tag{16.2.1}$$

where $f(x, m)$ is the pdf of X.

Since the true value m is unknown, naturally we may think of finding an estimator for m on the basis of the random observations $x_1, x_2, \ldots, x_n$. The best estimate of m, according to Gauss, is the most probable value of m. Let $m(x_1, x_2, \ldots, x_n)$ be the best estimate of m on the basis of n independent observations $x_1, x_2, \ldots, x_n$. The corresponding estimator or statistic $m(X_1, X_2, \ldots, X_n)$, according to Gauss, will be independently distributed and consistent, and satisfies the following axioms.

Axiom 1: $m(x_1, x_2, \ldots, x_n)$ will be independent of origin, i.e. for all h, $m(x_1 + h, x_2 + h, \ldots, x_n + h) = m(x_1, x_2, \ldots, x_n) + h$.

Axiom 2: $m(x_1, x_2, \ldots, x_n)$ will be independent of the unit of measurement, i.e. for every real number k, $m(kx_1, kx_2, \ldots, kx_n) = km(x_1, x_2, \ldots, x_n)$.

Axiom 3: $m(x_1, x_2, \ldots, x_n)$ will be independent of the order of the observations $x_1, x_2, \ldots, x_n$, i.e. $m(x_1, x_2, \ldots, x_n)$ is a symmetric function of $x_1, x_2, \ldots, x_n$.

Axiom 4: The function $m(x_1, x_2, \ldots, x_n)$ must have a single valued continuous derivative at every point.

We may now see that the most plausible estimator of m, satisfying the above four axioms will be the arithmetic mean of the observations

$$\bar{x} = \frac{1}{n}\sum_{i=1}^{n} x_i$$

Because,

$$m(kx_1, kx_2, \ldots, kx_n) = m(0, 0, \ldots, 0) + k\sum x_i \left[\frac{\partial m}{\partial x_i}\right]_{\theta k x_i} \quad (0 < \theta < 1) \text{ [axiom 1]} \tag{16.2.2}$$

Making $k \to 0$, we have

$$m(0, 0 \ldots, 0) = 0 \tag{16.2.3}$$

$$\therefore m(kx_1, kx_2, \ldots, kx_n) = k\sum_{i=1}^{n} x_i \left[\frac{\partial m}{\partial x_i}\right]_{\theta kx_i} + m(0, 0, \ldots, 0)$$

$$\Rightarrow km(x_1, x_2, \ldots, x_n) = k\sum_{i=1}^{n} x_i \left[\frac{\partial m}{\partial x_i}\right]_{\theta kx_i} + m(0, 0, \ldots, 0)$$

Dividing by k and making $k \to 0$, we have

$$m(x_1, x_2, \ldots, x_n) = \sum_{i=1}^{n} x_i \left[\frac{\partial m}{\partial x_i}\right]_{(0,0,\ldots,0)} \quad (0 < \theta < 1) \tag{16.2.4}$$

By axiom (3), the constants

$$\left[\frac{\partial m}{\partial x_i}\right]_{(0,0,\ldots,0)}$$

must all be the same, say C.

$$\therefore \hat{m}(x_1, x_2, \ldots, x_n) = C\sum_{i=1}^{n} x_i \tag{16.2.5}$$

By axiom (1),

$$C\sum_{i=1}^{n}(x_i + h) = C\sum_{i=1}^{n} x_i + h \Rightarrow \sum_{i=1}^{n}(x_i + h) = \sum_{i=1}^{n} x_i + \frac{h}{C}$$

$$\Rightarrow \sum x_i x + h = \sum x_i + \frac{h}{C} \Rightarrow C = \frac{1}{n} \tag{16.2.6}$$

Substituting C in (16.2.6)

$$\hat{m}(x_1, x_2, \ldots, x_n) = \frac{1}{n}\sum_{i=1}^{n} x_i = \bar{x} \text{ (= sample mean).}$$

We now refer to our result in (16.2.1). It was obtained on the basis of the hypothesis that the probability distribution of error is independent of the true value m.

From (16.2.1), we have

$$f(x, m) = g(x - m)$$

Then the likelihood function L is given by

$$L = g(x_1 - m)g(x_2 - m)\cdots g(x_n - m)$$

$$\text{or, } \log L = \sum_{i=1}^{n} \log g(x_i - m) \tag{16.2.7}$$

To obtain the maximum likelihood estimator, we take

$$\frac{\partial}{\partial m}\log L = 0 \Rightarrow \sum_{i=1}^{n} \frac{g'(x_i - m)}{g(x - m)} \tag{16.2.8}$$

Setting for convenience,

$$G(x - m) = \frac{g'(x_i - m)}{g(x - m)}$$

we have

$$\sum_{i=1}^{n} G(x_i - m) = 0, \quad \text{i.e.} \quad \sum G(e_i) = 0 \tag{16.2.9}$$

Since $\hat{x} = \bar{x}$ is the maximum likelihood estimator of m, therefore we obtain

$$\sum_{i=1}^{n}(x_i - m) = 0, \quad \text{i.e.} \quad \sum_{i=1}^{n} e_i = 0$$

Then

$$\sum_{i=1}^{n} G(x_i - m) = 0, \quad \text{and} \quad \sum (x_i - m) = 0$$

and we have

$$\sum_{i=1}^{n} \{G'(x_i - m) + \lambda\} \cdot d(x_i - m) = 0$$

where λ is Lagrange's multiplier (constant).

$\therefore G'(x_i - m) + \lambda = 0$

Integrating,

$$G(x_i - m) + \lambda(x_i - m) + A = 0 \tag{16.2.10}$$

where A is an unknown constant.

$\because \sum G(x_i - m) = 0$ and $\sum (x_i - m) = 0. \quad \therefore A = 0$

$$\therefore G(x_i - m) = -\lambda(x_i - m) \Rightarrow \frac{g'(x_i - m)}{g(x_i - m)} = -\lambda(x_i - m)$$

or generally, we have

$$\frac{g'(x - m)}{g(x - m)} = -\lambda(x - m), \ \lambda > 0$$

Integrating,

$$\log g(x - m) = B - \frac{\lambda}{2}(x - m)^2$$

where B is a constant.

$$\therefore g(x - m) = Ce^{-\lambda/2(x-m)^2}, \quad \text{i.e.} \quad f(x, m) = Ce^{-\lambda/2(x-m)^2} \tag{16.2.11}$$

The constant C is so chosen as to make

$$C \int_{-\infty}^{+\infty} e^{-\lambda/2(x-m)^2} dx = 1 \ [\because f(x, m) \text{ is the pdf of } X]$$

Writing $\lambda = 2h^2$, we get

$$C \cdot \int_{-\infty}^{+\infty} e^{-h^2(x-m)^2} dx = 1$$

Putting $h(x-m) = Z$ $(-\infty < Z < +\infty)$, we have

$$\frac{C}{h}\int_{-\infty}^{+\infty} e^{-Z^2}dz = 1 \Rightarrow \frac{2C}{h}\int_{0}^{+\infty} e^{-Z^2}dz = 1 \Rightarrow \frac{2C}{h}\cdot\frac{\sqrt{\pi}}{2} = 1 \Rightarrow C = \frac{h}{\sqrt{\pi}}$$

$$\therefore f(x, m) = g(x-m) = \frac{h}{\sqrt{\pi}}e^{-h^2(x-m)^2} \tag{16.2.12}$$

This shows that the probability distribution of X is normal with mean = m (*true value*) and variance $\sigma^2 = 1/2h^2$.

Since $(x-m)$ is the error of measurement, the above result shows that the distribution of E is normal with zero mean and variance = $1/2h^2$.

Note: If variance of E is small, it implies that the observations closely cluster round the true value m. In other words, the measurements have good precision.

Therefore, h is called the index of precision, and $h = 1/\sigma\sqrt{2}$.

16.3 The law of error of a linear function of independent quantities

We next state a fundamental theorem of great importance.

THEOREM 16.3.1. If $M_1, M_2, \ldots, M_n$ are independent observed quantities whose laws of error are

$$y = \frac{h_1}{\sqrt{\pi}}e^{-h_1^2x^2},\quad y = \frac{h_2}{\sqrt{\pi}}e^{-h_2^2x^2},\quad \ldots,\quad y = \frac{h_n}{\sqrt{\pi}}e^{-h_n^2x^2}$$

then any linear function of these quantities obeys a similar law of error.

Proof. Let the linear function be

$$F = L_1M_1 + L_2M_2 + \cdots + L_nM_n \tag{16.3.1}$$

where $l_1, l_2, \ldots, l_n$ are arbitrary constants.

If $x_1, x_2, \ldots, x_n$ be the errors of $M_1, M_2, \ldots, M_n$ respectively and ξ denotes the corresponding error in F, we have

$$F + \xi = l_1(M_1 + x_1) + l_2(M_2 + x_2) + \cdots + l_n(M_n + x_n)$$
$$= (l_1M_1 + \cdots + l_nM_n) + l_1x_1 + \cdots + l_nx_n$$
$$\therefore \xi = l_1x_1 + l_2x_2 + \cdots + l_nx_n$$

Thus, the error in F will be a linear function of the corresponding errors in $M_1, M_2, \ldots,$ M_n. Since $x_1, x_2, \ldots, x_n$ are independent errors, we may say that they are independently and normally distributed, each with mean zero and variance

$$\frac{1}{2h_1^2}, \frac{1}{2h_2^2}, \ldots, \frac{1}{2h_n^2}. \qquad \therefore \xi = l_1x_1 + l_2x_2 + \cdots + l_nx_n$$

will be normally distributed with mean zero and variance

$$\frac{a_1^2}{2h_1^2} + \frac{a_2^2}{2h_2^2} + \cdots + \frac{a_n^2}{2h_n^2} \equiv \frac{1}{2H^2}, \text{ (say)},$$

where

$$\frac{1}{H^2} = \frac{a_1^2}{h_1^2} + \frac{a_2^2}{h_2^2} + \cdots + \frac{a_n^2}{h_n^2}$$

$\therefore$ probability distribution of ξ will be

$$g(\xi) = \frac{H}{\sqrt{\pi}} e^{-H^2\xi^2}, \quad -\infty < \xi < +\infty, \; H > 0$$

Note: H is the index of Precision of ξ.

16.4 The principle of least squares

Let $x_1, x_2, \ldots, x_n$ be a set of n measurement of a quantity, whose true value is m, and the measurements be made using the same care, methods and instruments. Whatever care one may take, it will be seen that the results of these measurements do not agree exactly among themselves. The question is: how to decide upon the best result obtainable from the given set of measurements.

This problem can be tackled by using the principle of Least Squares, which states that the most probable value of the unknown true value m, is that value for which the sum of squares of errors is least. This is quite in agreement with common sense reasoning. The requirement that the sum of squares of the errors shall be minimum implies that the arithmetic mean of the measurements is the best value which may be obtained from any sėt of equally trustworthy direct measurements.

The principle of least squares also follows from the Normal Law of errors, as will be shown now.

Let $m_1, m_2, \ldots, m_n$ be equally trustworthy direct measurements of a physical quantity, whose true value is m and let $x_1, x_2, \ldots, x_n$ be the errors of this measurements. The precision constant h of the probability distribution will be same for all the measurements.

Since the separate measurements are independent, the probability that a set of errors $x_1, x_2, \ldots, x_n$ will be made is

$$P = \left(\frac{h}{\sqrt{\pi}}\right) e^{-h^2(x_1^2+x_2^2+\cdots+x_n^2)} dx_1\, dx_2 \cdots dx_n, \quad -\infty < x_i < +\infty, \quad i = 1, 2, \ldots, n \tag{16.4.1}$$

The set of errors for which the above probability is maximum will lead to the solution of the most plausible value of the quantity measured.

Now P in (16.4.1) will be maximum, when the exponent of e is the least, i.e. when $x_1^2 + x_2^2 + \cdots + x_n^2$ is a minimum.

Thus, from the joint probability distribution of errors, we obtain the *Principle of Least Squares.*

16.5 Residual

In the preceding section we have defined error as the difference between the observed value (measured) and the true value. Thus, $E = X \sim m$.

i.e. Error = −True Value + Measured Value. (16.5.1)

Since m is not exactly known, we do not get an exact value of the error committed in the measurement. We, however, find the most probable value of m, say m_0, and define the difference between the most probable value and the measured value as the Residual. Thus,

Residual = Most Probable Value ~ Measured Value. (16.5.2)

Let $x_1, x_2, \ldots, x_n$ denote the values of n separate measurements and let m_0 be the most probable value of the measured quantity, whose true value is m. Then the errors are

$$e_1 = x_1 - m,$$
$$e_2 = x_2 - m,$$
$$\cdots\cdots\cdots\cdots$$
$$e_n = x_n - m$$

The residuals $v_1, v_2, \ldots, v_n$ are

$$v_1 = x_1 - m_0$$
$$v_2 = x_2 - m_0$$
$$\cdots\cdots\cdots\cdots$$
$$v_n = x_n - m_0 \tag{16.5.3}$$

The true value m is unknown and cannot be exactly found. The principle of least squares says that the most probable value is that which makes the sum of the weighted squares of the errors a minimum, i.e.

$$\phi(m) = \omega_1(x_1 - m)^2 + \omega_2(x_2 - m)^2 + \cdots + \omega_n(x_n - m)^2 \tag{16.5.4}$$

must be minimum.

The solution for m will be obtained by taking

$$\frac{d}{dm}\phi(m) = 0$$

$$\sum_{i=1}^{n} \omega_i(x_i - m) = 0 \Rightarrow \sum \omega_i x_i = m \sum \omega_i$$

$$\therefore\ m_0 = \sum_{i=1}^{n} \omega_i x_i \Big/ \sum_{i=1}^{n} \omega_i \tag{16.5.5}$$

This value m_0 is called the weighted average of the several measurements.

If the measurements have equal weights $\omega_1 = \omega_2 = \cdots = \omega_n$, then (16.5.5) reduces to

$$m_0 = \frac{1}{n} \sum x_i \tag{16.5.6}$$

Equations (16.5.5) and (16.5.6) immediately suggest the following theorem of importance:

In any set of measurements of equal weight, the algebraic sum of the residuals is zero, and in a set of measurements of unequal weights, the algebraic sum of the weighted residuals is zero.

16.6 Law of error for residuals

We shall now show that when errors of a set of measurements follow a Normal Law of Error, the residuals will follow a similar law. With notations used in section (16.5), we may write

(v)	(e)
$v_1 = x_1 - m_0$	$e_1 = x_1 - m$
$v_2 = x_2 - m_0$	$e_2 = x_2 - m$
....	....
$v_n = x_n - m_0$	$e_n = x_n - m$

and let $\omega_1, \omega_2, \ldots, \omega_n$ be the weights.

For the measurements of equal weight, we have from column (v),

$$\sum_{i=1}^{n} v_i = \sum x_i - nm_0 = 0, \quad \text{or,} \quad m_0 = \frac{1}{n}\sum x_i \tag{16.6.1}$$

and from column (e), we may similarly get

$$\sum_{i=1}^{n} e_i = \sum_{i=1}^{n} x_i - nm, \quad \text{or,} \quad \sum x_i = nm + \sum e_i$$

$$\Rightarrow nm_0 = nm + \sum e_i \Rightarrow m_0 = m + \frac{1}{n}\sum e_i \tag{16.6.2}$$

Substituting the value of m_0 in the equations of column (v), we have

$$v_1 = x_1 - m_0 = x_1 - m - \frac{1}{n}\sum e_i$$

$$= e_1 - \frac{1}{n}\sum_{i=1}^{n} e_i = \left(1 - \frac{1}{n}\right) e_1 - \frac{1}{n} e_2 - \frac{1}{n} e_3 \cdots - \frac{1}{n} e_n$$

Similarly,

$$\left.\begin{aligned} v_2 &= -\frac{1}{n} e_1 + \left(1 - \frac{1}{n}\right) e_2 - \frac{1}{n} e_3 - \cdots - \frac{1}{n} e_n \\ &\cdots\cdots\cdots\cdots\cdots\cdots\cdots\cdots \\ v_n &= -\frac{1}{n} e_1 - \frac{1}{n} e_2 - \cdots - \frac{1}{n} e_{n-1} + \left(1 - \frac{1}{n}\right) e_n \end{aligned}\right\} \tag{16.6.3}$$

Thus, we have proved that the residuals are the linear functions of the errors. Hence they follow the Normal Law.

In particular, v_i will be normally distributed with mean zero and variance

$$\left[\frac{1}{n^2} + \frac{1}{n^2} + \cdots \text{ to } (n-1) \text{ terms} + \left(1 - \frac{1}{n}\right)^2\right] \sigma^2$$

where σ^2 = common variance of the e_i's.

$$\therefore \text{ Variance } = \left[(n-1) \cdot \frac{1}{n^2} + \frac{(n-1)^2}{n^2}\right] \cdot \frac{1}{2h^2}$$

$$= \left[\frac{1}{n^2}(n-1)\{1 + n - 1\}\right] \frac{1}{2h^2}$$

$$= \left(\frac{n-1}{n}\right) \cdot \frac{1}{2h^2} \quad [h = \text{precision index of } e_i\text{'s}].$$

If H_i is the Precision index of v_i's, then

$$\frac{1}{2H_i^2} = \left(\frac{n-1}{n}\right)\cdot\frac{1}{2h^2} \Rightarrow \frac{1}{H_i^2} = \left(\frac{n-1}{n}\right)\cdot\frac{1}{h^2}$$

$$\therefore H_i^2 = \left(\frac{n}{n-1}\right)\cdot h^2 \Rightarrow H_i = \sqrt{\frac{n}{n-1}}\cdot h \quad (i = 1, 2, \ldots, n)$$

$$\therefore \phi(v) = \frac{H}{\sqrt{\pi}} e^{-H^2 v^2}, \quad -\infty < v < +\infty$$

Thus, the precision index of v is always greater than h, the precision index of error.

When the measurements are of unequal weights, the weighted residuals and the weighted errors are as given in column (ωv) and (ωe) below:

(ωv)	(ωe)
$\omega_1 v_1 = \omega_1 x_1 - \omega_1 m_0$	$\omega_1 e_1 = \omega_1 x_1 - \omega_1 m$
$\omega_2 v_2 = \omega_2 x_2 - \omega_2 m_0$	$\omega_2 e_2 = \omega_2 x_2 - \omega_2 m$
....	....
$\omega_n v_n = \omega_n x_n - \omega_n m_0$	$\omega_n e_n = \omega_n x_n - \omega_n m$

Adding the equations in (ωv) column

$$\sum \omega_i v_i = \sum \omega_i x_i - m_0 \sum \omega_i = 0 \Rightarrow m_0 = \sum \omega_i x_i \Big/ \sum \omega_i$$

($\because$ sum of the weighted residuals is zero).

Again adding the equations in (ωe) column,

$$\Sigma\omega_i e_i = \Sigma\omega_i x_i - m\Sigma\omega_i$$
$$\Rightarrow \Sigma\omega_i x_i = \Sigma\omega_i e_i + m\Sigma\omega_i$$
$$\Rightarrow m_0\Sigma\omega_i = \Sigma\omega_i e_i + m\Sigma\omega_i$$
$$\therefore m_0 = \frac{\Sigma\omega_i e_i}{\Sigma\omega_i} + m$$

Substituting this value of m_0 in the equations of column (ωv)

$$v_1 = x_1 - m - \frac{\Sigma\omega_i e_i}{\Sigma\omega_i} = e_1 - \frac{\Sigma\omega_i e_i}{\Sigma\omega_i}$$
$$= e_1 - \frac{\omega_1 e_1}{\Sigma\omega_i} - \frac{\omega_2 e_2}{\Sigma\omega_i} - \cdots - \frac{\omega_n e_n}{\Sigma\omega_i}$$
$$= \left(1 - \frac{\omega_1}{\Sigma\omega_i}\right) e_1 - \frac{\omega_2}{\Sigma\omega_i} e_2 - \cdots - \frac{\omega_n}{\Sigma\omega_n} e_n$$

In general, we may write

$$v_i = -\frac{\omega_1}{\Sigma\omega_i} e_1 - \frac{\omega_2}{\Sigma\omega_i} e_2 - \cdots + \left(1 - \frac{\omega_i}{\Sigma\omega_i}\right) e_i - \cdots - \frac{\omega_n}{\Sigma\omega_i} e_n$$

v_i, being linear function of independently distributed e_i's, will be normally distributed with mean zero, and index of precision H_i is given by

$$\frac{1}{H_i^2} = \frac{\omega_1^2/(\Sigma\omega_i)^2}{h_1^2} + \frac{\omega_2^2/(\Sigma\omega_i)^2}{h_2^2} + \cdots + \frac{\left(1 - \dfrac{\omega_i}{\Sigma\omega_i}\right)^2}{h_i^2} + \cdots + \frac{\omega_n^2/(\Sigma\omega_i)^2}{h_n^2}$$

$$= \frac{1}{(\Sigma\omega_i)^2}\left[\frac{\omega_1^2}{h_1^2} + \frac{\omega_2^2}{h_2^2} + \cdots + \frac{(\Sigma\omega_i - \omega_i)^2}{h_i^2} + \cdots + \frac{\omega_n^2}{h_n^2}\right]$$

We may have similar results for other residuals.

16.7 Measures of precision

Direct measurements with instruments always involve some error which is inherent in every system of measurements. In our present discussion, by error we shall always mean accidental error. A precise measurement is therefore one which is free from this accidental error. The precision of a measurement can be estimated in several ways.

There are three measures of precision in common use.

- The mean square error (MSE), denoted by μ.
- The probable error (PE), denoted by r.
- The average error, denoted by η.

We shall try to express μ, r and η in terms of the precision index h.

If X be random variable, denoting the accidental error for a single measurement, then

$$\text{Prob}\,(x < X \le x + dx) \equiv p = \frac{h}{\sqrt{\pi}} e^{-h^2x^2}\, dx$$

h being the index of precision.

1. MSE ($= \mu$) : Let n measurements of an object have accidental errors, represented by the random variables $X_1, X_2, \ldots, X_n$. The probability

$$\text{Prob}\,(x_i < X_i \le x_i + dx_i) \equiv p_i = \frac{h}{\sqrt{\pi}} e^{-h^2x_i^2}\, dx_i \quad (i = 1, 2, \ldots, n)$$

Errors being independently distributed, the probability of simultaneous occurrence of errors $x_1, x_2, \ldots, x_n$ is

$$P = p_1, p_2, \ldots, p_n = \left(\frac{h}{\sqrt{\pi}}\right)^n e^{-h^2(x_1^2+x_2^2+\cdots+x_n^2)}\, dx_1\, dx_2 \cdots dx_n$$

Let us now find the best or the most probable value of the index of precision h, which will make P maximum. For this, we make

$$\frac{dP}{dh} = 0 = \left(\frac{h}{\sqrt{\pi}}\right)^n \cdot e^{-h^2(x_1^2+\cdots+x_n^2)}\left\{-2h(x_1^2 + \cdots + x_n^2)\right\}$$

$$+ n \cdot e^{-h^2(x_1^2+\cdots+x_n^2)}\left(\frac{h}{\sqrt{\pi}}\right)^{n-1} \cdot \frac{1}{\sqrt{\pi}}$$

$$\text{or, } e^{-h^2(x_1^2+x_2^2+\cdots+x_n^2)}\left(\frac{h}{\sqrt{\pi}}\right)^{n-1}\frac{1}{\sqrt{\pi}}\left[-2h^2(x_1^2+x_2^2+\cdots+x_n^2)+n\right]=0$$

$$\text{or, } 2h^2(x_1^2+x_2^2+\cdots+x_n^2)=n$$

$$\therefore \frac{1}{2h^2}=\frac{x_1^2+x_2^2+\cdots+x_n^2}{n} \Rightarrow \frac{1}{\sqrt{2}h}=\sqrt{\frac{x_1^2+x_2^2+\cdots+x_n^2}{n}} \tag{16.7.1}$$

The quantity on the right of (16.7.1) is known as the mean square error (MSE) of a single observation and is denoted by the Greek letter μ.

$$\therefore \mu=\frac{1}{\sqrt{2}h}=\sqrt{\frac{x_1^2+x_2^2+\cdots+x_n^2}{n}} \tag{16.7.2}$$

2. The probable error (PE). The probable error, r, of a particular measurement of a set of measurements, is a quantity ($= r$) such that one-half the errors of the series are greater than it and the other half is less than it. In other words,

   $$\text{Prob}(-r \le x \le +r)=\frac{1}{2}$$

   Thus,

   $$\int_{-r}^{r}\frac{h}{\sqrt{\pi}}e^{-h^2x^2}\,dx=\frac{1}{2}$$

   Putting $t = hx$, we have

   $$2\int_0^r\frac{h}{\sqrt{\pi}}e^{-h^2x^2}\,dx=\frac{1}{2}, \quad \text{or,} \quad \int_0^{hr}\frac{1}{\sqrt{\pi}}e^{-t^2}\,dt=\frac{1}{4}$$

   $$\text{or, } \int_0^{hr}e^{-t^2}\,dt=\frac{\sqrt{\pi}}{4}, \quad \text{or, } \int_0^{\rho}e^{-t^2}\,dt=0.4431135, \ (\text{where } \rho=hr)$$

   $$\Rightarrow \int_0^{\rho}\left(1-t^2+\frac{t^4}{2}-\frac{t^6}{6}+\frac{t^8}{24}-\frac{t^{10}}{120}+\cdots\right)dt=0.4431135$$

   $$\text{or, } \rho-\frac{\rho^3}{3}+\frac{\rho^5}{10}-\frac{\rho^7}{42}+\frac{\rho^9}{216}-\frac{\rho^{11}}{1320}-0.4431135=0$$

   This can be solved by iteration method and we find

   $$\rho=0.4769363. \quad \therefore r=\frac{\rho}{h}=\frac{0.4769}{h} \tag{16.7.3}$$

   $$\because \frac{1}{\sqrt{2}h}=\sqrt{\frac{x_1^2+x_2^2+\cdots+x_n^2}{n}}$$

   $$\therefore r=0.4769\cdot\sqrt{2}\cdot\sqrt{\frac{x_1^2+x_2^2+\cdots+x_n^2}{n}}$$

   $$=0.6745\sqrt{\frac{x_1^2+x_2^2+\cdots+x_n^2}{n}} \quad (\sqrt{2}=1.414) \tag{16.7.4}$$

3. The average error is the arithmetic mean of all the errors of a set, without regard to signs. Thus, we write

   $$\eta=\frac{|x_1|+|x_2|+|\cdots|+|x_n|}{n} \tag{16.7.5}$$

We shall now find an expression for η in terms of h. Let us suppose that we have a set of n measurement, each of which is affected with an error of the same size. In case of a single measurement, the probability of an error of magnitude x to $x + dx$ is

$$\frac{h}{\sqrt{\pi}} e^{-h^2x^2}\, dx$$

$\therefore$ probable number of errors of this size in n measurements is

$$\frac{nh}{\sqrt{\pi}} e^{-h^2x^2}\, dx$$

The sum of these errors is therefore

$$S = \int_{+\infty}^{+\infty} x\frac{nh}{\sqrt{\pi}} e^{-h^2x^2}\, dx = \frac{2nh}{\sqrt{\pi}} \int_0^{\infty} xe^{-h^2x^2}\, dx$$

$$= -\frac{n}{h\sqrt{\pi}} \int_0^{\infty} e^{-h^2x^2}(-2h^2xdx) = -\frac{n}{h\sqrt{\pi}}\left[e^{-h^2x^2}\right]_0^{\infty}. \quad \therefore S = \frac{n}{h\sqrt{\pi}}$$

Hence

$$\eta = \frac{S}{n} = \frac{1}{h\sqrt{\pi}} = \frac{1}{\sqrt{\pi}} \cdot \sqrt{2} \cdot \sqrt{\frac{x_1^2 + x_2^2 + \cdots + x_n^2}{n}} \tag{16.7.6}$$

16.8 Relations between the different measures of precision

We have seen from (16.7.4)

$$r = 0.6745\mu \simeq \frac{2}{3}\mu \text{ (approximately)}$$

and $$\mu \simeq \frac{r}{0.6745} = 1.4826r \tag{16.8.1}$$

Also

$$\mu = \frac{1}{h} \cdot \frac{1}{\sqrt{\pi}} = \mu\frac{\sqrt{2}}{\pi} = 0.7979\mu \simeq \frac{4}{5}\mu \text{ (approximate).}$$

Thus,

$$\mu \simeq \frac{3}{2}r, \quad \mu \simeq \frac{5}{4}\eta. \quad \therefore \mu : r : \eta = 1 : 2/3 : 4/5 = 15 : 10 : 12 \tag{16.8.2}$$

Appendix A

Miscellaneous Problems

1. Show that

$$\sum_{x=0}^{\infty} u_{2x} = \frac{1}{2}\sum_{x=0}^{\infty} u_x + \frac{1}{4}\left(1 - \frac{\Delta}{2} + \frac{\Delta^2}{4} - \cdots\right) u_0$$

[*Hints:*

$$\begin{aligned}
\text{RHS} &= \frac{1}{2}(u_0 + u_1 + u_2 + \cdots) + \frac{1}{4}\left(1 + \frac{\Delta}{2}\right)^{-1} u_0 \\
&= \frac{1}{2}(1 + E + E^2 + \cdots)u_0 + \frac{1}{4}\cdot 2(2 + \Delta)^{-1} u_0 \\
&= \frac{1}{2}(1 - E)^{-1} u_0 + \frac{1}{2}(1 + E)^{-1} u_0 \\
&= \frac{1}{2}\left\{(1 - E)^{-1} + (1 + E)^{-1}\right\} u_0 \\
&= \frac{1}{2}\cdot 2\left(1 + E^2 + E^4 + \cdots\right) u_0 \\
&= u_0 + u_2 + u_4 + \cdots \\
&= \sum_{x=0}^{\infty} u_{2x} = \text{LHS}]
\end{aligned}$$

2. $\Delta x^m - \frac{1}{2}\Delta^2 x^m + \frac{1\cdot 3}{2\cdot 4}\Delta^3 x^m - \frac{1\cdot 3\cdot 5}{2\cdot 4\cdot 6}\Delta^4 x^m + \cdots$ to m terms $= \left(x + \frac{1}{2}\right)^m - (x - 12)^m$.

[*Hints:* Since $\Delta^r x^m = 0$ for $r > m$, the given series may be written as

$\Delta x^m - \frac{1}{2}\Delta^2 x^m + \frac{1\cdot 3}{2\cdot 4}\Delta^3 x^m - \frac{1\cdot 3\cdot 5}{2\cdot 4\cdot 6}\Delta^4 x^m + \cdots$ to infinity.]

3. Find the sum of the series: $x^3 + (x+1)^3 + (x+2)^3 + \cdots + (x+m-1)^3$.

[*Hints:* The series $= \left(1 + E + E^2 + \cdots + E^{m-1}\right) x^3 = \frac{E^m - 1}{E - 1}x^3 = \frac{(1+\Delta)^m - 1}{\Delta}x^3$

$$= \left\{\binom{m}{1} + \binom{m}{2}\Delta + \binom{m}{3}\Delta^2 + \binom{m}{4}\Delta^3\right\} x^3$$

since higher order differences of x^3 will be zero.]

4. Show that

$$\Delta^n y_x = y_{x+n} - \binom{n}{1} y_{x+n-1} + \binom{n}{2} y_{x+n-2} - \cdots + (-1)^n y_x$$

5. Find the value of

$$f(2n) - \binom{n}{1} \cdot 2f(2n-1) + \binom{n}{2} \cdot 2^2 f(2n-2) - \cdots + (-2)^n f(n)$$

if $f(x) = x^3$.

[*Hints:* Experiment $= \left[E^n + \binom{n}{1}(-2)E^{n-1} + \binom{n}{2}(-2)^2 E^{n-2} + \cdots + \binom{n}{n}(-2)^n\right] f(n)$

$= (E-2)^n f(n) = (\Delta - 1)^n f(n) = (-1)^n (1-\Delta)^n f(n)$, etc.]

6. Prove that $u_x = u_{x-1} + \Delta u_{x-2} + \Delta^2 u_{x-3} + \cdots + \Delta^{n-1} u_{x-n} + \Delta^n u_{x-n}$.
7. Obtain $\Delta^{10}\left[(1-ax)(1-bx^2)(1-cx^3)(1-dx^4)\right]$.

   [*Hints:* $\Delta^r x^n > 0, \forall r > n$, hence the above $= \Delta^{10}(abcdx^{10})$.]
8. Prove that $(x\Delta)^n u_x = (x+n-1)^{(n)} \Delta^n u_x$.

   [*Hints:* Use the method of mathematical induction. Remember $(x\Delta)^{(n)} = (x\Delta)(x\Delta - 1)(x\Delta - 2) \cdots (x\Delta - n + 1)$.]
9. Show that

$$x^n - \binom{n}{1}(x-y)^n + \binom{n}{2}(x-2y)^n + \cdots + (-1)^n \cdot (x-ny)^n = y^n \cdot n!$$

[*Hints:* Take interval of differencing $= y$.

$$\therefore \text{ LHS} = \left[1 - \binom{n}{1}E^{-1} + \binom{n}{2}E^{-2} - \cdots + (-1)^n \cdot \binom{n}{n} E^{-n}\right] x^n$$

$$= (1 - E^{-1})^n x^n = \frac{\Delta^n}{E^n} x^n = \Delta^n (x - ny)^n$$

$$= (ny)(\overline{n-1}y)(\overline{n-2}y) \cdots (1y) = n! \cdot y^n = \text{RHS}]$$

10. (a) Prove that the divided differences are symmetric function of their arguments.

    (b) Show that the nth order divided differences $(x_0, x_1, \ldots, x_n)$ for the function $u_x = 1/x$ is equal to $\dfrac{(-1)^n}{x_0, x_1, \ldots, x_n}$.
11. Show that the nth divided difference can be expressed as the quotient of two determinants each of order $(n+1)$.

    [*Hints:* For four arguments x_0, x_1, x_2, x_3 divided difference of order 3 is given by

$$\delta[x_0, x_1, x_2, x_3] = \sum \frac{f(x_0)}{(x_0 - x_1)(x_0 - x_2)(x_0 - x_3)}$$

$$= \sum \frac{(x_1 - x_2)(x_1 - x_3)(x_2 - x_3)}{(x_0 - x_1)(x_0 - x_2)(x_0 - x_3)(x_1 - x_2)(x_1 - x_3)(x_2 - x_3)} f(x_0)$$

$$= \sum \left[\frac{\text{Difference product of } (x_1, x_2, x_3)}{\text{Difference product of } (x_0, x_1, x_2, x_3)} f(x_0)\right]$$

$$= \frac{\Sigma\{\text{Difference product of } (x_1, x_2, x_3)\} f(x_0)}{\text{Difference product of } (x_0, x_1, x_2, x_3)}$$

We shall now use the important result in determinants due to Vander-Monde as given below:

Difference product of

$$(x_1, x_2, x_3) = \begin{vmatrix} x_1^2 & x_2^2 & x_3^2 \\ x_1 & x_2 & x_3 \\ 1 & 1 & 1 \end{vmatrix}$$

and Difference Product of

$$(x_0, x_1, x_2, x_3) = \begin{vmatrix} x_0^3 & x_1^3 & x_2^3 & x_3^3 \\ x_0^2 & x_1^2 & x_2^2 & x_3^2 \\ x_0 & x_1 & x_2 & x_3 \\ 1 & 1 & 1 & 1 \end{vmatrix}$$

∴ we write

$$\delta[x_0, x_1, x_2, x_3] = \frac{\Sigma\left\{\begin{vmatrix} x_1^2 & x_2^2 & x_3^2 \\ x_1 & x_2 & x_3 \\ 1 & 1 & 1 \end{vmatrix} f(x_0)\right\}}{\begin{vmatrix} x_0^3 & x_1^3 & x_2^3 & x_3^3 \\ x_0^2 & x_1^2 & x_2^2 & x_3^2 \\ x_0 & x_1 & x_2 & x_3 \\ 1 & 1 & 1 & 1 \end{vmatrix}} = \frac{\begin{vmatrix} f(x_0) & f(x_1) & f(x_2) & f(x_3) \\ x_0^2 & x_1^2 & x_2^2 & x_3^2 \\ x_0 & x_1 & x_2 & x_3 \\ 1 & 1 & 1 & 1 \end{vmatrix}}{\begin{vmatrix} x_0^3 & x_1^3 & x_2^3 & x_3^3 \\ x_0^2 & x_1^2 & x_2^2 & x_3^2 \\ x_0 & x_1 & x_2 & x_3 \\ 1 & 1 & 1 & 1 \end{vmatrix}}$$

Thus, the third order divided difference is expressed as a ratio of two determinants, each of order four. By induction, we may show that

$$\delta[x_0, x_1, x_2, \ldots, x_n] = \frac{\begin{vmatrix} f(x_0) & f(x_1) & \cdots & f(x_n) \\ x_0^{n-1} & x_1^{n-1} & \cdots & x_n^{n-1} \\ x_0^{n-2} & x_1^{n-2} & \cdots & x_n^{n-2} \\ \cdots & \cdots & \cdots & \cdots \\ x_0 & x_1 & \cdots & x_n \\ 1 & 1 & \cdots & 1 \end{vmatrix}}{\begin{vmatrix} x_0^{n} & x_1^{n} & \cdots & x_n^{n} \\ x_0^{n-1} & x_1^{n-1} & \cdots & x_n^{n-1} \\ \cdots & \cdots & \cdots & \cdots \\ x_0 & x_1 & \cdots & x_n \\ 1 & 1 & \cdots & 1 \end{vmatrix}}$$

which is the ratio of two determinants, each of order $(n + 1)$.]

12. By means of divided differences find the polynomial $f(x)$ given that

x	0	1	2	4	5
$f(x)$	0	16	48	88	0

[*Hints:* The divided difference table is

x	$f(x)$	δ^1	δ^2	δ^3	δ^4
0	0				
		$\frac{16-0}{1}=16$			
1	16		$\frac{32-16}{2-0}=8$		
		$\frac{48-16}{2-1}=32$		$\frac{-4-8}{4-0}=-3$	
2	48		$\frac{20-32}{4-1}=-4$		$\frac{-8+3}{5}=-1$
		$\frac{88-48}{4-2}=20$		$\frac{-36+4}{5-1}=-8$	
4	88		$\frac{-88-20}{5-2}=-36$		
		$\frac{0-88}{5-4}=-88$			
5	0				

The interpolating polynomial

$$
\begin{aligned}
f(x) &= 0 + 16x + 8x(x-1) - 3x(x-1)(x-2) - 1x(x-1)(x-2)(x-4) \\
&= 16x + 8x^2 - 8x - 3x^3 + 9x^2 - 6x - x^4 + 7x^3 - 14x^2 + 8x \\
&= -x^4 + 4x^3 + 3x^2 + 10x.]
\end{aligned}
$$

13. Calculate f' (7.50) from the values

x	7.47	7.48	7.49	7.50	7.51	7.52	7.53
$f(x)$	0.193	0.195	0.198	0.201	0.203	0.206	0.208

[*Hints:* Taking origin at $x = 7.50(= x_0)$ and $h = 0.01$, we may write $f(x)$ is approximated by the interpolating polynomial by Stirling's interpolation formula and then find

$$\frac{d}{dx}f(x)\Big|_{x=7.5} = \frac{d}{dx}\cdot\frac{du}{dx}f(x)\Big|_{x=7.5} = \frac{1}{h}\cdot\frac{d}{du}f(x)\Big|_{u=0}.]$$

14. Find the value of $f'(0.04)$ from the following table:

x	u	$y = f(u)$	Δy	$\Delta^2 y$	$\Delta^3 y$	$\Delta^4 y$
−3	0.01	0.1023				
			0.0024			
−2	0.02	0.1047		0		
			0.0024		0.0001	
−1	0.03	0.1071		0.0001		−0.0001
			0.0025		0	
0	0.04	0.1096		0.0001		−0.0001
			0.0026		−0.0001	
1	0.05	0.1122		0.0000		
			0.0026			
2	0.06	0.1148				

[*Hints:* Taking origin at $u = 0.04$ and $h = 0.01$, we write $x = (u - 0.04)/0.01$.

We replace $f(x)$ by Gauss interpolating polynomial

$$y_x = y_0 + x\Delta y_0 + \frac{x^2 - x}{2}\Delta^2 y_{-1} + \frac{x^3 - x}{6}\Delta^3 y_{-1}$$

$$= \frac{x^4 - 2x^3 - x^2 + 2x}{24}\Delta^4 y_{-2} + \cdots$$

$$h\left[\frac{d}{dx}y_x\right]_{x=0} = \Delta y_0 - \frac{1}{2}\Delta^2 y_{-1} - \frac{1}{6}\Delta^3 y_{-1} + \frac{1}{12}\Delta^4 y_{-2} + \cdots$$

$$\therefore 0.01f'(0.04) = 0.0026 - \frac{1}{2} \times 0.0001 - \frac{1}{6} \times 0 + \frac{1}{12} \times (-0.0001) = 0.00254$$

$$\therefore f'(.04) = 0.254.]$$

15. Deduce the following approximations

(a) $\frac{d}{dx}y_x = \Delta y_{x-(1/2)} - \frac{1}{24}\Delta^3 y_{x-(3/2)}$;

(b) $\frac{d^2}{dx^2}y_x = \frac{1}{2}\left[\Delta^2 y_{x-(3/2)} + \Delta^2 y_{x-(1/2)}\right]$;

(c) $\frac{d^3}{dx^3}y_x = \Delta^3 y_{x-(3/2)}$.

[*Hints:* (a) Use Bessel's formula with origin = 0 and $h = 1$.

$$\therefore v = x - \frac{1}{2}, \quad \text{i.e.} \quad x = v + \frac{1}{2}$$

By Bessel's formula

$$y_{v+(1/2)} = \frac{y_0 + y_1}{2} + v\Delta y_0 + \frac{\left(v^2 - \frac{1}{4}\right)}{2!}\cdot\frac{\Delta^2 y_{-1} + \Delta^2 y_0}{2} + \frac{v\left(v^2 - \frac{1}{4}\right)}{2}\Delta^3 y_{-1}$$

$$\frac{dy_{v+(1/2)}}{dx} = \Delta y_0 + v\cdot\frac{\Delta^2 y_{-1} + \Delta^2 y_0}{3!} + \left(\frac{v^2}{2} - \frac{1}{24}\right)\Delta^3 y_{-1}$$

Putting $v = 0$,

$$\frac{dy_{1/2}}{dx} = \Delta y_0 - \frac{1}{24}\Delta^3 y_{-1}$$

Shift the orign to $x - \frac{1}{2}$. Then

$$\frac{dy_x}{dx} = \Delta y_{x-(1/2)} - \frac{1}{24}\Delta^3 y_{x-(3/2)}$$

For (b) and (c) use Bessel's formula.]

16. Compute the value of the definite integral

$$\int_{0.2}^{1.4} (\sin x - \log_e x + e^x)dx$$

by (a) the trapezoidal rule, (b) Simpson's one-third rule, (c) Simpson's three-eighths rule and (d) Weddle's rule. After finding the true value of the integral, compare the errors in the four cases.

17. Prove that

$$\int_{-1}^{1} f(x)dx = \frac{13}{12}\{f(1) + f(-1) - f(3) - f(-3)\} \text{ approximately.}$$

18. Prove that if $f(x)$ is a continuous function whose fifth differences are constant, then

$$\int_{-1}^{+1} f(x)dx = \frac{8}{9}f(0) + \frac{5}{9}\{f(\sqrt{0.6}) + f(-\sqrt{0.6})\}$$

19. Solve the equation $u_{x+2} - 7u_{x+1} + 10u_x = 12.4^x$.
20. Solve by the method of differences $u_{x+2} - 5u_{x+1} + 6u_x = 5^x$.
21. Solve the following difference equation $u_{x+3} - 5u_{x+2} + 8u_{x+1} - 4u_x = x\cdot 2^x$.

[*Hints:*

$$\text{P.I.} = \frac{1}{E^3 - 5E^2 + 8E - 4}x\cdot 2^x = \frac{1}{(E-1)(E-2)^2}x\cdot 2^x = 2^x\cdot\frac{1}{(2E-2)^2}x$$

$$= \frac{2^x}{\Delta}\frac{1}{(E-1)^2}x = \frac{2^x}{\Delta}\cdot\frac{1}{\Delta^2}x = \frac{2^x}{\Delta}\cdot\frac{x^{(3)}}{6} = \frac{2^x}{\Delta}x(x-1)(x-2).$$

Rule: $\frac{1}{f(E)}a^x\phi(x) = a^x\cdot\frac{1}{f(aE)}\phi(x),\ f(aE) \neq 0.$]

22. Solve: $u_{x+2} - 2mu_{x+1} + (m^2 + n^2)u_x = m^x$.
23. Solve: $y_{x+2} - 5y_{x+1} + 6y_x = 5^x$.
24. Solve: $y_{x+2} - 4y_{x+1} + 4y_x = 3\cdot 2^x + 5\cdot 4^x$.
25. Solve: $u_{x+2} - 4u_x = 9x^2$.
26. From the equation $x^5 - a = 0$, deduce the Newtonian iterative procedure

$$x_{n+1} = \frac{1}{5}\left(4x_n + \frac{a}{x_n^4}\right) \text{ for } \sqrt[5]{a}$$

Use this method to find $\sqrt[5]{3}$.

[*Hints:* $f(x) = x^5 - a$, $f'(x) = 5x^4$.

From Newton-Raphson's iterative formula

$$x_{n+1} = x_n - \frac{f(x_n)}{f'(x_n)} = x_n - \frac{x_n^5 - a}{5x_n^4} = x_n - \frac{x_n}{5} + \frac{a}{5x_n^4} = \frac{1}{5}\left(4x_n + \frac{a}{x_n^4}\right) \text{ for } \sqrt[5]{a}$$

This is a relation connecting the $(n+1)$th approximation and the nth approximation of the solution.

For $a = 3$, we have $\sqrt[5]{3}$.

$$x^{(1)} = \frac{1}{5}\left(4x_0 + \frac{3}{x_0^4}\right)$$

Taking $x_0 = 1.2$

$$x^{(1)} = \frac{1}{5}\left(4 \times 1.2 + \frac{3}{2.0736}\right) = 1.249$$

$$x^{(2)} = \frac{1}{5}\left(4 \times 1.249 + \frac{3}{2.4336}\right) = 1.2457$$

(correct to 4 decimal places).]